# Tracing M<

Walter Benjamin famously defined modernity as 'the world dominated by its phantasmagorias'. The chapters in this book will focus on one such phantasmagoria, that of 'modernity' itself. From the late seventeenth century until today, the 'modern' has served as a key category by which to understand an ever-changing present. Art and architecture have played a key role in this pursuit as the means by which the modern was to manifest itself. The aim of this anthology is to trace the modern project through its multifarious manifestations in order to understand contemporary culture in a deeper sense than facile discussions of modernism and postmodernism often grant. Drawing on architectural and urban history as well as philosophy and sociology, the chapters outline the complex and conflicting roots of modernity by tracing its manifestations in architecture and the city.

The book is divided into three parts, each exploring a distinct aspect of modernity. While part one scrutinises the much-abused concepts of modernity, modernism and the modern, parts two and three look at the manifestations of the modern in architecture and the city respectively. Focusing particularly on the transition between historicism and modernism, the chapters offer a reinterpretation of early modern architectural and urban culture as it came to expression through people such as Cerdá, Semper, Bötticher, Scott, Baudelaire, the Goncourt brothers, Benjamin, Warburg, Kracauer, Mackintosh, Behrens, Taut and Le Corbusier. For all their differences, these were thinkers and practitioners whose undisputed modernity arose from a deep preoccupation with history. A re-reading of their legacy may throw light on the neglected reciprocity between modernity and its historical conditions of becoming.

**Mari Hvattum** is an architect specialising in nineteenth century architectural thinking. She is Senior Lecturer in architectural history and theory at the Oslo School of Architecture, Norway, and Lecturer at Strathclyde University.

**Christian Hermansen** is Guest Professor of Architecture at the Oslo School of Architecture, Norway, and Senior Lecturer at the Mackintosh School of Architecture.

# Tracing Modernity

## Manifestations of the modern
## in architecture and the city

**Edited by Mari Hvattum
and Christian Hermansen**

Routledge
Taylor & Francis Group

LONDON AND NEW YORK

First published 2004
by Routledge
11 New Fetter Lane, London EC4P 4EE

Simultaneously published in the USA and Canada
by Routledge
29 West 35th Street, New York, NY 10001

*Routledge is an imprint of the Taylor & Francis Group*

© 2004 Mari Hvattum and Christian Hermansen (selection and editorial
matter); contributors (individual chapters)

Typeset in Charter by Bookcraft Ltd, Stroud, Gloucestershire
Printed and bound in Great Britain by TJ International, Padstow, Cornwall

*British Library Cataloguing in Publication Data*
A catalogue record for this book is available from the British Library

*Library of Congress Cataloging in Publication Data*
Tracing modernity : manifestations of the modem in architecture and the city /
edited by Mari Hvattum and Christian Hermansen – 1st ed.
    p. cm.
Includes bibliographical references and index.
1. Modern movement (Architecture)   2. Architecture and society
3. Modernism (Aesthetics)   I. Hvattum, Mari, 1966–  II. Hermansen,
Christian.
 NA682. M63T73 2004
 724' . 6–dc22                    2003022333

ISBN 0-415-30511-X hb
ISBN 0-415-30512-8 pb

# Contents

Illustration credits                                                   vii

Notes on contributors                                                   ix

Introduction                                                            xi

## PART 1  MODERNITY

**1**  Analysing modernity: some issues                                 3
       *David Frisby*

**2**  What modernism was: art, progress and the avant-garde           23
       *Arnfinn Bø-Rygg*

**3**  Modernity and architecture                                      42
       *Iain Boyd Whyte*

**4**  Modernity and the uses of history: understanding classical
       architecture from Bötticher to Warburg                          56
       *Caroline van Eck*

**5**  Projecting modern culture: 'aesthetic fundamentalism' and
       modern architecture                                             68
       *Gabriele Bryant*

**6**  Modernity and the question of representation                    81
       *Dalibor Vesely*

## PART 2  ARCHITECTURE

**7**  'How is it that there is no modern style of architecture?'
       'Greek' Thomson versus Gilbert Scott                           103
       *Gavin Stamp*

# Contents

**8** 'A complete and universal collection': Gottfried Semper and
the Great Exhibition    124
*Mari Hvattum*

**9** The interior as aesthetic refuge: Edmond de Goncourt's
*La maison d'un artiste*    137
*Diana Periton*

**10** Timely untimeliness: architectural modernism and the idea
of the *Gesamtkunstwerk*    156
*Gabriele Bryant*

**11** Le Corbusier and the restorative fragment at the Swiss Pavilion    173
*Dagmar Motycka Weston*

**12** The concrete memory of modernity: excerpts from a
Moscow diary    195
*Jonathan Charley*

## PART 3   THE CITY

**13** Ildefonso Cerdá and modernity    217
*Christian Hermansen*

**14** 'To knock fire out of men': forging modernity in Glasgow    232
*Juliet Kinchin*

**15** The expressionist utopia    256
*Iain Boyd Whyte*

**16** Walter Benjamin's Arcades Project: a prehistory of modernity    271
*David Frisby*

**17** Impromptus of a great city: Siegfried Kracauer's *Strassen in
Berlin und Anderswo*    291
*Graeme Gilloch*

**18** Orpheus in Hollywood: Siegfried Kracauer's Offenbach film    307
*Graeme Gilloch*

Bibliography    325

Index    341

# Illustration credits

Architekturmuseums der TU, München ©,10.5

Ateneo Barcelones ©, 13.1

Bjørn Sandaker ©, 6.1

Christopher Schulte ©, 9.5

Commission of the Great Exhibition, *Art Journal Illustrated Catalogue of the Industry of All Nations* ©,8.1

Commission of the Great Exhibition, *Art Journal Illustrated Catalogue of the Industry of All Nations* ©,8.5

Commission of the Great Exhibition, *The Official Descriptive and Illustrated Catalogue of the Great Exhibition* ©, 8.2

Commission of the Great Exhibition, *The Official Descriptive and Illustrated Catalogue of the Great Exhibition* ©,8.6

Fernand Lochard, Bibliothèque Nationale de France ©,9.1, 9.2, 9.3, 9.4

Fondation Le Corbusier, Paris, © ADAGP, Paris and DACS, London 2004 ©, 11.3

Gavin Stamp ©, 7.5, 7.6, 7.9, 7.10, 14.2

Hedrich Blessing, Chicago Historical Society ©, 6.5

*Histoire naturelle*, 1925, © DACS, London 2004 ©, 11.9

Instituto de Estudios Territoriales, Archivo Cerdá ©, 13.2, 13.3

*L'Architecture vivante*, 1933, © ADAGP, Paris and DACS, London 2004 ©, 11.4

*L'Architecture vivante*, 1933, © FLC L2 (8)21/ADAGP, Paris and DACS, London 2004 ©, 11.1

Mitchell Library ©, 7.7

Oeuvre complète II, © ADAGP, Paris and DACS, London 2004 ©, 11.8, 11.10

Oeuvre complète II, p. 83, © FLC L2(8)50/ADAGP, Paris and DACS, London 2004 ©, 11.5

Peter Blundell Jones ©, 6.4

Robert Gibbs ©, 14.4

Royal Incorporation of Architects in Scotland ©, 7.11

*Studying the Victorians at the Victoria and Albert Museum*, Victoria and
    Albert Museum 1994 ©,8.1
Sylvester Bone ©, 14.1The Arts Council of Great Britain, Le Corbusier:
    Architect of the Century, © ADAGP, Paris and DACS, London 2004
    ©,11.2, 11.6
The Arts Council of Great Britain, *Le Corbusier: Architect of the Century*, ©
    FLC n.342/ADAGP, Paris and DACS, London 2004 ©, 11.7
University of Glasgow ©, 14.3
Victoria and Albert Museum, © V&A Picture Library 8.4W. Stewart (ed.),
    *University of Glasgow Old and New*, Glasgow 1891 ©, 7.8

# Contributors

**Gabriele Bryant** is researcher at Linacre College, University of Oxford.

**Arnfinn Bø-Rygg** is Professor in Aesthetics at Oslo University. His recent publications include *Modernisme, Antimodernisme, Postmodernisme*.

**Jonathan Charley** is Senior Lecturer at the Department of Architecture, University of Strathclyde.

**David Frisby** is Professor of Sociology at Glasgow University. His recent publications include *Simmel on Culture* and *Cityscapes of Modernity*.

**Graeme Gilloch** is Lecturer in Sociology at the University of Salford. His recent publications include *Myth and Metropolis: Walter Benjamin and the City*.

**Christian Hermansen** is Senior Lecturer at the Mackintosh School of Architecture, Glasgow School of Art. He is currently guest professor at the Oslo School of Architecture, and is has recently published *(theorizing) History in Architecture*.

**Mari Hvattum** is Senior Lecturer at Oslo School of Architecture and a lecturer at the University of Strathclyde. Her recent publications include *Gottfried Semper and the Problem of Historicism*.

**Juliet Kinchin** is Lecturer at the University of Glasgow. Her recent publications include *E.W Godwin: Aesthetic Movement, Architect and Designer*.

**Diana Periton** is Assistant Director of Histories and Theories at the Architectural Association, London.

**Gavin Stamp** was Senior Lecturer at the Mackintosh School of Architecture at the Glasgow School of Art, and is currently a Mellon Senior Fellow at the University of Cambridge. His recent publications include *An Architect of Promise: George Gilbert Scott Junior and the Late Gothic Revival.*

**Dalibor Vesely** has recently retired as Senior Lecturer at the University of Cambridge. His recent publications include *The Poetics of Architecture.*

**Caroline van Eck** is guest Professor in the History and Theory of Architecture at Ghent University and teaches Architectural History at the University of Groningen. Her recent publications include *Germain Boffrand's Livre d'Architecture.*

**Dagmar Motycka Weston** is Lecturer at the University of Edinburgh.

**Iain Boyd Whyte** is Professor of Architectural History at the University of Edinburgh. His recent publications include *Modernism and the Spirit of the City.*

# Introduction
## Tracing modernity

This anthology examines the heterogeneous modern culture of the late nineteenth and early twentieth centuries, and traces its manifestations in architecture and the city. The choice of the hackneyed epithet 'modern' as our theme was not done out of a *fin-de-siècle* preoccupation with labels, though the constant desire for branding and rebranding one's own time is an unmistakable characteristic of the modern. Both the careful differentiation of a 'modernity' to stand in wilful opposition to a non-modern past, and the war waged on the modern under the prefix banners of 'post-' or 'super-', are profoundly modern impulses. The insistence that we have somehow overcome the modern is perhaps the most modern assertion of all – presupposing the very linearity that the postmodern purported to question. Rather than projecting the end of modernity, it may be prudent to scrutinise some of its beginnings, so better to understand the nature of the 'inescapable' modern condition. This is the ambition that inspired this collection of essays: to understand modern architecture and urban culture in a deeper sense than the facile discussion of modernism and postmodernism often grants.

David Frisby, in his exploration of the concept of modernity, evokes Benjamin's definition of the modern: 'The world dominated by its phantasmagorias … this is modernity'. The essays that follow focus – in different ways and by different means – on one such phantasmagoria, that of modernity itself. From the late seventeenth century until today, the notion of the modern has served as a key category by which to understand an ever-changing present. Part One investigates this construct and looks at the concepts of modernity and modernism in different intellectual traditions. More than a theoretical construct, however, modernity is embodied in the very fabric of society. Architecture and the city perform this embodying function in two ways. They form the inert background against which modernity can be grasped and discerned, and they also constitute the active means by which the modern project came to manifestation. Both roles are examined by the essays that follow, with Parts Two and Three focusing respectively on architecture and the city.

The essays present a wide variety of approaches and materials, a heterogeneity which in itself constitutes an apt image of the multiplicity of viewpoints characterising modernity. Yet the anthology's aim is more specific than simply to display diversity. Focusing on the ambiguous relationship between history and modernity, we aim to question the neat linearity outlined in much modernist historiography, and to uncover the complex ways in which the modern incorporates and transforms history as a part of itself. Furthermore, if modernity rests on a reinterpretation of history then modernism as its aesthetic equivalent is intrinsically bound up with its self-proclaimed antithesis, historicism. The supposed break between a corrupt historicism and a heroic modernism – presupposed by modernist historiography – is an idea which itself rests on powerful historicist presuppositions. The essays trace this relationship from two contrasting points of view. On the one hand they trace the inherent modernity of historicist preoccupation with phenomena such as style, method, the collection and the interior. On the other hand they examine the lurking historicism of the modernist utopia, expressed by its redemptive obsession with *Zeitgeist* and the *Gesamtkunstwerk*. The anthology focuses deliberately on figures who transcend a narrowly construed idea of the modern: Cerdá, Semper, Bötticher, Scott, Baudelaire, the Goncourt brothers, Benjamin, Warburg, Kracauer, Mackintosh, Behrens, Taut and Le Corbusier. For all their differences these were thinkers and practitioners whose undisputed modernity arose from a deep preoccupation with history, and a re-reading of their legacy throws light on the neglected reciprocity between modernity and its historical conditions of becoming.

An anthology requires the contribution of many people who deserve our thanks. We are grateful to Caroline Mallinder at Routledge for her help and patience, and also to Michelle Green and the rest of the Routledge staff for their unfailing support. We thank institutions and individuals who have granted copyright for the use of illustrations and texts. Our deepest gratitude, however, goes to the contributing authors for sharing their ideas and insights so generously. Working with them has been a privilege.

Christian Hermansen and Mari Hvattum
Oslo, August 2003

Part One

# Modernity

# Chapter 1

# Analysing modernity
## Some issues

*David Frisby*

# I

Just over a century ago, in 1896, Otto Wagner published in Vienna what is probably the first modernist architectural manifesto – his *Modern Architecture*.[1] Despite its theoretical weaknesses, it was read by his contemporaries as a rejection of the historicism of the recent past and a plea to create an architecture appropriate to modern life. Indeed, in architecture 'the sole starting point of our artistic endeavours should be *modern life*'.[2] Architecture, like other modern arts, 'must represent our modernity, our capabilities and our actions through forms created by ourselves'.[3] And Wagner's answer as to where this modern life, this modernity, is most visibly located is unequivocal: 'the most modern of that which is modern in architecture are indeed our metropolitan cities'.[4] Yet the identification of modern life and modernity with the physical location of the metropolis is only one of the possible sites for the origins of modernity.

But if a modern architecture is to represent, reflect or mirror modern life and modernity, even in the somewhat naive positivistic manner in which Wagner stated it, then it must presuppose a reading of modernity that can be given architectural form. Unlike many of his contemporaries in *fin-de-siècle* Vienna, Wagner's reading of modernity focuses not on the fragmentation and disintegration of modern experience that we associate with the city's other modernist movements, but rather on unlimited practical progress and the possibility of an unbounded metropolis. The features of modern life which Wagner chose to highlight can be subsumed under the processes of abstraction and levelling (quantitative expansion of the metropolis, a quantitative conception of progress, democratisation as abstract political participation, the levelling of forms of life, the rented apartment block as a 'conglomerate of cells', the significance of money in time calculation and purposive action, and increasingly abstract ornamentation in street facades), movement and circulation (the acceleration of circulation of individuals and commodities in traffic systems, including the straight – as opposed to crooked – street, the circulation of money and capital in apartment block building and investment), and the monumental (the modern continuous street facade as monumental, the demand for a modern monumentalism). What disturbed many of his contemporaries was Wagner's emphasis on a close connection between modernity and fashion, not merely in terms of the cycles of fashion but also in relation to fashion's role in the creation of the new in architecture.

Wagner indicates the problematic connection between architecture and intelligibility all too briefly. For him, the lack of intelligibility of much contemporary (historicist) architecture lies in the fact that it does not reflect modern needs and modern life. Even more briefly, he suggests that the 'language of forms' created by engineers is also unintelligible to the mass of the population. The issue of intelligibility is, of course, related to that of the legibility of architecture and the modern metropolis. The language of modern architecture must therefore express modern life for modern people. With few exceptions, Wagner's reading of the forms of modern life or modernity is unproblematical, a reading of modernity without contradictions. Nonetheless, contradictions were also present within the successful Wagner School (1894–1911); for instance between the tendency towards a universal modernism in Wagner's own work and attempts by some of his students, especially after 1918, to reconcile modernism with nationalism and to develop a modern national architecture.[5]

The call for a modern architecture to reflect modern needs and uses coincides with the much wider debate about modernity at the end of the nineteenth century. Only a decade before the publication of Wagner's *Modern Architecture*, the German concept of modernity/modernism (*die Moderne*) had appeared in a Berlin literary journal in 1886.[6] The explosion in avant-garde literary modernisms around 1890 was followed some years later by similar developments in the architectural field. Wagner's 1894 inaugural lecture at the Vienna Academy of Fine Arts, his *Modern Architecture*, and the completion of the Secession Building in 1898 by one of his students, Josef Olbrich, initiated a debate on the modern which, in diverse forms, continued intermittently over subsequent decades.[7]

The Viennese response to the modern indicates the diversity and contradiction within the concept of modernity. Wagner's concept of modernity revealed one aspect of the attempt to capture the forms of modern life – through order and abstraction. Many of the contemporary modernist movements in Vienna explored metropolitan modernity as the fragmentation and disintegration of experience. It has been argued by Zygmunt Bauman, Marshall Berman and others that modernity is experience of the tension and contradiction in modern social formations;[8] in particular the tension between modernity as dynamic, discontinuous and unregulated movement, and modernity as a process of rationalisation whose consequence is the regulation of movement, a dynamic evident in much of the discourse on modernity in the nineteenth and twentieth centuries. The notion that everything is in motion is a disturbing one, and efforts to capture the 'labyrinth of movement' (von Stein) through regulation (whether statistical, social, political or monumental rationalisation) helped contribute to powerful strategies of containment.

Thus there is a tension between the desire to give the modern world new modes of structure, order and regulation on the one hand, and the recognition of the disintegration of modern experience of that world (seeing time as transitory, space as fleeting, and causality replaced by the fortuitous and the arbitrary) on the other; a tension between the old and the ever-new. These tensions and contradictions are apparent in many aspects of modern experience, including our experience of the social spaces and built environment of modernity.

Many theories of modernity can be distinguished by the way they analyse the contrast between the structuring, rationalising dimensions of modernity and the discontinuity and destruction of modern life. Not all were able to capture both dimensions. A schematic overview of prominent social theories of modernity broadly contemporary with Wagner's reflections, together with some later theories, illustrates some of the issues and problems involved in the analysis. The diversity of theories of modernity demonstrates both the qualitative differentiation of concepts of modernity and their contested nature, and suggests that a modern architecture that would reflect them would be equally diverse, calling into question undifferentiated readings of a 'modernist movement' and of modernism as a unified project. This brief overview of theories of modernity is followed by a reflection on one of Wagner's main concerns – the intelligibility and legibility of architecture and the modern metropolis.

# II

The social sciences abound in theories of modernisation – social, economic, political, psychological and cultural explorations of how and through what processes modern societies emerged. Such accounts often rest on a juxtaposition between traditional and modern societies, between static and dynamic socioeconomic formations. Yet an account of modernity, understood as modes of experiencing that which is new in modern society, would presuppose an account of the transitions to modern society but without itself being reduced to a theory of modernisation. Similarly, the aesthetic representations of transitions to modern society and modernity since the second half of the nineteenth century have given rise to a series of aesthetic modernisms, often accompanied by avant-garde manifestos announcing the arrival of new, modernist movements and exploring 'the shock of the new'. The closer the concept of modernity is to that of modernisation, the more it is likely to become a conceptualistation of historical periodisation. Where the concept is closer to aesthetic modernisms it is more likely to become a conceptualisation of modes or qualities of modern social experience. A third, more recent, conceptualisation of modernity is modernity as an historical project (Habermas).[9] None of these concepts of modernity is without analytical and methodological problems.[10]

There has been considerable uncertainty surrounding the concepts of the modern, modernity, modernisation and modernism in some historical periods, such as the turn of the nineteenth century and also in recent decades, when the concept of modernity has come to encompass or be fused with all these related concepts. Indeed, the common associations of modernity with changes in historical consciousness, an emphasis on accelerating change and an identification of the present as modernity, raise the issue of historical periodisation.

The historical periodisation of modernity often relies on abstract chronologies and temporalities, and on uncontextualised stages of presentness. Modernity conceived by Marshall Berman as emergent in the late Renaissance around 1500 and its successive phases – 1500–1789, 1789–1900 and 1900 to the present – relies on an abstract conception of historical epochs, and certainly contrasts with his broad definition of modernity as

> a mode of vital experience – experience of space and time, of the self and others, of life's possibilities and perils – that is shared by men and women all over the world today … To be modern is to find ourselves in an environment that promises us adventure, power, joy, growth, transformation of ourselves and the world – and, at the same time, that threatens to destroy everything we have, everything we know, everything we are. Modern environments and experiences cut across all boundaries of geography and ethnicity, of class and nationality, of religion and ideology.[11]

Modernity as co-terminous with the development of the capitalist mode of production makes sense only if the processes by which capitalism as a socioeconomic formation transforms social relations and experience into modernity can be delineated. An association of modernity with late capitalism would have to confront not only the demarcation and justification of lateness, but also the possibility that capitalism as a socioeconomic phenomenon may be in its infancy. Modernity as a project co-terminous with the enlightenment and autonomous reason can be said to rest on a demonstration of the continuity of this intellectual project since Kant.[12] The most ambitious attempt to abandon the connection between modernity and periodisation and turn to modernity as process in the past and present – Walter Benjamin's 'prehistory of modernity' – itself retains elements of periodisation of capitalism (Baudelaire is viewed by Benjamin as a poet of high capitalism).[13]

Accounts of the transition to modernity and the contemporary analysis of the present modernity were always associated with a critique of modernity, rather than with its celebration. The focus on the present in earlier theories of modernity, and their claims to be an analysis of the present, were often framed in the context of a sense of crisis that problematised the present. Even then, and more commonly in the early decades of the twentieth century, this did not preclude the development of theories of an 'anti-modern' modernity,[14] and mythological and post-historical political projects associated with fascism. The 'prehistory' project of the road to communism was also problematical.

When Baudelaire introduced the notion of *modernité* as a new concept in his essay 'The painter of modern life' (1863) he defined it as 'the transitory, the fleeting, the fortuitous, the half of art whose other half is the eternal and the immutable'.[15] The emphasis on the transitory, fleeting and fortuitous dimensions of modern experience was located in the modern metropolis. This modernity was both a 'quality' of modern life and a new aesthetic object, grounded in the 'ephemeral, contingent newness of the present' and in 'the daily metamorphosis of external things' on the surface of everyday existence. Baudelaire's concept of modernity thus emphasises the experience of newness, everyday metropolitan existence, and the dynamic movement of fields of metropolitan signifiers.

Although Baudelaire was concerned largely with modes of aesthetic representation of modernity, it is possible to transpose these transitory, fleeting and fortuitous dimensions of modern existence onto a more general level. If modernity is conceived as the discontinuous and disintegrating experience of time as transitory (moments of presentness), space as fleeting (disintegrating, variable space), and causality as replaced by fortuitous or arbitrary constellations, then such a concept also has significant consequences for human individuality and subjectivity that coalesce around more recent discussions of the relationship between modernity and self-identity.[16] This reading of Baudelaire's concept of modernity takes it beyond what Habermas terms 'aesthetic modernity', and provides a framework within which the relationship between modernity and historical consciousness, social space, the conflation of time and space, and the challenge to fixed forms of causality and historical and natural necessity can be examined.

The identification of modernity and capitalism is nowhere more powerful than in the work of Marx. The *Communist Manifesto*, described by Berman as the first modernist manifesto, highlights both the dynamic and destructive features of capitalism that shape modernity, encapsulated in Marx's assertion that

> Constant revolutionising of production, uninterrupted disturbance of all social relations, everlasting uncertainty and agitation distinguish

the bourgeois epoch from all earlier ones. All fixed, fast-frozen relationships, with their train of venerable ideas and opinions, are swept
away, all new-formed ones became obsolete before they can ossify.
All that is solid melts into air, all that is holy is profaned.[17]

Here and elsewhere Marx draws attention to the permanently
dynamic 'revolutionising of production' within capitalist socioeconomic structures, and the accompanying 'uninterrupted disturbance of all social relations'
and 'everlasting uncertainty and agitation'. This revolutionary new destruction
of the past (thereby destroying historical specificity) is accompanied by a second
dimension of modernity, the ever-new destruction of the present (all newly
formed social relations 'become obsolete before they can ossify'). But whereas in
these earlier writings Marx assumed that this destructive dynamic in which 'all
that is solid melts into air' would reveal the 'real conditions' of social life, in his
mature works a third dimension of modernity is introduced with his theory of
commodity fetishism, announcing the ever-same reproduction of the 'socially
necessary illusion' of the commodity form as a barrier to a qualitatively different
future. Berman puts forward a different argument concerning the persistence of
capitalism, namely that the instability of permanent 'everlasting uncertainty and
change' serves to maintain capitalism as a socioeconomic phenomenon.

The identification of modernity with endless movement that is
present in Marx's exploration led him to search for 'the laws of motion' of
capitalism. Although he wrote little about the built environment, the dimensions
of modernity which he identified as the revolutionary new destruction of the
past, the ever-new destruction of the present, and the ever-same reproduction of
the 'socially necessary illusion' as commodity, are all relevant to reading metropolitan modernity. The destruction of the past as built environment is both a
feature of modernisation and a central theme in the debate surrounding the new
discipline of city planning. The destruction of the present is manifested in the
need for capital accumulation in the urban context to facilitate the maximisation
and cheapening of commodity production and an acceleration in the circulation
of commodities. The commodification of urban forms bestows on them characteristics of the commodity form, especially the transformation of 'every product
of labour into a social hieroglyphic [which] human beings try to decipher'.[18]
This implies that modern architecture displays constellations of hieroglyphics
that must be deciphered.

In examining the commodity form as a feature of modernity, Marx's
analysis of commodity fetishism and the illusions of the commodity form in the
spheres of circulation and exchange suggests a reenchantment or reification of
the 'movement which proceeds on the surface of the bourgeois world'. This
implies the creation of new illusions within 'the daily traffic of bourgeois life' as
it appears in the circulation of commodities.[19] The revolutionary movement of

the capitalist socioeconomic formation and experience of its everlasting destructive dimension could be viewed as functional to the continuation of capitalist social relations (Berman) and the features of modernity associated with them. Similarly, the socially necessary illusions within the phenomenal forms in which capitalist society appears to its members masks its transitory present as eternal, its economy as natural rather than historical, and its social relations as harmonious rather than contradictory. Yet Marx does not devote a great deal of attention to the everyday experience of capitalist modernity. Rather, he is largely concerned to search for the 'laws of motion' of the capitalist mode of production, and it is this socioeconomic formation which receives the greatest attention.

Other attempts to delineate what is new and modern in modern society are conceptualised as the juxtaposition of modernity with its opposite. Ferdinand Tönnies (1887) provided an opposition between *Gemeinschaft* and *Gesellschaft*, and Emile Durkheim (1893) an opposition between traditional societies based on mechanical social solidarity and modern, complex societies based on organic solidarity.[20] Tönnies' concepts are intended to reveal constituent elements of modern experience, with the transition to society as a shift to contractual, conventional social relations epitomised in capitalist exchange relations and the metropolis. Society as a 'a strange country', a transitional phenomenon, is contrasted with community as the location of creative formative forces.[21] And although not intended by Tönnies, the normative and ideological import of these two concepts made them accessible to anti-modern ideologies that denounce modern society. The ostensibly positive features of community could also contribute to variants of communitarianism and to the establishment of model and utopian communities in the late nineteenth and early twentieth centuries.

In a different manner, the disintegration of community and collective consciousness constitutes a central theme in Durkheim's explorations of the moral order of modernity in the context of modern societies based on a complex division of labour and its abnormal consequences. Specialised divisions of labour increasingly emphasise individual specialisation and the development of individualism. The collective consciousness of societies based on mechanical solidarity (segmental, low interdependence, common sentiments and beliefs) gives way to a more problematical collective consciousness grounded in individualism and emerging out of modern complex societies based on organic solidarity (differentiated, high interdependence, the cult of the individual). In this context Durkheim maintained that the collective consciousness of a society manifested itself in collective representations. In traditional societies this may have taken the form of totems, but in modern society it would be possible to see monumental architectural representations of modernity in this light.[22]

The weakening of collective consciousness and moral integration and regulation of the individual produces a crisis in the individual's relationship with group and society characterised as the predominance of egoism (weak integration) and anomie (weak regulation). Egoism results in an overnourished intelligence, thought without an object, and an individual dream world; anomie in unconstrained emotion, passion without a goal, and unlimited desires. Such individual pathologies result from a breakdown in moral regulation that has its origins either in economic and social disturbances, or in the breakdown of social relations. Insofar as modern capitalist society requires excessive consumption in order to keep the economy in motion, and to the extent that excessive individualism is encouraged, the negative currents which Durkheim detected have become endemic to capitalist society and are not merely pathological deviations. Further, insofar as individuals seek to invest at least part of their identity in acquisitions, this investment will never be permanently realised. Individualism, and certainly not excessive individualism, can secure individuality. Such an account of modernity renders problematic the relationship between personal identity, individuality and modern society.

Unlike those theories of modernity which identified many of its central features with the transformations of production and the industrial enterprise, Georg Simmel's delineation of modernity focuses on two different but interrelated topics: the metropolis and the mature money economy. The strength of Simmel's analysis of modernity lies in his exploration of the transformations of social relations, and of their experiential and emotional consequences on these two sites.

This focus manifests itself in his only 'definition' of modernity, to be found in his positive appreciation of Auguste Rodin's work as expressing the tensions of modernity, where Simmel states that

> The essence of modernity as such is psychologism, the experiencing and interpretation of the world in terms of the reactions of our inner life and indeed as an inner world, the dissolution of fixed contents in the fluid element of the soul, from which all that is substantive is filtered and whose forms are merely forms of motion.[23]

This experience of the world as an inner world in flux has affinities not only with Walter Benjamin's later characterisation of the dominant mode of experienced modernity as inner, lived experience (*Erlebnis*), but also with the fluidity, flux and dissolution in metropolitan modernity and in the sphere of money exchange. The dissolving of fixed contents into fragments and the recurrent immediacy of modernity, its presentness, anticipates dimensions of modernity highlighted in its later characterisations.

In their different but related ways, the metropolis as the site of concentration and intensification of modernity and the mature money economy as the site of the diffusion and extensification of modernity, both focus on the spheres of circulation (of commodities and individuals), exchange and consumption, the increase in social differentiation (and dedifferentiation in the case of the levelling effect of money as universal equivalent of all values), the increase in the functionalisation (abstraction) of social relations, and the widening gap between what Simmel terms subjective and objective culture.

Simmel's delineation of modernity focuses on modes of experiencing the immediate present in modern society as differentiated and discontinuous (fragmented). Both within the metropolis and the money economy there is a tendency for human culture to be transformed into a culture of things.[24] This reification is at the heart of Simmel's theory of the dialectical relationship between subjective and objective culture, with objective culture becoming an autonomous sphere with its own laws of development that confront individual, subjective culture. Confronted by this objective culture of material objectifications (including architecture), the individual in the metropolis faces reification in motion, the shocks of abstract confrontation. The tumult of the metropolis, its abstract existence, is responded to by individuals through social distance and dissociation from the continuous shocks to their nervous life. Personal reserve, hostility, and the blasé attitude that characterise metropolitan existence (represented dramatically in German expressionism's depictions of the city) also constitute the elementary forms of socialisation in the money economy where individuals must respond to the reification of the social relations of exchange and the dynamic abstract flux of commodity circulation. In his *Philosophy of Money*, Simmel shows how value as substance is transformed into a relational concept, the teleology of means and ends becomes the elevation of money to the absolute means and the reduction of quality to quantity, individual freedom is paid for by increasing functionalisation of social relations, personal values are reduced to money values, and the style of life presents itself to us as an objective totality but is in fact composed of the fragmentary.

By taking seriously the fragmentation of experience in modern society and by investigating its inner consequences for the individual, Simmel provides the most sustained study in social theory of the everyday world of modernity around 1900. His concern for the 'fortuitous fragments of reality', for 'the delicate, invisible social threads' that bind individuals together, provides a theory of modern culture to which his students such as Martin Wagner and Adolf Behne, and others such as Erich Mendelsohn who were impressed with his work, were drawn.[25] The study of the fragmentation and flux of the world of appearances in modernity requires an appropriate methodological approach that can focus on the relations between individuals, groups and things.

The development of modern western rationalism and its consequences, the most important of which is modern western rational capitalism, is the key to Max Weber's theory of modernisation and modernity outlined in the first two decades of the twentieth century. The distinctive form of rationality, whose origins for Weber lay in the application and practical transformation of aspects of protestant reformation doctrine, and whose internal dynamic enabled it to dominate all the major spheres of social, economic, artistic, legal, administrative and religious life, obtained its superiority through its lack of attachment to any end or goals of these institutional arrangements. It was a purely formal rationality concerned with the precise calculability of means and the most efficient procedures or means to achieve a given end. The (formally) rational organisation of wage labour, enterprises, systems of administration, legal systems, systems of state legitimation and religion, and the domination of purposive rational action above all other orientations to action, could readily be construed as a universal totalising process (extendible to, and influencing, Georg Lukács' theory of reification). The rationalisation of building practices that was favoured by some architects and planners in the Weimar Republic, such as Martin Wagner as chief city planner of Berlin after 1926, can be illuminated by reference to Weber's earlier discussion of rationalisation, as can the more general and diverse treatments of this process in modern architecture.[26]

The creation of systems and subsystems of purposive–rational action in all social spheres was accompanied by a progressive disenchantment (*Entzauberung*) of the world, in which formal rationality asserted its superiority over other forms of meaning attached to the world, by the creation of the irrationality of the sphere of human valuation and value systems (and the accompanying 'irreconcilable conflict' between value systems), and by the loss of individual freedom in those spheres (bureaucratic, economic, legal, political) in which rational organisation predominated. In particular, the consequences for the individual lay in a loss of meaning and a loss of individual freedom. Where the domination of formal rationality produces a situation in which the world's processes merely 'happen' or 'are', then the response to this may be a search for a reassuring world view – a return to mythology and the irrational. If the domination of formal rationality is universalised and read as a totalising tendency, then the response to the 'iron cage' of rationalisation and lack of dynamic forces, especially in the growing bureaucratisation of organisations, may be to resort to charismatic leadership which might transcend this ossification. In this context, Weber's theory of modernity posits a transition from an inner dynamic of this new rationality to a situation in which the domination of formal rationality is characterised by external constraint and compulsion, as well as by ossification.

There is evidence that Max Weber was, in part, confronting the radical critique of totalisation provided by the work of Nietzsche. Indeed, the thesis of the disintegration of totalities takes a more radical turn in Nietzsche's critique of modernity. The shattering of all foundations, their dissolution into a continuous flow, society's disintegration into a conglomerate of components, the dissolution of cultural forms, a pervasive decadence, a present saturated with historicist illusions, the permanent duration of the eternal recurrence – all

these are elements of Nietzsche's critique of modernity. The f.
longer lives in totalities' [27] means that he takes seriously the mea
tiniest fragments, the smallest forms. The focus on the fragmentary, the explicit
acknowledgement of the end of totalities, the dissolution of foundations, are all
explicit elements of Nietzsche's critique of modernity. They are all dimensions
claimed by many theorists of postmodernity as constituents of the postmodern
condition.[28]

# III

Later critiques of modernity found more radical expression in the tradition of
critical theory. Its earliest sustained, though incomplete, instance is Walter
Benjamin's 'Arcades Project'.[29] Two of Benjamin's 'definitions' of modernity – 'The
world dominated by its phantasmagorias ... this is modernity' and 'The new in the
context of what has always been there' – indicate two possible affinities with
earlier delineations of modernity.[30] The first statement draws on Marx's notion of
'socially necessary illusion', an illusory world created by the commodity form and
its fetishisms. The second can also be related to the commodity form, to the
commodity's ever-new face, but it can also be fused with Nietzsche's doctrine of
the ever-same, such that the ever-new masks the ever-same.

What is different in Benjamin's prehistory of modernity is that his
exploration of Paris as capital of the nineteenth century – with its architecture, figures, representations and media – is intended as illumination of our
modernity today. Further, the phantasies, their representation, the sources of
newness, are explored through such dialectical images as those of modernity
and antiquity, the masses and the city, the new and the ever-same, in order to
build up a constellation of interrelated dimensions of modernity. There is a
conscious abandonment of linear progression such that antiquity is recognised
as embedded within modernity itself.

In the course of this ambitious construction of Paris in the mid
nineteenth century – commencing with the arcade and extending through
panoramas, department stores, railway stations and cafes – Benjamin seeks to
capture the constitutive transformations of experience and perception in metropolitan capitalist modernity, the representations of modernity, the significance
of means of representation (panoramas, mirrors, fashions), the architecture and
streets of the modern metropolis, the figures who exemplify key dimensions of
modernity (*flâneur*, gambler, prostitute, idler) and what Benjamin termed 'the
extinct world of things'. At the centre of the later versions of his arcades project
was to be the phenomenal life of the commodity as a thing.

Benjamin's approach to his prehistory of modernity is one that
commences with the fragments, the images, the ruins of modernity. The world

of metropolitan modernity is a not yet deciphered text, a text that can be a dream (requiring awakening), a picture puzzle (requiring a solution), or hiero-glyphics (requiring deciphering). Modernity is also explored as the techniques for the reproduction of the ever-new and the continuous shock of the new (hiding both the ever-same and the old). The commodity, on which Benjamin increasingly focuses, signifies both the 'phantasmagoria' of modernity (its allegorical effect) and 'the new in the context of what has always been there'.

The analysis of modernity as a critical confrontation with the mundane everyday world and the profane world of things, an analysis that also confronts contemporary developments in modern architecture, was developed and continued within the critical theory tradition by Theodor Adorno, Siegfried Kracauer and Ernst Bloch.[31] The dialectic of rationalisation and mythology was explored notably in Kracauer's 'The Mass Ornament' and in Bloch's *Heritage of Our Times*, an interpretation of non-contemporaneous experience, the hiero-glyphics of the nineteenth century, and the contradictory inheritance of the past in early Nazi Germany.[32] It was Kracauer in particular, trained as an architect, who in his many explorations of modernity in the Germany of the Weimar Republic developed another trajectory into the hieroglyphics of urban modernity. Such explorations were undertaken on the basis of guiding insights such as 'Spatial images are the dreams of society. Wherever the hieroglyphics of any spatial image is deciphered, there the basis of reality presents itself'; and 'knowledge of cities is bound up with the deciphering of their dreamlike expressive images'.[33]

The problematic relationship between rationality, mythology and modernity had already appeared at various points within the German philosoph-ical tradition, notably in Kant, Hegel and Nietzsche, and was given one of its most devastating treatments in Max Horkheimer and Theodor Adorno's *Dialectic of Enlightenment*.[34] In a discursive philosophy of history and the human subject, the authors engage in a critique of the 'totalitarian' nature of enlightenment reason, the entanglement of enlightenment and myth, the progressive alienation of the bourgeois human subject (often presented from a male-centred perspec-tive), and the illusory nature of scientific (positivist) progress.

Dimensions of this critique both draw on critiques of Hegel and Nietzsche, and in turn inform subsequent explorations of modernity by Adorno and, more recently, Jürgen Habermas. The identification of enlightenment and modernity draws on an interpretation of Kant's philosophy as the site of the formal dissociation of reason. The critiques of pure reason (cognitive, instru-mental) and scientific truth; practical reason (moral, practical) and normative rightness; and judgement (aesthetic, expressive) and authenticity and beauty, already announce the fragmentation of reason and possibility for the develop-ment of autonomous spheres of objectifying science, moral concepts and aesthetic judgement. Such a dissociation was radicalised later in the neo-Kantian theory of value spheres. Earlier however for Hegel, the existence of these autonomous spheres was seen not only as the symptom of the disintegration of the totality, but also as connected with the development of subjectivism and a crisis in the identity of the human subject. The critical defence of modernity as a project has been undertaken most notably by Habermas, who has highlighted the philosophical discourse on modernity (with regard to the present) and of modernity (with regard to philosophy's own task), as well as the historical development of modernity (largely in terms of systems and subsystems of purposive–rational action). Habermas's delineation of the role of rationalisation in modernity has strong Weberian overtones with its assumption of rationalisation as a universal process extending beyond the rational organisation of production, administration, technology and other systems of purposive–rational action that are themselves remote from the hermeneutics of everyday communication into the life world. The concomitant separation and autonomous specialisation of the scientific, moral and artistic spheres may well lack sufficient differentiation with regard to the development of these spheres. The colonisation of the life world by instrumental rationality also threatens the maintenance of the public bourgeois sphere, insofar as accumulators of capital also seek to accumulate and control the spheres of cultural signification.[35] In architectural discourse the nature of the separation of the public and private spheres in modernity, not least the gendered nature of this separation, has become a significant issue.[36] With regard to what Habermas terms aesthetic modernity, his focus is on the transformation of time consciousness. The four dimensions relate to newness, the future, the present and the past. That which is modern constitutes an expression of contemporaneity, a manifestation of what is new – albeit a newness that will be destroyed. Modernity also implies the transformation of time consciousness, especially within artistic avant-gardes as ventures into unknown realms with an orientation towards a not yet realised future. The overvaluation of the fleeting, contingent and ephemeral and the celebration of their dynamic also expresses the desire for an untarnished, coherent present. This secret desire for a harmonious present coincides with an abstract opposition to history, and thereby favours a present that no longer anchors itself in the past.

The connection between modernity and the enlightenment has been the subject of considerable debate. Lyotard has claimed that Habermas is creating another 'grand narrative' of history that has been rendered obsolete with the radical proliferation of language games and the undermining of all forms of foundational thought. This has become a central strand in the debate between modern and postmodern theorists.[37]

In recent decades discussions about modernity have often built upon and amplified themes present in earlier delineations of modernity. The fundamental ambiguity and contradictory nature of modernity as a project[38] has important implications for the possibility of a continued critique of modernity and self-identity.[39] It is also appropriate to examine the extent to which the reconstruction of cities in Europe from 1945 to the 1960s and beyond was driven by a 'project' of modernity understood as 'progress'. Against the totalising tendencies of some theories of modernity such as total rationalisation and one-dimensionality, the critique of such generalisations has been undertaken for the political sphere and its spaces, notably in Foucault's exploration of power.[40] Similarly, the universalistic inevitability of modernisation and modernity that has been presumed in versions of globalisation theory overlooks the differentiated nature of the paths to modernity, the diversity of experience of modernisation even within capitalist socioeconomic formations,[41] and the disjunction between experience of modernity in so-called 'north–south' social formations. Stated simply, experience of modernity is structured differently at the centre of an empire than in its colonies. A similar need for greater differentiation and attention to the variety of experiences of modernity should also take gender, ethnicity and social class into account, all of which require specific historical investigation.

The transition to modernity and the discontinuity between modern and traditional societies was explored by sociology at its inception. The concept of modernity as discontinuous experience of time, space and causality has been developed in new directions with respect to time–space distanciation. Every exploration of the crucial sites of modernity – the metropolis, the state, the industrial enterprise – relied on the transformation of time–space relations, and this transformation remains central within the context of globalisation. It involves the rethinking of territoriality and the notion of the temporally bounded society, beyond the reflections on time–space transcendence outlined by Marx and Simmel in relation to media such as money and the creation of a transspatial community.[42] Although the spatial dimensions of modernity – in contrast with the focus on the temporal origins and developments of modernity – were analysed by earlier theorists, notably by Simmel, Kracauer and Benjamin, their study has been significantly advanced by Henri Lefebvre's analysis of the production of space, modernity and everyday life.[43] His tripartite division of the production of space, representations of space, and spatial practices, has been influential in the subsequent analysis of the modern metropolis and its architecture. His theory of alienation and proliferation of abstract space also had a significant impact on an avant-garde movement that could conceive of architecture itself as a critique of modernity – situationism.[44]

The proliferation of discussions on modernity has been stimulated by the reassessment of theorists of modernity at the turn of the nineteenth century; by the discovery and reappropriation of key works in the critical theory tradition, notably those of Benjamin; by the defence of modernity as an unfinished project by Habermas; and the ensuing debate and by the proliferation of theories of postmodernity. Although these theories have produced no serious prehistory of postmodernity in Benjamin's sense, its history must be located within modernity itself. Indeed, it has been argued by Calinescu that postmodernity should be viewed as another new 'face' of modernity itself, an accentuation or acceleration of dimensions of existence already found within modernity.[45] Such a location for postmodernity is rejected by those who see it as a radical reaction to and break with modernity.

Insofar as there is a strong correlation between modernity and capitalism, the radical break thesis rests on fundamental qualitative shifts in the nature of capitalism in the last quarter of the twentieth century. However, the assertion of a shift to postcapitalism, postproduction or postindustrial socio-economic formations, and at all events away from production towards the spheres of circulation, exchange and consumption, runs counter to the globalisation-of-capitalism thesis – including the universalisation of the commodity form – that has been prominent in more recent discussions of modernity. Furthermore, since the discourses on modernity and postmodernity are grounded in specific concepts of the nature, development and transformation of capitalism, much rests on the delineation of its development – as mature capitalism (Marx), high capitalism (Benjamin) or late capitalism.[46] The end of modernity might then be associated with the demise of late capitalism. But if we assume that capitalism as a socioeconomic phenomenon is only at its beginning, then it might be appropriate to view modernity in the same light. The same rethinking may be necessary with respect to assumptions about postindustrial or postproduction societies. These often ethnocentric assumptions ignore the possibility that capitalism as a global system can function perfectly well where a major part of the production of commodities takes place outside the metropolitan core.

Even if the connection between capitalism and modernity remains problematical, delineations of modernity by earlier social theorists focused in many instances on the sphere of circulation, exchange and consumption; the problematical shifts from differentiation to dedifferentiation (above all in the cultural sphere); the problematisation of the relationship between signifier and referent; the shift from discourse to figuration (and the proliferation of images), the shift to self-referential systems (language games, but also the sphere of commodity circulation); the end of history and society (already announced in Nietzsche's critique of modernity); and the centrality of the aesthetic sphere – all of these being key features of postmodernity. The possibility of postmodernity as an absolutely 'new' condition remains susceptible to Benjamin's (and Nietzsche's) dialectic of the ever-new as the ever-same, or at least of the absolutely-new as illusory.

# IV

However schematic the presentation of the theorisations of modernity outlined above, their diverse foundations and methodological implications indicate that the study of modernity as a qualitative rather than as a chronological concept opens up a range of possibilities. The notion of a unilinear modernity process ignores the differentiated modes of modernisation with which it is associated, together with the impact of resistances to modernity. It also overlooks different ways of experiencing the modern between those on the core and those on the periphery of modernisation processes and between the imperial and the colonised, together with the related issues of gender, social class and ethnicity. The architectural reflection of modernity which Wagner sought to achieve is not as unproblematic as he and many others thought. The tendency to conceive of modernism as a representation of modernity and as a single homogeneous movement or project of the twentieth century ignores the diversity and plurality of multiple modernisms. At the same time, many postmodern critiques of modernism in architecture contrast undifferentiated concepts of modernism (such as the universal functional glass box) with a differentiated, vernacular postmodernism (itself of course subject to a multiplicity of postmodernisms), thus leaving the caricature of architectural modernism undifferentiated and unexplored.

Over a century ago Wagner raised another significant issue which remains relevant to any exploration of modernity today. He not only argued for a modern architecture that would be capable of reflecting modern life – metropolitan modernity, including its suburban villa developments that feature so largely in the architecture journals of the early twentieth century – but also for an architecture that would be intelligible, and by implication legible. Although Wagner raised this issue in the context of his critique of contemporary historicism, on the grounds that the latter did not speak to contemporary metropolitan dwellers, the problems of intelligibility and legibility applied directly to Wagner's own work. How to read the city, and how to make that reading intelligible to others as well as creating an architecture that is intelligible to urban dwellers, are issues that go beyond Wagner's own framing of them. We have ask for *which* urban dwellers this architecture is legible taking into account the gender, class and ethnic differentiation of its 'readers'; this is particularly important when considering the capital of an empire with complex class, gender and status hierarchies and a plurality of 'language games'.[47]

Experiencing the city through reading and using it (and not just walking though it, but also dwelling, working and playing in it) must be constituent elements both of building the city (*Städtebau*) and building in the city. Yet each of these practices in themselves can be rendered problematic, as can the relationship between them in modernity. In a number of contexts, Benjamin for example explored the transformation of experience in modernity from individual lived experience (*Erlebnis*) to concrete historical experience (*Erfahrung*), a development he took to be essential for any critical confrontation with the phantasmagorias of modernity. Although experience of the city comes from 'reading' it, Benjamin pointed out that reading the city does not exclude erroneous readings where the city's text 'is riddled with error'. Kracauer maintained that the point at which one learned to love a city coincided with the search for its defects. In contrast with contemporary reportage in the Weimar Republic and its orientation towards 'descriptions' of reality, Kracauer insisted that we should 'seek out the traces of its errors of construction'.[48] More radically, when Benjamin was investigating the traces of 'Paris, capital of the nineteenth century', he spoke of 'the coming to legibility' of the nineteenth century only in the twentieth.

In reading the city we must be sceptical of pre-existing mappings of its text – not merely in tourist guides but in official statements and reports on the city. Both Franz Hessel and Siegfried Kracauer were aware of the dangers of pre-framed mappings of the city and neglect of what Hessel, in his *flanerie* through Berlin in the late 1920s, referred to as 'the other Berlin', for 'whoever wishes to get to know a city cannot leave out the quarters of the poor'. Viewed merely from its facades, 'at first glance there is not much to be seen from the outside'.[49] Rather, as in Kracauer's exemplary sociospatial analysis of the employment agency in the same period in Berlin, it is 'in the back courts of society [that] human entrails are hung out like pieces of washing'.[50]

Though Hessel recognised some of the problems in reading the city, not least in his essay 'On the Difficult Art of Going Strolling',[51] he nonetheless saw strolling as 'a kind of reading of the street, in which human faces, shop windows, café terraces, automobiles and trees become a wealth of equally valid letters of the alphabet that together result in words, sentences and pages of an ever-new book'.[52] Transitory texts were also not excluded from this reading, such as

> the group of temporary buildings, the demolition scaffolding, new building hoardings, the wooden partitions ... Between the slats and visible through holes is a battlefield of stone.[53]

Knowledge of the destructive–constructive moments of the city require that we 'get to know thresholds'. The stroller should also attend to 'advertising's momentary architectures', of which Hessel observes that

> No newspaper reads so exciting as the illuminated wall text gliding along the roof above billboards. And the disappearance of this text,

that one cannot flick back over as in a book, is a conspicuous symbol of transitoriness.[54]

Its reader is exhorted 'to retain the timeless moment in his consciousness'.

Decades later it was Roland Barthes who expressed some scepticism in the ease of reading the city as a text. He argued that

The world is full of signs, but these signs do not all have the fine simplicity of the letters of the alphabet, of highway signs, or of military uniforms: they are infinitely more complex.[55]

Barthes elsewhere applauds Kevin Lynch's work on the image of the city, which identifies its 'basic rhythm of signification which is opposition, alternation and juxtaposition of marked and non-marked elements'.[56] But there is a limitation in such research:

The city is a discourse, and this discourse is actually a language: the city speaks to its inhabitants, we speak our city, the city where we are, simply by inhabiting it, by traversing it, by looking at it. Yet, the problem is to extract an expression like 'language of the city' from the purely metaphorical stage.[57]

However, the meanings generated in the city as discourse emerge out of our use of the city, its streets, its spaces, its architecture. Inhabiting, walking and viewing the city and its spaces are all practices whose meanings are related to their uses.

Wittgenstein's notion that the meaning of a term is the use to which it is put suggests a link between meaning, use and intelligibility.[58] Otto Wagner was deeply concerned with the uses to which the modern city, its spaces and its architecture were to be put. Against the historicist elaborations of facades and front elevations, Wagner juxtaposed the ground plan (*Grundriss*) to the elevation (*Aufriss*). The ground plan was the starting point in the creation of new spaces. The detailed attention to aspects of modern usage in his architectural projects – be it the city railway or the post office savings bank – was already announced at the very end of the 1880s by his striving for a style for usage (*Nutzstil*). In this respect at least, Wagner's somewhat uncritical reading of a modernity without contradictions (in contrast to the readings of modernity produced by many of his Viennese modernist contemporaries in other fields) did not prevent him from producing both useful and aesthetically innovative modern architecture.[59]

# Notes

1   Wagner (1988).
2   Wagner (1988:60); translation amended.
3   Ibid., p.75; translation amended.
4   Ibid., p.103.
5   On Wagner and modernity see Frisby (forthcoming). On the contradictions within the Austro-Hungarian Empire and the role of architecture see Moravánszky (1996). For a useful and different discussion of architecture and modernity see Heynen (1999).
6   On the development of the concept of modernity see Gumbrecht (1978:93–131).
7   The formal context for the debate on architectural modernity in Vienna was the Association of Austrian Engineers and Architects debate in 1898–9, initiated by Franz von Neumann's (1899) lecture 'Die Moderne in der Architektur und im Kunstgewerbe'. A more general controversy about city planning had already commenced around 1890 with the publication of Camillo Sitte's book on city planning in 1889 and that of Joseph Stübben in 1890. On the parameters and significance of this debate – which was fought out in the leading Berlin architecture journal *Deutsche Bauzeitung* – see my 'Straight or Crooked Streets: The contested rational spirit of the modern metropolis' in Frisby (2003:57–64)
8   See for example Bauman (1989) and Berman (1982).
9   Habermas (1979).
10  For a critical discussion see Osborne (1992), more fully outlined in Osborne (1995).
11  See Berman (1982:15).
12  See Habermas (1987).
13  Benjamin (1973); Benjamin's prehistory of modernity is available in Benjamin (1999).
14  See Herf (1984).
15  Baudelaire (1969:1–40).
16  On the broader conceptualisation of Baudelaire's definition see Frisby (1985) and Frisby (2001). On modernity and self-identity see Giddens (1992).
17  See Berman (1982:21 and Ch. 1), Frisby (1985:20–27), and Sayer (1992).
18  Cited in Frisby (1985:21).
19  For Marx the sphere of circulation of commodities is experienced as total alienation. Its analysis in his *Grundrisse* contains a wealth of insights into circulation and consumption.
20  See Tönnies (1955); Durkheim (1984).
21  For a discussion of Tönnies' distinction see Frisby (1992: Ch. 3). On Tönnies' relevance for the architecture of community see Welter (2002:138–9).
22  Durkheim does not make the connection with architecture, but his discussion of religion and collective representations opens up neglected possibilities; see Durkheim (1995).
23  For Simmel's 'definitions' see Simmel (1909). The two key Simmel texts exploring the sites of modernity are 'The Metropolis and Mental Life' (1903) in Frisby and Featherstone (eds) (1997:174–85); and Simmel (2003). For Simmel's impact on avant-garde movements see Leck (2000). On the metropolis in detail see Frisby (2001: Ch. 3). On Simmel's metropolis analysis and modern architecture see Cacciari (1993) and Vidler (2000). On the metropolis see also Borden (1997).
24  See Frisby (1992: Ch. 8).
25  On the 'delicate invisible threads' see Simmel, 'Sociology of the Senses' (Frisby and Featherstone 1992:109–20).
26  On Martin Wagner see Frisby (2001: Ch. 7). More generally, see Scaff (1995).
27  Nietzsche (1980:27). There are already many references to modernity in Nietzsche's writings in the 1870s and 80s.

28  For an exploration of Nietzsche on modernity, compared with Marx, see Love (1986).

29  See Benjamin (1999). For a critical account see Buck-Morss (1989). For Benjamin on the city see Gilloch (1996); more generally Gilloch (2002).

30  These 'definitions' are also discussed in Frisby (1985: Ch. 4).

31  Some of their contributions to the analysis of modern architecture are available in Leach (1997), including Adorno on functionalism, Kracauer on employment agencies and the hotel lobby, and Bloch on ornament.

32  See Kracauer, 'The Mass Ornament', in Kracauer (1995:75–88); Bloch (1990).

33  Cited in Frisby, 'Deciphering the Hieroglyphics of Weimar Berlin', in Haxthausen and Suhr (eds) (1990:152–65).

34  Horkheimer and Adorno (1973).

35  For Habermas's account see Habermas (1989).

36  See Colomina (1994); on gender and space see Durning and Wrigley (2000).

37  Lyotard (1984); Bernstein (1985); Kelly (1996).

38  Bauman (1991); Bauman (2000); Cascardi (1992).

39  Lash and Friedman (1992).

40  M. Foucault, 'What is Enlightenment?', in Rabinow (1984:35–50).

41  Wagner (1993).

42  On time–space distanciation see Giddens (1990); Friedland and Boden (1994). On money in this context see Harvey (1985: Ch. 1).

43  See Lefebvre (1984); Lefebvre (1991); Lefebvre (1995).

44  See Heynen (1999:148–74). Lefebvre's work has also had a major impact on the work of Harvey and Soja, amongst many others.

45  Calinescu (1987).

46  See Harvey (1989); Jameson (2002); Jameson (1991). To take but one example, with few exceptions such as Hubert Gessner, Wagner's students did not design working class housing in the period of the Wagner School (1894–1911), yet after 1918 the majority of the social housing blocks of Red Vienna were designed by Wagner's students. See Blau (1999).

47  More generally on reading the city see Donald (1992); Donald (1999); Fritsche (1996); Frisby (2002).

48  Cited in Frisby (1985:150).

49  Hessel (1999:274).

50  S. Kracauer, 'On Employment Agencies' (Leach 1997:64).

51  Hessel (1999).

52  Hessel (1984:145).

53  Hessel (1999:70).

54  Hessel (1999:69).

55  Barthes (1994:158).

56  Barthes (1994:159); Lynch (1960).

57  Lynch (1960).

58  For a brief discussion see Frisby (2001: Conclusion).

59  On Wagner's architecture see Mallgrave (1993).

**Chapter 2**

# What modernism was
## Art, progress and the avant-garde

*Arnfinn Bø-Rygg*

'... the memory of the present ...'

Charles Baudelaire

'To be modern is to know what is no longer possible'

Roland Barthes

A justified scepticism has spread in aesthetic discourse regarding concepts of universality and historical evolution, linear chronology and homogenous periodisation. This scepticism is directed in equal measure against artists' own self-understanding and against historical description relying on such vocabulary. This is particularly true in the case of the modern avant-garde, insofar as it contained, in its very concept, ideas of the superiority of the new and the notion that it is possible today to determine which art is 'ahead', thus to help map out the future. But does this partly justified criticism mean that history – what Hegel called 'the fury of disappearance' – has outplayed its role?

History is itself a modern product. History, modernity and art are contemporary and intertwined concepts, and we cannot have one without the others. The very history that made art possible also generated modernism in art, in which the past itself became a problem. When today – from an allegedly postmodern vantage point – we historicise modernity or declare ourselves to have reached a postmodern state, this itself is a modern impulse. The cunning of history is possibly even greater than the cunning of reason.

## Modernism as historical construction

In general terms 'modernism' refers to an international tendency that came to expression in western literature, theatre, music, visual arts and architecture in the latter half of the nineteenth century, and continued to dominate twentieth century art. Its exact timing is debatable: should it be located around 1850 with the novel understanding of 'the modern' as modernity (Baudelaire), or towards the end of the nineteenth century with tendencies like literary symbolism and impressionism in painting and music? I will argue here for the former, seeing the concepts of modernity and the new as keys to understanding the radical transformation in the role and status of art under high capitalism. Despite the fact that we are dealing with several different movements, certain stylistic features characterising modernism as a whole may be identified: the dissolving of classical notions of space and the object, the deconstruction of tonality in music, and fragmentation and stream-of-consciousness writing in the novel. An emphasis on the formal aesthetic is often also seen as a distinct feature.

The concept of modernism is by no means without problems. It is highly complex, fraught with paradoxes and inherent contradictions. One such difficulty stems from the fact that the concept of modernism is used, and must be used, in both a descriptive and emphatic-normative manner. Furthermore, the concept is insinuated into a wider conceptual landscape by its close association with concepts such as the modern and modernity, far transcending the aesthetic domain. I will try to clarify some of these difficulties by tracing the history of these related concepts, and then moving on to a discussion of the characteristics and intentions of modernism.

All concepts in the field of aesthetics are historical constructions, and as such more or less adequate. Yet such concepts are necessary: in a world of total nominalism, individual facts would remain silent. The real question is whether such concepts are fruitful. Throughout the history of art there are many examples of unfruitful concepts, like 'baroque' in music or 'postimpressionism' in art history. At times, concepts that are difficult to define are fruitful, because they are flexible and can be given new content as the knowledge and understanding of a style or epoch is expanded and refined. The concept of modernism is valuable not least because it was itself an integral part – a hermeneutic constituent – of the modern artwork. The concept of modernism was itself a part of making history, not simply a *post facto* creation.

The paradox of 'the modern' and other concepts deriving from it is that they are relatively old. *Modernité* was first used in Chateaubriand's *Memoire d'Outre Tombe* from 1848, hinting at a new world and heightened self-consciousness. Baudelaire, often seen as the precursor of modernism, made the term a key concept in his program for a new aesthetics. Yet the concept of *modernité* has a history. In the late fifth century the term 'modern' was used to differentiate the Christian age from a classical–pagan past. Until the enlightenment, according to H. R. Jauss, the concept was used every time one consciously invoked antiquity as a means to understand oneself as the product of a transition from the old to the new.[1] With renaissance humanism the term modern was used in visual art, music and poetry. A dispute between 'ancients' and 'moderns' took place in virtually every generation, climaxing in the famous literary feud of late seventeenth century France, *la querelle des anciens et des modernes*. This dispute resulted in the final dispelling of the renaissance idea of ideal antiquity in favour of the belief that art should express the specific *Zeitgeist* of its epoch. Thus, the philosopher Hans Blumenberg sees *la querelle* as a key source for the modern notion of progress.[2] With the enlightenment idea of humanity's unbroken progress towards rational and moral advancement, the concept of modernity lost its connection with antiquity. For the romantics modernity was intimately linked with the new philosophy of history, and the notions of a 'modern world' (Schelling) and a 'modern consciousness' can be traced back to the historical consciousness of the romantic movement. Hegel called historical experience as well as the category of experience grounding it 'the modern principle', and declared its realisation to be 'in the interest of modern philosophy'.[3] A radically new development emerged around 1850 with the extraordinary emphasis placed on the newly constructed noun *modernité*. The timing of this construction is significant, for it was at this time that society began construing itself as modern, and thus defining itself.

From the renaissance onwards, technological and scientific progress as well as the rational division of labour resulting from the industrial revolution gradually altered social life and eroded traditional culture. At the same time the division of labour caused deep political rifts, manifesting themselves as social and political struggles. Added to all this came a dramatic increase in population, an unprecedented urban expansion, and a rapid development of communication and infrastructure. These transformations resulted in a society and a way of life bent on change and innovation, but also in instability, continual movement, and crisis. A new feeling for time is characteristic of the period, with history itself often seen as a universal and continuous crisis. A curious notion of being part of a critical epoch prevailed, sometimes even taking the form of an eschatological experience of witnessing the last days. History, it seemed, had collapsed into a 'before' and 'after'. Stendhal saw in the revolution of 1789 just such a moment: a decisive incision in time, making subsequent history stand wilfully opposed to all earlier history. According to Stendhal's bourgeois self-consciousness, modern society was not separated from *l'ancien regime* only by a new constitution, but also by a new morality, a new way of life, new ideas and – perhaps most importantly – by a new relationship with beauty. Stendhal gives his definition of modern anti-platonic beauty thus: 'que le beau n'est que la promesse de bonheur' – beauty is nothing but the promise of happiness.

# Baudelaire and modernity

In *Le peintre de la vie moderne* Baudelaire – the poet of modernity *par excellence* and a man with a profoundly ambivalent relationship with bourgeois self-consciousness – criticised Stendhal's definition of modern beauty because it depended on the ever-changing ideal of happiness. Baudelaire's correction of Stendhal, as well as his own definition of modern beauty, displays the ambiguity that pervades modernity in art. Modernity in art is constituted, according to Baudelaire, on the one hand by a moment of relativism – to do with an epoch's fashion, passion and morals – and on the other hand by something eternal and unvarying. Modern beauty has a dual character, it is 'the ephemeral, the fugitive, the contingent, the half of art whose other half is the eternal and the immutable'.[4] To despise the 'transitory, fugitive element', we are warned, inevitably leads to the return of an empty and abstract concept of beauty. Baudelaire emphasised this notion of the contemporary and the present so strongly that he even subsumed memory under it: 'Woe to him who studies the antique for anything else but pure art, logic and general method! By steeping himself too thoroughly in it, he will lose all memory of the present; he will renounce the rights and privileges offered by circumstance – for almost all our originality comes from the seal which Time imprints on our sensations'.[5]

The contemporary as experienced present becomes fundamental to aesthetic experience to such an extent that 'The pleasure which we derive from the representation of the present is due not only to the beauty with which it can be invested, but also to its essential quality of being present'.[6] Yet at the same time Baudelaire also stressed the other aspect of modern beauty: the eternal, understood as the immutable presence (antiquity) to be extracted by the artist from a fleeting modern life. The task of the artist, according to Baudelaire, is that of 'distilling from it [the age] the mysterious element of beauty that it may contain, however slight or minimal that element may be'.[7] The eternal is not privileged over and above the ephemeral and fugitive, but rather extracted from it; a product of what Baudelaire called an *idéalisation forcée*.[8] What we are dealing with here is not the traditional notion that poetry captures the moment, but rather that modernity itself becomes 'antique': 'In short, for any "modernity" to be worthy of one day taking its place as "antiquity", it is necessary for the mysterious beauty which human life accidentally puts into it to be distilled from it'.[9] Jauss stresses that Baudelaire's use of the term 'antique' here does not refer to antiquity as an epoch but rather to its role as a functional opposite to modernity: the fact that modernity itself is made 'antique' by extracting its fleeting spirit.

In Baudelaire modernity loses practically all of its descriptive aspect. It cannot be compared to the past because the past, for Baudelaire, appears as a series of unique events with no inner connection. To be modern, for Baudelaire, means to step out of the continuity of history. Yet precisely by being lifted out of history the present may begin to correspond to a remote past. Baudelaire sees the creative imagination – *l'imagination* – as the artist's particular power to create such correspondences.[10]

The emphatic concept of modernity is in Baudelaire linked to the normative concept of the new. We may view the idea of the new as axiomatic in an art increasingly construing itself as modern. The new, as Baudelaire sees it, is that which cannot be determined. Its position in modernism becomes consequently quite different from the demand for novelty in earlier periods. When, for instance, the medieval troubadour singers claimed to sing 'a new song', the novelty consisted of variations within a determined genre.[11] For Baudelaire, it is the undetermined and undeterminable that constitutes the new. This 'indeterminability' – Hugo Friedrich called it 'the indeterminate other' or 'the empty ideal'[12] – is invoked in Baudelaire's *Les fleurs du mal*. The last lines in the cycle *La mort* read: 'Plonger au fond du gouffre, Enfer ou Ciel, qu'importe? / Au fond de l'Inconnu pour trouver du *nouveau!*'[13] Adorno aptly characterises this equation between the new and the unknown as 'the cryptogram of modernity'.[14]

# Fashion and the new

The concepts of modernity and the new, as they were used by Baudelaire and as they came to constitute modernism, are complex and even contradictory. Baudelaire is as remote from the hermeneutic concept of tradition as he is from historicism. His notion of the modern encompassed both the contemporary, that for which it 'is time', and that which breaks with the past. But where Baudelaire overemphasised break and ignored continuity, the new lost its novelty. The genuinely new must contain tradition. This becomes clear if we compare the novelty of the new with that of fashion. While the novelty of genuine innovation constitutes something aesthetically striking – the aging Hugo said of Baudelaire's poems that they had created a 'new shudder' – the simulated novelty of fashion, necessary for it to differentiate itself from previous fashion, is forced to construe itself as convention.[15] The novelty of fashion is abstract; a discontinuous shift to something ever different. 'Even if fashion copies something from the day before yesterday, it has no tradition, whereas, in what is truly and substantially new, tradition is always contained and transcended, even if in the form of explicit negation'.[16] Schönberg's atonality may, for instance, be seen as a 'determinate negation' of tonality.

The simulated novelty of fashion has threatened modernism ever since Baudelaire; threatened, that is, to put an end to it. For with fashion's change for change's sake any substantial value and any real break are lost. Change as formal game cancels the very idea of progress inherent in the concept of modernism. There are good reasons to see artistic tendencies from the last decades, particularly within painting, in this light: the aesthetically modern converges with fashion. Here change has become something cyclical, where forms from the past as well as from various subcultures may appear. Emptied of substance, they function as signs that do not differentiate between new and old: a 'timeless fashion' (Adorno). This relationship with history and tradition is a fundamental feature of postmodernity.

The emphatically new in modernism often corresponds to the 'antique' in Baudelaire's sense, to the archaic and primitive. This is noticeable not only in literature but also in painting such as cubism and music such as Stravinsky's. It relates back to the notion of the 'aesthetically striking': the moment of shudder associated with aesthetic experience since the beginning of modernity. As Dahlhaus has pointed out, the genuinely remote, just like the genuinely new, has the capacity to unsettle, not simply to be observed. The remote represents a new beginning, not simply continuity.[17] This may at least partly explain Baudelaire's wish to step out of the continuity of tradition. When the new no longer has the genuinely old as its enemy but rather the common-place, the *juste-milieu* art, it suddenly becomes possible to differentiate the new from the merely contemporary.

A radical concept of the new also becomes entangled in other contra-dictions. A demand for radical change and upheaval, hinted at by Baudelaire and made into a programme by later avant-garde movements, is hard to combine with a demand for general acceptance. When it does get accepted the new easily turns into tradition, yet as tradition its emphatic and normative aspect dissolves. Modernism thus turns into what Harold Rosenberg called 'the tradition of the new'. The emphatically modern must represent itself as that which Octavio Paz calls 'tradition against itself'. Paz poses the question, 'Even if we had to concede that negation of tradition ... may itself constitute tradi-tion; how could it be tradition without negating itself, i.e. without ascribing continuity to a certain moment. The tradition of disruption implies not only a negation of tradition, but a negation of disruption itself.'[18] Modernity, for Paz, is therefore never really itself, always another. Stated differently, the modern, in order to be emphatically modern, must continuously die. It must be in a state of continuous crisis, in which the crisis itself becomes a value. This is a dilemma which essentially cannot be solved, yet which in practice results in the rich experimentation so characteristic of modernism. As such, it is not simply a *contradictio in terminis*.

# The aporia of the avant-garde

As an art-historical category the modern has something paradoxical about it. The genuinely new is a unique moment, and as such hardly appropriate as a characterisation of an epoch. Moreover, the new can only be identified once it has already had its effect. The related concept of the avant-garde is entangled in a similar aporia. As Enzensberger points out, one can never determine, outside the avant-garde itself, what lies ahead. 'One can only say with certainty what *was* ahead, not what *is* ahead'.[19] Yet I would be the last to reject modernism's project and the idea of the new on the grounds of such philosophical and art-historical paradoxes. When the idea of the new is pushed so far in Baudelaire, it may be because, as pointed out by Adorno in *Minima Moralia*, 'dass es nicht Neues mehr gebe'.[20] With Adorno we may view the aporias in the modern and the new as necessary aporias in modern art itself, which, compared with the harmonious works of traditional art, is fundamentally imperfect. This necessary aporia consists in the fact that modern art vouches to say what has not been said before, yet it cannot say this as long as its material is intrinsically linked to the past. What art must but cannot say; that is Baudelaire's unknown. Adorno pinpoints the quest of the modern in this way: 'Dinge machen, von denen wir nicht wissen was sie sind.'[21] Or to quote Beckett: 'Dire cela, sans savoir quoi'. According to this aesthetic conception, the task of art is constantly to reveal this aporia, constantly, as Adorno said in *Ästhetische Theorie*, to attempt 'the Münchhausen act, to identify the nonidentical.'[22] Moreover, and this is perhaps the key point, in order to do this art must be in command of its 'formal level or 'material level' as Adorno would say. Rimbaud's imperative of being absolutely modern is transformed into a specific charge regarding the aesthetic material.

# The two concepts of modernity

Despite hinting at a break with history and tradition, Baudelaire maintained an exemplary continuity in his own poetry, using tradition as the means by which new experiences of the Parisian metropolis could be articulated. While he ridiculed contemporary faith in the advancement of reason and technology, he did not for a moment doubt the aesthetic progress of art. In fact, the project of modernism is intrinsically linked to the belief in change, progress, hope and utopia. In *Faces of Modernity* Matei Calinescu points out how the nineteenth century developed two concepts of the modern. One was the bourgeois idea of modernity as a stage in the evolution of western civilisation. Related to this was the belief in progress, faith in the possibilities of science and technology, the cult of reason, the quantification of time into a calculable entity equivalent to money, and a general emphasis on pragmatism, action and success. Against this, however, stood the idea of a cultural or aesthetic modernity with strong anti-bourgeois undercurrents. The extreme emphasis on the autonomy of art expressed by the slogan 'l'art pour l'art' belonged to the latter, and was intimately linked to another slogan: 'épater le bourgeois' – make the bourgeois speechless.[23]

Even if the aesthetic concept of modernity stood as a direct antithesis to the bourgeois–rational, it could not avoid relying, at least in part, on its bourgeois counterpart. This was particularly true for the bourgeois concept of progress. The stronger the idea that history is static and that any social progress is an illusion, the stronger is the emphasis on aesthetic innovation and advancement. Rimbaud, who launched the call for absolute modernity, repeated Baudelaire's demand that the poet should aim for the new and the unknown, saying explicitly that in this way poetry will remain ahead (*en avant*).[24] Implicit here is the concept of an avant-garde, an artistic forefront constantly venturing into the unknown. True, Baudelaire rejected the term 'avant-garde', seeing it as yet another example of the French penchant for military terms. However, the expression had not yet, as it would later, come to signify artistic innovation and experimentation.

## Gombrich on the core of modernism

The aesthetic concepts of modernity and the avant-garde were conceived within a general notion of progress, described by Hegel as the very logic of history and conspicuously displayed by the new technical and scientific innovations. Ernst Gombrich sees this notion of progress as the very core of modernism.[25] With increasing aesthetic autonomy and the advent of an anonymous art market, traditional norms and frameworks governing artistic activity began to dissolve. Gombrich believes that the notion of progress established a new framework for the work of the artist, serving as an inner directive that structured the situation and provided the criteria for evaluation. Gombrich draws on a famous quote from Richard Wagner's *Das Kunstwerk der Zukunft*, illustrating the artist's frame of mind:

> After Haydn and Mozart, a Beethoven not only could, but *must* come; the genie of Music claimed him of Necessity, and without a moment's lingering – he was there. Who now will be to Beethoven what *he* was to Mozart and Haydn, in the realm of absolute music?[26]

Wagner's profound influence on artists as well as art critics is well known. One such, who incidentally defended him during the many scandals of the early 1860s, was Baudelaire himself, who saw Wagner as an ally. It may be argued that Wagner's 'Tristan und Isolde' was for the new music what *Les fleurs du mal*, published in the same year, was for literary modernism. This may even be true for the works themselves, where similarities have been pointed out regarding for instance the principle of ambiguity. Because of the avant-garde character of his work Wagner was compared, not without reason, to the realists. The conservative critic F.-J. Fétis even called him 'the Courbet of music'.

## Criteria for the avant-garde

Gombrich situates the notion of the avant-garde within the discussion of art and progress. I will not here trace the concept of the avant-garde itself, which has its own history, but it is worth noting that the concept had a double significance in the nineteenth century, referring both to politics and aesthetics.[27] Gradually, however, the concept went from being a general term for artistic frontiers to signify specific historical avant-garde movements, all characterised by an insistence on aesthetic autonomy. As already mentioned, we can find early traces of the avant-garde concept in Baudelaire, with his notion of 'finding the new'. Calinescu, however, argues that we cannot really speak of an avant-garde consciousness before Rimbaud in the 1870s, when the political and aesthetic aspects of the term coalesced. Linda Nochlin similarly argues that we cannot strictly call Courbet's painting avant-garde.[28] Following Renato Poggioli,[29] she identifies psychological and social alientation as the key criteria for the avant-garde and argues that the avant-garde in painting only truly began with Manet. Though she believes that Baudelaire and Flaubert were alienated, she argues that Courbet 'was never an alienated man, that is, in conflict with himself internally or distanced from his true social situation externally.'[30] Flaubert's *Madame Bovary* of 1857 was the clear precursor to the modern novel, and if the aim is to identify the exact moment of the avant-garde's breakthrough we might as well go for the remarkable year of its publication. However, I will refrain from constructing a history based purely on dates and return to the key point: that Baudelaire, Flaubert and Courbet – even Wagner – must be included in the twin concepts of the modern and the avant-garde, insofar as the work of all these artists reflects the idea of aesthetic autonomy. The concepts are indispensable insofar as they were themselves hermeneutic constituents of the artworks; the concepts themselves made history.

# Modernism, realism and the avant-garde

We find support for this view in an essay by Charles Rosen and Henri Zerner entitled 'What is, and is not, realism?'[31] They start with the many attempts made from the beginning of the 1980s onwards to revise the idea of a split between avant-garde and official art in the latter half of the nineteenth century. Often accompanying such attempts is a desire to show that the so-called avant-garde is incoherent and inconsistent, and consequently that the modern has no continuity. The strategy of these attempts – sharply criticised by Rosen and Zerner – is to trace antimodern aspects of nineteenth century art and to abandon a distinction between avant-garde realism and other forms of representations of modern life, such as the academy realists. Rosen and Zerner's assertion is that impressionism must be seen as an expansion and radicalisation of realism and that there is no definitive break between the two, as is often assumed when construing impressionism as the first modern style. In a wider perspective, they argue, realism and *l'art pour l'art* are not polar opposites but rather two sides of a coin minted by the avant-garde. Through analyses of Courbet and Flaubert, Rosen and Zerner reach the conclusion that despite moral and political content in their work, a shift of interest takes place away from content, morals and meaning to forms and means of representation. Opposed to categories like 'the sentimental', 'the picturesque' and 'the anecdotal' (attributed to the petit realists whose aim lies beyond the work itself), the avant-garde breaks down any reference to topic, establishing in the process a new relationship between motif and execution.

It is the absence of topic that guarantees the truth of the content of the picture and gives the depiction its objectivity. To be sure it was the realists' achievement to accept the trivial and banal and resist the temptation to idealise it or make it picturesque. Yet the point is that the more emphasis that is placed on the banal and everyday as aesthetic material, the more important style becomes as an all-encompassing value. Avant-garde realism places the emphasis firmly on the means of representation:

> The rhythm of the prose or the patterns of brush strokes are always obtrusively in evidence. We are always acutely conscious of the surface of the picture, the texture of the prose. Neither novel nor picture effaces itself modestly before the scene represented. A work of avant-garde Realism proclaims itself first as a solid, material art object, and only this allows us access to the contemporary world it portrays.[32]

Finding support in Flaubert's own writing (his letters to Louise Colet), Rosen and Zerner see his determinism as that which allows facts to speak for themselves: the inevitable becomes an aesthetic quality. It is not a substitute for beauty, but appears beautiful in itself. 'This is the key to a style that is both abstractly beautiful and that can exist only as a perfect representation of events which are absolutely not beautiful: the style represents the inevitability of events, and is beautiful only insofar as it succeeds in making what happens seem inexorable.'[33] In Courbet's mature work, likewise, it is made clear to us that we stand in front of a work of art, a representation. Referring to the 'Funeral in Ornans', Rosen and Zerner write:

> The insistence is entirely on representation, on painting as a transcription of the experience of things. Not that the things represented are of no interest or importance but they preserve what might be called the ordinary indifference of their being. The burial that Courbet represents is strongly individualized and characterized because only the particular event has real existence; but it is not a special burial. To make it beautiful would make it special: it is the picture which is beautiful, not the burial, and the picture in no way embellishes the burial.[34]

The method of Courbet and Flaubert was not romantic alienation like that of Novalis, 'to make the familiar unknown and the unknown familiar'. Flaubert let the familiar be familiar and did not care for picturesque and exotic effect. 'What place was left then for art?' ask Rosen and Zerner, and answer:

> Only the technique and the virtuosity of the means of representation. If contemporary life was to be represented with all its banality, ugliness, and mediocrity undistorted, unromanticized, then the aesthetic interest had to be shifted from the objects represented to the means of representation. This is the justification of the indissoluble tie of mid-nineteenth century Realism to art-for-art's-sake; and although it is sometimes seen as an odd contradiction in Realism, it is, in fact, the condition of its existence.[35]

From such a vantage point Rosen and Zerner can see the realist movement from Courbet to impressionism as the first step towards abstract art, and while the authors do not recognise it explicitly, the logic of this reading is warranted by the increased mastery over the artistic means granted by aesthetic autonomy.

# Autonomy and institution

The exposure of this indissoluble tie between realism and *l'art pour l'art* is, I believe, of great importance. The principle of *l'art pour l'art* implied not only that art no longer needed a religious, political or practical legitimacy, but also that the idea of beauty itself had become superfluous. Realism shared this conviction with *l'art pour l'art*, and the pivotal point for both was the autonomy of art. For realism too it was this autonomy that guaranteed the truth content in what is being represented. Furthermore, it was the principle of autonomy that made evolutionary logic and the idea of progress in art possible. As long as art is dependent on the demands of a patron or the implicit expectations of a benefi-ciary, progress, understood as immanent development of the artistic material, is limited by the artwork's reception. It is only with the advent of the aesthetically modern, when art frees itself from its reception, that the idea of evolution becomes feasible. It is also only from this point onwards that art can make claim to insight; make claim to be not only beautiful, but also to be true. The tradi-tional harmonious artwork – Benjamin called it 'auratic' – did not recognise such insight but allowed cognition to withdraw within it. A sign of this increasing demand for insight is the new art's integration of objects from outside the aesthetic realm. Baudelaire thus incorporated into his poetry not only the non-natural metropolis with its crowds and its asphalt, but hideousness as such. Benjamin even argued that Baudelaire's 'empathy into the exchange value' and the inorganic was one of his sources of inspiration.[36] We may add that for art to integrate this alien material, a great effort on behalf of form is required; hence the modernist preoccupation with formal problems.

The demand for insight or cognition is a paradigm for emphatically modern art, as expressed by Schönberg's motto: 'Art shall not adorn but be true'. For Schönberg this demand concerned inner, subjective truths, but it can also be applied with respect to perceptions of the physical or social world, or concern the nature of the aesthetic material, artistic means, down to the 'flatness', as Clement Greenberg puts it, and to the total reduction of abstract painting. Based on notions of an alienated society and a completely administered world, some like Adorno would argue that if art is to be authentic it cannot free itself from dissonance, fragmentation, negativity. The guideline of modernism is F.H. Bradley's 'Where everything is bad it must be good to know the worst',[37] for only in this way can art uncover the monstrosity of reality and negatively reveal utopia. Adorno would add that because communication is conditioned by an alienated society, emphatically modern art – the art that wants to be ahead – must break with communication and make itself incomprehensible. Further-more, he argues that the social *Gehalt* of great artworks from the very beginning of the aesthetically modern lies in its protest against social reception. However, we need not accept these assertions to conclude that heightened reflectivity – so prone to irritate in modern art – is integral to the new autonomy and the corresponding demand for insight.

Of course art did not become autonomous overnight, and autonomy is in no way a trait exclusive to the aesthetically modern. A Mozart symphony, a Brahms quartet, Goethe's *Werther* and a poem by Mallarmé are all autonomous artworks. The autonomy of art is linked to the sectorisation of different areas of life following in the wake of capitalist modes of production, a process which resulted in religion, politics, economy and culture becoming distinct realms with relative independence. When art is severed from other areas of society, an institution develops which establishes the framework for the production and reception of art.[38] A precondition for the emergence of the art institution is the market, defining art's character as a consumer good. It is only when art has turned into a consumer good that it can be understood as art and nothing else, and only then can it be discussed and evaluated. With the emergence of the art institution, a number of 'arts' were lumped together that had never previously been conceived collectively: painting, music and literature. This newly unified concept of 'art' lost its immediate connection with society at large. Salons, exhibitions and concerts are social occasions constructed for the public to experience and appraise what is art and only art. Aesthetics emerged as a distinct philosophical discipline whose role it was to determine the nature of art. On the social scene the critic appeared as a new figure, charged with judging art according to the rules of good taste. Philosophical aesthetics was called on to give the new autonomous art its theoretical basis. In the wake of Kant's *Kritik der Urteilskraft* (1890), firmly establishing the autonomy and universality of the judgments of taste, Schiller developed the idea that art, because it stands apart from the life world, is the very place where the lost unity can be restored.

The principle of autonomy constitutes the very core of the institution of art. This principle warrants both the critical function of art and the fact that art remains without consequences. From the point of view of production the principle of autonomy frees art from a normative code, setting itself so to speak in the place of aesthetic rules.[39] It is only when art imposes its own rules on itself that an autonomous development of the means of art can begin, and the idea of progress as inherent evolution can take shape. From the point of view of reception a gradual differentiation of aesthetic experience takes place, coming to full expression in the purely aesthetic approach of aestheticism, for instance in Mallarmé. It is important to note that the principle of autonomy is a *sine qua non* for the later ideal of progress or advancement in art. This does not, of course, prevent a more general belief in progress – transferred from natural science – from playing its part, as Gombrich believed.[40] The point is, however, that the principle of autonomy is what allows for ideas of immanent progress in art; progress, that is, understood as a progressing consequence. To be sure, art's actual development is not independent from the general notion of history held by the artist. For instance, an attitude to history maintaining that progress does *not* take place, that history does *not* have a goal, that all events are unique events without inner connection and consequence, will inevitably affect the resulting works, as it has done in much postmodern art. However, the modern principle of autonomy goes beyond such notions, warranting, independent of our conception of history, that progress may (again) take place. Interestingly enough it was during the enlightenment, when the institutional autonomy of art first emerged, that the idea of progress without limits and without telos emerges. Condorcet, for instance, no longer sees perfection as the realisation of a *telos* intrinsic in the object, but rather as a limitless process. The progress of the human spirit is not governed by an immanent *telos*; it is free, yet reliant on external circumstances.[41] When this notion of progress is gradually put to effect in art, the classical conception of perfection and progress breaks down.[42]

The autonomy of art is understood as a development of a tendency that does not follow a straight line or progress without questioning. Art's autonomous status may be questioned, for instance, when other social interests so demand. Distinguishing between the institutional autonomy of art and the individual artwork's potential realisation of autonomy, Peter Bürger, in his book on the historical avant-garde movements, has outlined the evolutionary aspect of aesthetic autonomy.[43] Particular works of art may well have a moral or political content; in fact, this is the norm in the early stages of autonomous art. Here art is a medium for bourgeois subjectivity – a fact brought out not least by the phenomenon of art criticism as a necessary correlate to art itself. Only in its reflection on art does bourgeois subjectivity reach a full understanding of itself.[44] In bourgeois public life content is primary, opposed to the old formal criteria for beauty. This new emphasis comes to expression in the early novel's emphasis on morality and sentimentality, or eighteenth century sensationalism in music and painting. Art, in bourgeois public life, exists in a tension between institutional autonomy and the individual work's content; an unstable tension which is gradually effaced.[45] Even if aspects of institutional autonomy may be traced as far back as the eighteenth century, the individual work holds content which resists such autonomy. It is only from the mid-nineteenth century, with the advent of the aesthetically modern, that the tension between institutional framework and the particular work's content tend to fade. Bürger's fruitful distinction makes it possible to distinguish between a formal, institutional autonomy in the mid-eighteenth century and the autonomy of the individual work by the mid-nineteenth. Since *l'art pour l'art*, the dialectic between form and content in art shifts in favour of form. In terms of production this shift manifests itself as an increasing emphasis on means, in terms of reception, through a refinement of the aesthetic observer. Bürger uses the term 'aestheticism' to mean artworks that have taken as their content the tension between their own autonomy and social praxis, a situation in which institutional framework and individual content merge.[46] With aestheticism, art becomes its own content, and self-reflection its dominant feature.

## Cultural rationalisation

The emergence of the art institution in the eighteenth century is not without its own history. In general terms we may say that it has to do with what Max Weber called cultural rationalisation: a part of an overall process of rationalisation in the western world, influencing all areas of human practice. When the 'substantial reason' (Weber) – previously manifested in a single religious–metaphysical image of the world – in the modern age divides into domains of knowledge, moral and taste, a differentiation of value takes place in which science, morals, and art become distinct spheres defined and handled by professionals.[47] The professionalisation of these spheres makes their specificity apparent: 'Art now constitutes itself as a cosmos, in which its autonomous value is more and more consciously recognised'.[48] This rationalisation occurs first and foremost on a technical level. Weber considered the development of polyphonic music based on the organisation of tonal systems, as well as the use of linear perspective in painting, in this light. It is interesting to note that the renaissance artist had already developed a strong notion of progressing, a notion which shifted the emphasis from the commissioning patron to the work itself. As Gombrich aptly says: 'The artist who believes in the progress of the arts is automatically excluded from the social context of buying and selling. His duty lies less towards the buyer than towards art itself.'[49]

The process of rationalisation, which includes specialisation and the increasing division of labour, causes art to follow much the same path as science.[50] In both spheres it is urgent to be *au fait* with the 'state of the art' at any given point, and to drive it further. As far as modernism is concerned, this is a specific state and understood as an international standard. Anything deviating from this mainstream becomes peripheral and provincial. With the new emphasis on the means of art, artistic production can be seen as a form of problem-solving and thus as a form of cognition. This is what is properly established with the aesthetically modern and what makes it possible to regard modernism as a gradually increasing mastery of the aesthetic material, understood first and foremost as mounting formalisation, construction and abstraction. Adorno looks at the development of music in this manner, while Clement Greenberg has used a similar model for the development of painting since Manet.[51] Both scholars emphasise the increased self-reflection at stake when the mastery of materials becomes the very theme for the representation itself. Especially after the first phase of modernism, art becomes, to an increasing degree, about art: art reflects on the making of art. The heightened self-reflection and problem-solving brings with it the formal nominalism so characteristic of modernism: every work creates its own form, stubbornly resisting concepts of style or genre. In other words every work desires to do what has never been done before, to be unique. Here lies the similarity between formal nominalism and the notion of the new, both of which harbour a humanistic–utopian aspect.

We have touched on the fact that autonomy paves the way for the critical function of art. With the advent of the aesthetically modern, criticism can turn against the very role capitalist society has granted art: a kingdom of beauty beyond reality, to be enjoyed with innocent abandon. To adopt current jargon, art refuses to represent 'soft values'. The modernist demand for insight and cognition comes to expression not so much as a matter of content, but rather as a criticism *qua* aesthetic form: discordant, fragmented form or hermetic, difficult, forthright, incomprehensible form. Aestheticism may be seen as this kind of protest: a protest *qua* hermetic form. Far from expressing the self-conception of bourgeois subjectivity, the modernist work now escapes communication at large, insofar as this communication takes place within the confines of society. It was these issues that exploded with the historical avant-garde movements at the beginning of the twentieth century when the protest turned inwards against the autonomous institution of art itself.

# From autonomy to isolation

These aspects of modernism – heightened reflectivity, formal nominalism, breakdown of communication – inevitably led to art's (and the artist's) isolation. Concurrent with this tendency was an erosion of the critical awareness of the late-nineteenth-century bourgeois public. Until this time the rules of aesthetic reception were in many ways given in a communal sphere shared by the artist and his spectator / listener. The educated public knew what to look at or listen for. This aesthetic competence and knowing, which made art comprehensible, broke down with the fall of the public sphere and the advent of a modernism reluctant to communicate. Habermas has shown how a critical bourgeois public disintegrated with the waning of the liberal era. When the modern state emerged in the latter half of the nineteenth century – characterised by direct interventions into economy and other areas of society, and for the first time detaching economy from the private sphere, the critical – and political – public was squeezed between state and economy.[52] When art no longer matches the level of its audience it loses its impact. This development was reinforced further by the breakdown of the public sphere and the recurring distinction between the aesthetic value of art and its utilitarian or entertainment value. Propelled by the invention of new means of representation and media, entertainment value manifested itself in a distinct category of products, the culture industry, filling the leisure sphere created by modern capitalism. Within the culture industry the autonomy of art is broken and the distinction between producer and recipient potentially cancelled: critical potential, creative individuality and independent reception are negated by standardised products. Culture, then, tends towards a split between avant-garde and kitsch.

Modernism and the avant-garde cannot entirely emancipate themselves from their audience. However, the unity of producer and recipient can only be realised in small circles of experts and connoisseurs like Mallarmé's *cercle* or Schönberg's *Verein für musikalische Privataufführungen* (1918). Such circles provided fora for both the performance of and reflection on the new art, and became crucial for art's further development. Schönberg's *Verein* is particularly interesting because it shows the kind of crisis avant-garde music was experiencing and the (necessary) naivity of the artist regarding the crisis. The *Verein* – characteristically labelled by its critics 'the Schönberg clique' – was formed to secure the adequate interpretation of the new music as well as the great works of the past, by providing sufficient rehearsal time, analyses, discussions on performance practice, and regular performances.[53] Schönberg wanted to save the new music from commercialisation and a failed public music world, but believed at the same time that as long as the new and difficult works were secured proper performances, they would reap commercial success. It was this belief that made it possible for Schönberg to concentrate on a steadfast development of his composition technique.

When reflections on art no longer have a home within the institution, they must attach to the artwork itself. Writing on modern painting, the sociologist Arnold Gehlen talks about its *Kommentarbedürfigkeit* (need for commentary), seeing the many modernist manifestos as part of this. Gehlen sees the history of painting as an increasing 'pictorial rationality', and modern painting in particular as 'reflective art'. With modern painting of the early twentieth century, he argues, 'readable' significance disappeared from the picture itself and was replaced with a necessary linked commentary.[54] Such a commentary is needed not only as reflective exegesis, but often guarantees the legitimacy of the picture. The role of the commentary is to answer the audience's foremost question of the artwork: what does it mean? Modern self-reflective art seems to require the aesthetic discourse as an integral part of itself.

# Notes

1   Jauss (1970:11).
2   Blumenberg (1974).
3   Anton (1965:7).
4   Baudelaire (1995:12).
5   Baudelaire (1995:12).
6   Baudelaire (1995:1).
7   Baudelaire (1995:12).
8   Baudelaire (1968b:553).
9   Baudelaire (1995:13).
10  In Walter Benjamin's work on Baudelaire this has become a central notion. Benjamin found his own idea of 'present time' (*Jetztzeit*) echoed in the notion of the modern. For Benjamin *Jetztzeit* enters into constellations with the most remote past. As such it breaks out of the continuity of history and is saved (Benjamin 1969b:253). Benjamin's conception of history runs counter to both historicism and the hermeneutic notion of tradition, and contains both messianic and anarchistic elements. The classless society was viewed by Benjamin as the end of history, equivalent to 'pulling the emergency brakes on the locomotive of history'. Both Benjamin and Baudelaire thus harbour a notion of posthistory.
11  See Bürger (1974:82), who also distinguishes several other contrasting conceptions of artistic novelty.
12  Friedrich (1956:33-37).
13  Baudelaire (1968b:124).
14  Adorno (1970:40).
15  Dahlhaus (1982:96).
16  Dahlhaus (1982:96).
17  Dahlhaus (1982:96).
18  Paz (1976:13).
19  Enzensberger (1962:299).
20  Adorno (1969:316).
21  Adorno (1963:437).
22  Adorno (1970:41).
23  Calinescu (1977:44).
24  'La Poésie ne rhythméra plus l'action; elle sera en avant … En attendant demandons au poète du nouveau … ' Quoted from Calinescu (1977:112). Calinescu does not, however, develop the general point that the aesthetic modernity takes its concept of progress from bourgeois modernity. In general, he draws the interesting distinction between aesthetic and bourgeois modernity rather too sharply, and emphasises aesthetic modernity as a tradition against itself. At times he even writes as if aesthetic modernity breaks completely with the idea of progress. The other – bourgeois – modernity created, according to Calinescu, its own aesthetic form, namely kitsch (1977:265).
25  Gombrich (1978a:76).
26  Gombrich (1978a:98). English translation from *The Artwork of the Future*, trans. W.A. Ellis, London 1892, Vol. I: *Richard Wagner's Prose Works*).
27  Calinescu (1977:95); Böhringer (1978:90–114).
28  Nochlin (1968).
29  Poggioli (1968).
30  Nochlin (1968:18).

31 Rosen and Zerner (1982). The essay is a review of the exhibition 'The Realist Tradition: French Painting and Drawing 1830–1900', and the viewpoints expressed in the accompanying catalogue published by G. P. Weisberg. The essay is later included in Rosen and Zerner, *Romanticism and Realism: the mythology of the nineteenth century*, London 1984.

32 Rosen and Zerner, 4 March 1982:29.

33 Rosen and Zerner, 4 March 1982:26.

34 Rosen and Zerner, 18 February 1982:26.

35 Rosen and Zerner, 18 February 1982:25.

36 Benjamin (1978:798).

37 Adorno uses this as his motto for the second part of *Minima Moralia*.

38 My use of the notion of institution coincides roughly with that of Søren Kjørup, Jacques Dubois and Pierre Bourdieu, as well as Peter Bürger in his *Theorie der Avantgarde* (Bürger 1974). I do not, however, corroborate the use of the term in Bürger's later writings, where the historical dimension of the concept is lost (Bürger 1978:260ff).

39 Bürger (1978:27).

40 Karl Löwith argues that the nineteenth century belief in progress must be seen as a secularised eschatology: history is secured a meaning only as it comes to an end (Löwith 1967). This notion has been rejected by Hans Blumenberg, who points out the opposing structures of the two concepts of progress. Blumenberg proposes a different origin for the idea of progress, two key features of early modernity: sixteenth-century advances in astronomy and the literary feud between 'ancients' and 'moderns' (*la querelle des anciens et les modernes*) which resulted in the gradual rejection of the renaissance idea of the classical ideal in favour of a notion of art as expressing the spirit of its time. What happened in the eighteenth century, according to Blumenberg, was that both these ideas were combined; the new scientific idea of combining the efforts of several scholars in an evolving totality, and the new aesthetic idea that nature rather than God is productive. The combination of these ideas formed general notion of progress where man creates history on all levels (science, technology and art), an idea of limitless self-affirmation (Blumenberg 1974).

41 Habermas (1981:211).

42 See Gombrich's presentation of this issue (Gombrich 1978a and 1978b:3). It may be argued, however, that Gombrich does not distinguish sufficiently between the classical and modern concepts of progress.

43 Bürger (1974:32).

44 Habermas (1962:46).

45 Bürger (1974:32).

46 Bürger 1974:35.

47 Habermas 1981:229.

48 Weber (1963:555).

49 Gombrich (1978b:4).

50 Gombrich (1978b:7).

51 Greenberg (1965 and 1977).

52 Habermas (1962:172).

53 For a thorough treatment of Schönberg's *Verein* see Stuckenschmidt (1974:227).

54 Gehlen (1965).

**Chapter 3**

# Modernity and architecture

*Iain Boyd Whyte*

> Ideology collapses and vanishes, utopianism atrophies, but something great is left behind: the memory of a hope.
>
> Henri Lefebvre

## The death of modernism

Architectural modernism – as everyone who has read a book on postmodernism knows – died in 1972 when an unprepossessing and hitherto utterly insignificant housing estate was blown up in St Louis. When the demolition contractors fired the detonator they flattened not only the Pruitt-Igoe housing but also, according to the postmodernist account, the final pretensions to authority of a modernism that was condemned as intellectually bankrupt and barren. The great reforming hopes of the 1920s, of Le Corbusier, Walter Gropius and Ludwig Mies van der Rohe, had run aground on the rocks of social pragmatism. In the process the dreams of an architecture that might improve the general lot of humanity were exposed as elitist and reductivist, with an unfashionable tinge of Calvinist dogma and asceticism. After the dust had settled and the twisted steelwork been cleared away, the site was cleared for the patricidal infant postmodernism, which offered pluralism in place of monotony, and joy, delight and wit in place of the purged white walls of a second reformation. As they swaggered their way on to the empty building site, with Serlio up their sleeves and styrofoam voussoirs under their arms, the apologists of postmodernism brought with them a simplified history that traduced the true complexity and inventiveness of modernism. The architectural revolution that had dominated the century was presented as the 'victory [of] the square, the crate, the box – the multipurpose case as universal packaging',[1] or as 'a Protestant Reformation putting faith in the liberating aspects of industrialisation and mass-democracy', led by the likes of

'John Calvin Corbusier', 'Martin Luther Gropius' and 'John Knox van der Rohe'.[2] In attacking modernism as reductionist, banal and monotone, the altar-boys of postmodernism invoked a history that was itself equally reductionist and banal. To a degree they had been invited to do this, as we shall see, by the polemicists of architectural modernism – both designers and historians. Yet modernism, both as a historical force and as an aesthetic project is too rich and complex to allow such easy dismissal. The brevity of the postmodernist interlude sharpens our sense of historical perspective and demands a more vigorous enquiry into the nature of architectural modernism.

# Modernity

'Modern' has many meanings. It means current and actual, as opposed to former, previous, or foregoing. It means self-consciously new in contrast to old. More negatively it is used to describe the passing, transient, and merely fashionable, in contrast to the eternal. With the insight that cultural production is transient comes the awareness that the modern age commenced at varying points for the different arts and sciences. Diderot, for example, in the article 'Moderne' in his *Encyclopédie ou dictionnaire raisonné des sciences, des art et des métiers* (1751–72), proposed that modern literature began with Boëthius in the fifth century, modern astronomy with Copernicus, modern philosophy with Descartes, and modern physics with Newton. It could be argued that modern architecture also began with Newton, or more exactly with the emergence in the later eighteenth century of rationalism as the dominant force in social, political, and scientific discourse. As Alexander Pope wryly observed in his epitaph for Sir Isaac Newton, penned in 1732,

> Nature and Nature's Laws lay hid in Night;
> GOD said, *Let Newton be!* and all was Light.[3]

In replacing the darkness, chaos, and mysticism of earlier history, enlightenment science brought order and illumination to the world.

If we are to believe the postmodernist account in general, and Jean-François Lyotard in particular, the enlightenment project was a single, holistic conspiracy of uncontrolled reason, intent on the domination of man and nature through the workings of technology, against the ultimate good of mankind. Critics of this position, most notably Jürgen Habermas, have responded that the enlightenment project of a world guided by reason was never a monolithic enterprise, but rapidly evolved into a separatist culture, with experts responsible for specific intellectual spheres, and fundamental divisions emerging between the three main forms of human thinking – science, morality and art, forms that before the enlightenment had been rolled together into a single worldview under the influence of religious or metaphysical principles.

Developing the more differentiated reading of Habermas, four themes might be identified as characteristic of modernity, which align themselves into two diametrically opposed groupings. Individualism and relativism on one hand – understood as the absence of any absolute values – are challenged by the authoritarian demands of instrumental reason and capitalism on the other – the demands of technological progress, cost-efficiency, and a docile labour market. Each of these four conditions is essentially modernist, yet in a state of total opposition to its inimical pair. Precisely this conflict led Karl Marx to the celebrated observation, made in 1856, that 'On the one hand there have started into life industrial and scientific forces which no epoch of human history has ever suspected. On the other hand, there exist symptoms of decay, for surpassing the horrors of the latter times of the Roman Empire. In our days everything is pregnant with its contrary.'[4] According to the Marxist account, the dialectical and confrontational nature of modernity simply reflected the demands of the ruling class, which had a vested interest not only in change but also in crisis and chaos, characterised by Marx himself as 'uninterrupted disturbance, everlasting uncertainty and agitation'. Stability or stasis in the world of the capitalist entrepreneur means slow death. Modernity is thus marked by the unprecedented pace and scope of change, and the emergence of new and unprecedented social institutions structured around the imperatives of the postfeudal society.

# Modernism

The creative response to modernity is modernism. Given the contested nature of modernity, it is unsurprising to find that modernism operates under similar conditions of ambiguity and irresolution. In the specific context of architecture any account of modernism must be grounded on a careful analysis of the intellectual parameters within which the architect is working. The analyst of architectural modernism must consider the relationship of architecture and of architects to three key epistemological positions: history, theology, and politics.

# History

Technical and material innovation, although central to understanding the forms taken by modernist architecture, cannot explain the modernist search for change, progress and novelty. Nor can it explain the motives of those conservative cultural forces that so staunchly resist these strivings. History offers stronger clues. In their understanding of history the conservative antimodernist and the radical modernist share similar convictions, albeit with opposing motives. For the architect working in the nineteenth and twentieth centuries, history was understood not as sequence of discrete events, but as an agency with its own irrefutable patterns, laws and logic. Replacing the Christian belief in progressive revelation, architecture turned to neo-Hegelian historicism to explain the significance of a particular building or constructional technique within the broader laws of development. This process is well expressed in Mandelbaum's definition of historicism as 'the belief that an adequate understanding of the nature of any phenomenon and an adequate assessment of its value are to be gained by considering it in terms of the place which it occupied and the role which it played within a process of development.'[5] Following the imperatives of historicism, architecture was understood as a significant expression of the Hegelian 'world spirit', the essence of which is movement, and thus history. The built fabric, accordingly, was seen as an expression of the life of previously held positions in the unfolding of history. The Hegelian position refutes relativism or the idea that one view is simply equivalent or relative to another, since all ideas are interconnected in the unfolding of the history of human culture and society. Even views that are sharply hostile to one another constitute, to the Hegelian, merely opposite determinations of the same spirit, the positive and negative of the same proposition. And these determinations, in turn, are the inevitable expressions of the *Zeitgeist*, the spirit of the age, expressing the concept of the sense or rationality in the order of things and the succession of one state of affairs after another according to some kind of lawful process.[6]

With architecture understood as a component in the Hegelian history of the spirit, architects became fired in the nineteenth century by a strong belief in the particularity of the present time and by the conviction that the revolutionary moment was imminent. For the conservative this is a moment of terror, when the values and traditions built up over the centuries will be brought tumbling down by the destructive forces of revolution. For the radical the revolutionary moment promises change, the redefinition of goals, and the banishment of historical prejudices and injustices. By the end of the twentieth century this all-dominating explicatory *Zeitgeist* fell from favour. As an expression of the *grand récit* of historical progress, it was rejected by the postmodern critique. At more or less the same time it was also attacked by conservative architectural historiography. David Watkin for example, in *Morality and Architecture* first published in 1977, saw it as evidence of a Germanic and Pevsnerian conspiracy: 'The underlying principle remains the same throughout Pevsner's work: art must "fit" into the Zeitgeist which is now a progressivist harbinger of the earthly new Jerusalem.'[7]

Far from being merely the invention of historians, the Hegelian and historicist account was being firmly prescribed by architects themselves, particularly in the gestatory decades of architectural modernism. Otto Wagner, for example, giving his inaugural lecture to the Academy of Fine Arts in Vienna in 1894, insisted that 'The starting point of every artistic creation must be the needs, ability, and achievements of our time.'[8] A few years later in 1898, his pupil Joseph Olbrich designed the Secession Building in Vienna, which carried above the door the inscription 'Der Zeit ihre Kunst, der Kunst ihre Freiheit' – to the age its art: to art its freedom. Similar sentiments are legion among the theorists of early modernism, stressing the inseparable bond between architecture and the *Zeitgeist*. In this relationship the architect functioned as a seismograph, highly and predictively responsive to the demands and the spirit of the age. Launching *L'esprit nouveau* in October 1920, Le Corbusier very predictably hailed the particularity of the moment: 'There is a new spirit: it is the spirit of construction and of synthesis, guided by a clear conception. Whatever may be thought of it, it animates to-day the great part of human society.'[9] In the Soviet Union too the search for the style to match the spirit of the age was equally strong, and informs the title of Moisei Ginzburg's seminal book *Style and Epoch*, which drew strongly on the Wöfflinian thesis that style in the visual arts and architecture was the direct expression of the spirit of a time and of a people.[10] This essentially Hegelian position dominated the architectural mindset for most of the twentieth century. When in the late 1960s the British group Archigram was proselytising for an architecture that was lightweight, technically sophisticated and highly mobile, they were responding self-confessedly to the spirit of the age, encapsulated at the time by the moon landing and 1960s pop culture. It is hard to imagine how it might have been otherwise, as the imperatives of modernity demand of the creative spirit a direct engagement with the technological inventions of the age. To bemoan the lack of mouldings or an indifference to historical models, as both conservative and postmodernist theoreticians did in the final decades of the century, denies these imperatives and wilfully misunderstands the nature of the modernist project. The inevitable result was the feeble, formalist posturing of postmodern architecture, devoid of any theoretical basis beyond a misinformed condemnation of modernism.

# Theology

Paradise on earth was the explicit goal of the early architectural modernism. Writing in 1902 the cultural sociologist Georg Simmel noted among his contemporaries a specific 'yearning after a final object' in a context that 'no longer renders possible its attainment'. This, he says, produces 'specifically modern feelings, that life has no meaning, that we are driven hither and thither in a mechanism built up out of mere preliminary stages and means, that the final and absolute wherein consists the reward of living, ever escapes our grasp.'[11] Both the brave new world of technology and the return to the wooden-framed cottage represent a search for certainty in a modern world that is categorically unable to provide such reassurance.

This search for certainty and spiritual reassurance is central to the dynamics of modernism in general and modernist architecture in particular. Following the triumph of enlightenment science in the eighteenth century and the decline of organised religion in the later nineteenth century when faced with the Darwinist challenge, the grand spiritual narratives were left in dysfunctional tatters. In the resulting vacuum countless specialist interests, ranging from teetotalism and vegetarianism to dance and neo-Buddhism, joined the fray, each offering itself as the holistic solution: the problems of the world would be solved if only the people would learn to dance or to shun the temptations of meat or alcohol. Architecture was another of these partial systems which offered itself as a total solution. Spiritual certainty and social harmony were to be achieved by good design, and the architect would take on a messianic role as the divinely gifted individual empowered to redirect and reconstruct the goals and ambitions of the industrial society. As Le Corbusier insisted in the *Charter of Athens*, 'Architecture holds the key to everything'.

In its spiritual ambitions modernist architecture fed on deeply ingrained Judeo-Christian ideas of progress, which understood history as a single, future-directed progression that would find its fulfilment in final events such as the coming of the messiah or the last judgement. As Karl Löwith has argued, the central modern idea of progress is simply a secularised version of ideas that derived from medieval Christianity: 'The ideal of modern science of mastering the forces of nature and the idea of progress emerged neither in the classical world nor in the East, but in the West. But what enabled us to remake the world in the image of man? Is it perhaps that the belief in being created in the image of a Creator-God, the hope in a future kingdom of God, and the Christian command to spread the gospel to all nations for the sake of salvation have turned into a secular presumption that we have to transform the world into a better world?'[12] This presumption is central to the mindset of architectural modernism, with its lingering belief in messianic leadership, its insistent emphasis on progress and goals, and its biblical patterns of guilt and expectation. As Colin Rowe concludes in *The Architecture of Good Intentions*, 'It is only this eminently dramatic and ultimately Hebraic conception of history in terms of architectural sin and architectural redemption which provides any real accommodation for the emotional preconditions of modern architecture's existence.'

# Politics

In his essay 'The Nature of Gothic' first published in the early 1850s in *The Stones of Venice*, John Ruskin called on a direct causality between political ideology and the process of designing and building. 'Go forth again', he advises, 'to gaze upon the old cathedral front, where you have smiled so often at the fantastic ignorance of the old sculptors: examine once more those ugly goblins, and formless monsters, and stern statues, anatomiless and rigid; but do not mock at them, for they are signs of the life and liberty of every workman who struck the stone; a freedom of thought, and rank in scale of being, such as no laws, no charters, no charities can secure; but which it must be the first aim of all Europe at this day to regain for her children.'[13] For Ruskin, social discontent and upheaval could be averted by freeing the operative from the dictate's machine and allowing the working force to invest its own energy and imagination in the process of production. For Ebenezer Howard the provision of good cheap housing, freed from the menace of speculation, would guarantee a similar outcome. His tract on the garden city was originally published in 1898 under the title *Tomorrow: A Peaceful Path to Real Reform*. A centrist, reformist socialism was the dominant political voice of the early decades of architectural modernism. As Bruno Taut sensed in 1919, 'A feeling exists, or at least slumbers in all of us, … that one should feel a sense of solidarity with all men. Socialism in the non-political, supra-political sense is the simple, straightforward relationship between men, far removed from any form of domination. It straddles the divide between warring classes and nations and binds all men together.'[14] In a similar tone, and echoing Ebenezer Howard, Le Corbusier concluded *Vers une architecture* in 1923 with the binary choice between good (which is to say modernist) Corbusian design and political revolution.

In accord with its revisionist socialist roots, the white modernism of the 1920s has traditionally been equated politically with social democracy and the liberal left. The great housing estates of Berlin and Frankfurt, built under the aegis of the respective chief city architects Martin Wagner and Ernst May, forged a concrete link between progressive architectural practice and the socialism of both city government and trades unionism. In contrast the architecture of the dictatorships, as it evolved in the 1930s in the Soviet Union, Germany and Italy, was seen as forging an inseparable link between monumental neoclassicism, stylistic conservatism and totalitarianism. This was the simple good/bad scenario argued right up to the 1960s and 70s, and paraphrased in Pevsner's celebrated admonition that any word devoted to Nazi architecture is a word too many. Only when the modernist project in architecture began to crumble in the 1960s under the weight of its own dullness did the historiography begin to question the simple, moralistic account of good modernism and bad neoclassicism.

This questioning operated at many levels, from the simple biographic through to more sophisticated theoretical enquiries. Biographically, such issues as Le Corbusier's relationship with the pro-Nazi Vichy government in France, or the initial willingness of the German avant-garde to enter Nazi-run design competitions, became the subject of scholarly enquiry. At the instrumental level, scholars like Boris Groys investigated the relationship between the avant-garde and the dictators not as one of opposites, but of similarity: 'totalitarian art was so unyielding towards the avant-garde, because it itself was inspired by an avant-garde purpose'.[15] The mutual purpose was a planned world, infatuated with technology and responsive to the dictates of an élite that promised a new world and a new social order, invented either on the architect's drawing board or in the corridors of political power. At an even more general level, revisionist sociologists like Zygmunt Bauman proposed that the entire enlightment project, far from being an emancipatory force for mankind, was a single, 'holistic' conspiracy of uncontrolled reason working ultimately against the good of mankind. Following this line of argument, the Holocaust was the inevitable offspring of enlightenment rationalism, only possible within an advanced technocratic society run by a powerful bureaucracy.

In this light the architecture of high modernism takes on a more sinister character. Were the broad avenues, regular blocks and broad expanses of glass favoured by the 1920s avant-garde dedicated to the emancipation of mankind in a new world of light, air and transparency, or was architecture being corrupted into yet another agency of surveillance and control? Were the inconoclastic rebuilding plans of the modernists – Le Corbusier's Voisin plan for Paris, or Ludwig Hilberseimer's Hochhausstadt, which demanded the destruction of large areas of the old city – essential preconditions and precursors for the dictatorial replanning of Moscow, Berlin and Rome in the 1930s?

Revisionist questions work on both sides of the equation. If architectural modernism was not as far removed from totalitarianism as its early histories suggested, then might monumental neoclassicism also have a more complicated status than simply the architectural expression of dictatorship? In their famous 1932 exhibition at the Museum of Modern Art in New York, Henry-Russell Hitchcock and Philip Johnson defined and illustrated the 'International Style' with the work of architects like Le Corbusier, Gropius, Mies and Oud. Yet in terms of its geographical distribution and the sheer volume of buildings, it would be equally possible to see the pared-down neoclassical revival as the international style of the 1930s. It was favoured not only for monumental building schemes in the dictators' capital cities, but also among the world's democracies – a shortlist of candidates from the 1930s includes the Parliament in Helsinki, Walthamstow Town Hall, the Brotherton Library at Leeds University, the Federal Triangle in Washington, the Hall of State in Dallas, and the State Capitol in Lincoln, Nebraska. The importance of the monumental gesture to civil society was a theme treated in the 1940s not only in the deliberations of dictators; it also featured in the 1943 paper 'Points on Monumentality' produced collaboratively by Josep Lluís Sert, Fernand Léger and Sigfried Giedion, which sought to reconcile architectural modernism and rhetorical monumentality. Their first point reads: 'Monuments are human landmarks which men have created as symbols for their ideals, and for their actions. They are intended to outlive the period which originated them, and constitute a heritage for future generations. As such, they form a link between the past and the future.'[16] This argument had a powerful impact on postwar discussions about urbanism and reconstruction, and marked a significant departure from the anti-monumental, cellular arguments put forward in the 1920s. By 1950, and charged with the symbolic rebuilding of democracy in Europe, the architectural avant-garde wrestled with the problem of freeing the monument from its burdensome association with dictatorship, a long and difficult process that found partial resolution with the construction of James Stirling's Staatsgalerie in Stuttgart in the mid-1980s, but which is still a source of conflict to this day, particularly in Germany.

The historian of modernist architecture not only has to acknowledge the dictatorial elements of high modernism in the 1920s and the universality of the neoclassical revival in the 1930s, but also the ambiguities endemic within the architectural policies of the dictatorships. These ambiguities are the inevitable result of the contradictions that lay at the heart of the totalitarian regimes: the desire to be both familiar and autocratic, to be technocratic yet wedded to the values of the soil, to be centralist yet alert to regional difference – to be all things to all people. Expressed in works of architecture, these contradictions allow the ultra-modernist *Rationalismo* of Terragni to coexist with the imperial Roman pastiche of the Piazzale Augusto Imperatore in Mussolini's Italy, or modernist factories, folksy Hitler Youth hostels and neo-romanesque autobahn bridges to be hailed by the party leadership in Germany as equivalent expressions of the spirit of National Socialism. This cocktail can be explained in the German context as an expression of the conservative revolution, which sought to emancipate technology from the world of means–ends rationalism, and ally it to the organic realm of *völkische Kultur*. The community of the people and the soil, and boundless technical optimism were fused together in an improbable alliance that once again challenges the simplistic assumptions of 'good' modernism and 'bad' monumental classicism.

# The city

As contradiction is fundamental to both modernity and modernism, it is hard to avoid the conclusion that the terms themselves are of relatively little use in establishing parameters or definitions. The power and fascination of modernity lie precisely in this dynamic irresolution, and modernism can only be investigated within the shifting parameters of modernity. Yet it is ultimately fruitless to indulge in limitless readings of modernity and modernism, and the theorist of twentieth century architecture needs to construct contexts or matrices within which the contradictions of modernity and modernist architecture can be located and studied. The most fruitful context of enquiry to date has been the city. As James Donald has noted, 'Modernity is inherently both rational and mythical. Nowhere is this more evident than in the modern city.'[17] The dialectic thus established is between rationality and enchantment, the city of the planner and engineer against the city of the artist, the poet and the *flâneur*. As described by Max Weber's in *The Protestant Ethic and the Spirit of Capitalism*, the planned city driven by formal, means–end rationality is an iron cage for the human spirit, from which all the magic has been driven out in a process of *Entzauberung* (disenchantment). Weber's position was grounded in a long tradition of negative critiques of the industrial city, which flourished most vigorously in nineteenth-century Britain, the first heavily industrialised society. In the mid-1830s, Thomas Carlyle was writing of the 'true sublimity', the incomprehensible extent of the power and horror of the city, which lay 'under that hideous coverlet of vapors, and putrefactions, and unimaginable gases.'[18] Developing the theme, Ruskin bemoaned in 'The Nature of Gothic' that 'The great cry that rises from all our manufacturing cities, louder than their furnace blast, is … that we manufacture everything there except men; we blanch cotton, strengthen steel, refine sugar, and shape pottery; but to brighten, to strengthen, to refine, or to form a single living spirit, never enters into our estimate of advantages.'[19] Summed up more pithily in 'Unto this Last', he writes: 'Men can neither drink steam nor eat stone.'[20]

In 1882 William Morris republished 'The Nature of Gothic', hailing it in his preface as 'one of the few necessary and inevitable utterances of the nineteenth century.'[21] Starting from this position it is a simple task to draw up the genealogy of architects and movements resistent to the cultural and social dictates of modernity, those who sought refuge from the industrial world in a regressive utopia. From Morris's Red House, built in 1859, it would move on to the British domestic revival of the 1890s and to Ebenezer Howard's new town impulse in the early years of the new century, seeking to combine the social advantages of the city with the primal innocence of the country. As Howard explained, 'The town is the symbol of society – of mutual help and friendly co-operation … of broad, expanding sympathies, of science, art, culture, religion. And the country! The country is the symbol of God's love and care for man. All that we are and all that we have comes from it … It is the source of all health, all wealth, all knowledge. But its fulness of joy and wisdom had not yet revealed itself to man.'[22] Very similar arguments resurfaced after World War I with the reactionary modernists in Weimar Germany, where the likes of Heinrich Tessenow, Paul Schmitthenner and Paul Schultze-Naumburg fell under the spell of the cultural pessimism typified by Oswald Spengler's *Decline of the West*. For Spengler the farmhouse and the village symbolise rootedness: 'The village, with its quiet hillocky roofs, its evening smoke, its wells, its hedges and its beasts, lies completely fused and embedded in the landscape. The country town *confirms* the country, it is an intensification of the picture of the country. It is the Late city that first defies the land, contradicts Nature in the lines of its silhouette, *denies* all Nature. It wants to be something different from and higher than Nature. These high-pitched gables, the Baroque cupolas, spires and pinnacles, neither are, nor desire to be related with anything in Nature. ... And then begins the gigantic megalopolis, the *city-as-world*, which suffers nothing beside itself and sets about *annihilating* the country picture.'[23] Alvar Aalto and the Scandinavian modernists would play a significant part in this anti-urbanist geneaology in the 1930s, and their influence resonated in the postwar new town movement in Britain, with its dispersed site planning, brick housing, and homey 'people's detailing'. In the 1960s the return to the soil took a more mystical turn towards the world of Paulo Soleri and Arcosanti, and by the end of the century an ever-expanding army of greens and advocates of sustainability was fighting to resist and reverse the powerful excesses of modernity and industrialisation.

The opposite reaction, at the other extreme of the dialectic, responds positively and affirmatively to the industrial city and to the technological, demographic and social forces that have created it. One of the earliest, and certainly the most celebrated affirmations of the metropolis as the defining site of modernity, was that of Charles Baudelaire. In *Le peintre de la vie moderne*, published in 1859, he offers the memorable sentence 'By modernity I mean the ephemeral, the fugitive, the contingent, the half of art whose other half is the eternal and the immutable'.[24] For Baudelaire the new metropolis, Hausmann's Paris, was both exciting and heroic, and this heroism lay precisely in the transient qualities of modern life, in the world of fashion, the chance encounter on the boulevard, the dazzling randomness of the great city. The nomadic sense of placelessness and flux that inevitably resulted from transience were also reasons for celebration in the eyes of the rationalists. Otto Wagner, in his plan for an ideal extension to Vienna, published in 1911 as *Die Großstadt*, praises anonymity as a particular quality of the metropolis. Such wholehearted commitment to the modernist city and to the technical and industrial forces that drive it was shared by Peter Behrens, Antonio Sant'Elia, Ludwig Hilberseimer, Martin Wagner, the high modernist Le Corbusier of the 1920s, the Soviet Constructivists, Adolf Meyer, Mart Stamm, Kenzo Tange, the British Archigram group of the late 1960s, and the prophets of high-tech in the 1980s and 90s: Norman Foster, Nicholas Grimshaw and Richard Rogers. Their message could be paraphrased in a passage from the *Manifesto of Futurist Architecture*, penned by Sant'Elia in 1914: 'Architecture now breaks with tradition. It must perforce make a new start. ... We must invent and rebuild the Futurist city like an immense and tumultuous shipyard, agile, mobile, and dynamic in every detail; and the futurist house must be like a gigantic machine.'[25]

Moving beyond the binary opposition of urbanites versus anti-urbanites, many sets of opposing poles offer themselves: rationalists versus enchanters, engineers versus organicists, machine versus handcrafts, the straight boulevard versus the picturesquely serpentine street, flat roof versus pitched roof. These in turn combine with the political, theological and historical positions previously touched on. As a result the absolute black-and-white differences that might allow confident definition and categorisation disappear as soon as each individual case is considered. Tessenow, for example, who was arguing for wooden architecture and traditional technology around 1920, was also a founder member of Bruno Taut's *Arbeitsrat für Kunst* (Working Council for Art), which was pressing for a radical architecture of coloured glass and steel. Both Taut and Tessenow were joined at that time in their advocacy of decentralisation and a return to the soil, yet Tessenow was politically conservative, Taut on the radical, anarcho-socialist left. Similarly Le Corbusier, in describing his visions of the city of the future in the 1920s, argued that the essence of his highly mechanistic grid-plan for the city was organic, in that it reflected the replication of cell-forms in nature. A recent commentator on Ludwig Mies van der Rohe has argued that the Mies of the 1920s might be understood as an organicist, resistant to predetermined forms and open to organic processes of becoming. According to this reading, Mies saw the city as 'a figure of totality and integration, symbol of the all-embracing but intangible structure, the unity of relationships and interdependences between things that he sought to express in material form and to make palpable for the beholder. Like many others of his generation, Mies in invoked the figure of the organicism to refer precisely to such a holistic and unified relational structure. Based on the organic priniciple of free relation among self-determined parts, the model of the organicism could be applied to machines and machine-like cities as well as to plants'[26] The contradictions pile up; the examples are endless.

# Conclusion

Modernism in architecture, as in all the arts, exists only as a response to the contradictory conditions of modernity. To understand the modernist statement it is essential in every single case to go back to the condition of modernity that prompted the statement. Stripped of this relationship, modernism becomes an empty formalism, just another chrome-steel chair. Yet paradoxically the working out anew of formal relationships is precisely the triumph of modernism, of an art that refuses to imitate the production and thus the conditions of any preceding generation. As Peter Bürger has noted, 'The category of artistic Modernism *par excellence* is form. In Modernism, form is not something pre-given which the artist must fulfil and whose fulfilment the critics and the educated public could check more or less closely against a canon of fixed rules. It is always an individual result, which the work represents.'[27] For the historian the task of understanding and description must investigate the real-world context of the modernist production and re-establish, if only at the theoretical level, the fusion of art and life to which the twentieth-century avant-garde constantly aspired, and which it invariably failed to achieve. For the architect the goal must be the affirmation of the essential categories of modernism, while reinvesting them with life and freeing them from modernist rigidity. Although penned in the context of literature, Malcolm Bradbury's conclusion is equally pertinent to architecture: 'Modernism is our art; it is the one art that responds to the scenario of our chaos'.[28]

# Notes

1   Klotz (1988:24).
2   Jencks (1989:27).
3   Pope (1921:461).
4   Marx (1978:577–8).
5   Mandelbaum (1971:42).
6   Hegel understood this rationality as a spirit that animated people, not as a creation of human labour. See Hegel (1991:373): 'The states, nations, and individuals involved in this business of the world spirit emerge with their own particular and determinate principles, which has its interpretation and actuality in their constitution and throughout the whole extent of their condition. In their consciousness of this actuality and in their preoccupation with its interests, they are at the same time the unconscious instruments and organs of that inner activity in which the shapes which they themselves assume pass away, while the spirit in and for itself prepares and works its way towards the transition to its next and higher stage. (Die Staaten, Völker und Individuen in diesem Geschäfte des Weltgeistes stehen in ihrem besonderen bestimmten Prinzipe auf, das an ihrer Verfassung und der ganzen Breite ihres Zustandes seine Auslegung und Wirklichkeit hat, deren sie sich bewußt und in deren Interesse vertieft sie zugleich bewußtlose Werkzeuge und Glieder jenes innern Geschäftes sind, worin diese Gestalten vergehen, der Geist an und für sich aber sich den Übergang in seine nächste höhere Stufe vorbereitet und erarbeitet.)' (Hegel 1972:296).
7   Watkin (1977:81).
8   Otto Wagner, Inaugural Address to the Academy of Fine Arts, Vienna, 15 October 1894, quoted in Wagner (1988:160).
9   Le Corbusier, Programme of *L'esprit nouveau*, no. 1 (October, 1920), quoted in Corbusier (1946:101)

10   For a discussion of the impact on Soviet constructivism of German notions of the *Zeitgeist* derived from Riegl, Wölfflin, Frankl and Spengler, see Senkevitch (1982:22–5).

11   Georg Simmel (1902:101); quoted in Frisby (1985:43).

12   Löwith (1949:203).

13   Ruskin (1907a:149).

14   Taut (1919:59–60).

15   Groys (1987:35).

16   Sert, Léger and Giedion (1993:29).

17   Donald (1992:437).

18   Carlyle (n.d.:17–18).

19   Ruskin (1907a, note 14:151).

20   Ruskin (1907b:195).

21   Morris (1892:i).

22   Howard (1945:48).

23   Spengler (1991:246).

24   Baudelaire (1970:13).

25   Sant'Elia (1973:169–70).

26   Mertins (2002:619).

27   Bürger (1992:44).

28   Bradbury and McFarlane (1976:27).

**Chapter 4**

# Modernity and the uses of history
## Understanding classical architecture from Bötticher to Warburg

*Caroline van Eck*

If one of the defining characteristics of modernity is the sense of an irrevocable separation between the present and the past, it is no longer evident why the past should be studied. In architectural history this breach became manifest in the last two decades of the nineteenth century when architects gave up looking to the styles of the past as models for a contemporary style that would accommodate the demands of their age and express its spirit. Instead they turned their energies to developing a truly modern style that would be free of the formal vocabularies of the past. In the study of architectural history and theory this change of attitude towards the past resulted in a transformation of the aims of writing architectural history or studying the theory of historical styles, particularly of classical architecture. History and theory were no longer the storehouse whose riches the architect could use to develop a style that would be both historically correct and an expression of the age. The aim and legitimation of their existence was no longer to be the foundation for present-day practice. In the case of the neogothic this led to an increasingly antiquarian approach once the gothic revival had ended, in the sense that the past was studied only for the sake of the increase of historical knowledge, and without consideration for the use of that knowledge except for conservation purposes. What happened to the study of the history and theory of classical architecture is the subject of this essay. It offers a reading of the three main theorists of classical art and architecture in the nineteenth century – Bötticher, Semper and Warburg – not to demonstrate the increasing irrelevance of their attempts to grasp the principles behind classical design, but to show how their emphasis on design was overtaken by a project that can best be termed hermeneutic: understanding the significance of classical art for the present day as a cultural phenomenon.

The nineteenth century saw both the establishment of architectural history as an academic discipline – Alois Hirt was appointed the first professor of architecture at the newly founded University of Berlin in 1809 – and a transformation of the relationship between the theory and history of classical architecture and its practice. Before the birth of modern classical archaeology in the 1750s the remains of Greek and Roman architecture had been studied mainly to establish firm principles of classical design. From the measuring excursions of Alberti, Brunelleschi and Palladio, through Antoine Desgodet's measuring and drawing of Roman buildings to solve the contradictions found in the treatises and former reconstructions, to the measuring, surveying, drawing and reconstructions by students at the Ecole des Beaux Arts who had won a Prix de Rome, historical investigation had been inseparable from design. In the nineteenth century that connection was significantly loosened, if not severed. The main point of historical investigation was no longer to illustrate, explain and support Vitruvian theory; it also became an academic subject in its own right, closely linked to archaeology and other disciplines in the humanities and social sciences, such as linguistics and anthropology.

Another way of describing this development, and a very familiar story, is to reconstruct the struggle of nineteenth-century architects to develop a style that would be based on historical precedent but at the same time expressive of the age and society in which they worked. After 1750 Vitruvianism lost its monopoly as a stylistic paradigm because of the criticism and new stylistic options offered by the gothic revival, the picturesque, and the rise of an archaeologically informed neoclassicism. For a while history appeared able to replace Vitruvian theory, offering both a repertoire of forms and clear examples of their use. Architects turned to history to solve their search for a style, and until the end of the century it was unthinkable to start designing a language of architectural forms from scratch. However, there was an increasing sense that the key to understanding the laws governing the use of these forms was irretrievably lost, and that the forms of the past could not be adapted indefinitely to accommodate new building types, materials and technical advances. As Karl Friedrich Schinkel wrote in 1835,

> I observed a great immeasurable treasure of forms, which had already come into being and had been recorded in the execution of works of building in the world through many centuries of development and among very different peoples. But at the same time I saw that our use of this accumulated treasury of often very

heterogeneous objects was arbitrary, because every individual form carries its own particular charm. ... It became particularly clear to me that the source of a great lack of character and style, of which so many new buildings seem to suffer, must be found in this arbitrariness of use.

It became the task of my life to gain clarity in this issue ... I continued my researches, but very soon I found myself trapped in a great labyrinth.[1]

Architects continued to write about the history of classical architecture, but during the course of the nineteenth century their writing changed to reflect how historicism as a viable and meaningful design option was overtaken by modernity's sense of the past as a foreign and fundamentally inaccessible territory. Of the three authors who figure in this essay, Bötticher and Semper were practicing architects, theorists and teachers. Aby Warburg was neither an architect nor a theorist, and seldom wrote directly about architecture, but the question that motivated all his work – what classical revivals mean for the culture in which they are revived – is the most radical formulation of the questions that exercised Bötticher and Semper.

# Bötticher

Carl Gottlieb Wilhelm Bötticher's *Die Tektonik der Hellenen* (first published in 1844–52 with a substantially revised edition in 1874) is one of the most misunderstood books on classical architecture ever written.[2] Its concerns were to discover the principles that ruled the forms of Greek architecture, to understand why these forms look the way they do in the light of these principles, and to offer general guidance on the creation of a new architectural style. It is not just a theory of architectural invention (what Bötticher called *eine Erfindungslehre* in a clear allusion to the rhetorical doctrine of invention, disposition and elocution as three stages of writing a speech), but also a systematic reflection on ways of understanding Greek temple architecture. In that sense Bötticher's project may also be called a hermeneutics of architecture.[3] The key concepts in his analysis of Greek architecture were tectonics – an approach to architecture in the technical and mechanical terms of statics and space creation – and the distinction between art forms and work forms, *Kunstformen* and *Werkformen*. Behind his project lies the scientific insight – quite recent at the time of writing the first edition – that mechanical and static forces are concepts rather than visible appearances; gravity, for example, is the force that causes an apple to fall rather than the phenomenon of the apple falling.

Work forms are those central parts of architecture which perform the material and mechanic tasks of building. He also called them 'schemes', to indicate that work forms as such are never found in a building since they are always 'dressed' in art forms. They are deduced by the observer from existing forms by a process of abstraction from the material appearance of a building. Art forms are the visual representations or characterisations of the static and mechanic functions performed by the work forms. Work forms are not easy to identify, because we arrive at an understanding of them only by means of abstraction from the art forms which cover them. We should rather think of them as constructional schemata such as the combination of columns and architrave, or of walls and arches. They are not inspired by existing historical models but deduced by the mind from the conditions of the material task these forms have to perform. Art forms, by contrast, are modelled after natural examples. Bötticher describes them, paraphrasing Aristotle's *Poetics*, as an imitation of appearances. They characterise and express the static function of the work forms, which in themselves cannot be perceived, through an analogical representation of forms borrowed from nature; for example the immovable solidity of the beam supporting the deadweight of the roof in the architrave of the Corinthian order is clothed in the visual form of the triple layer of the fascia. Although art forms do not contribute to the material solidity of a building, they do have another important role. Thanks to their expression of the invisible structural functioning of the building the dead stones are transformed into a living work of art: 'Lautlos und starr, verräth sich Gedanke und Begriff nur durch charaktervolle Zeichen.' Without such forms the way in which static forces work upon each other in a building would not be perceptible, and the building would seem dead: art forms are the visual language of tectonic forms.[4] They represent both the mechanic and static forces at work in a building and the connections between the parts of a building: in the classical orders of architecture astragals, thori, abaci, guttae, cymae rectae and reversae, coronae and so on. In calling such elements of the classical style *Juncturen*, Bötticher added a new chapter to the long history of attempts to define the essential but elusive element of the orders that articulates the transitions between base, shaft, capital and architrave, shaping the play of light and shadow on a building's façade.[5] Art forms imitate natural forms, but do not merely copy them. Their use is regulated by what Bötticher, again invoking Aristotle, called *anagkè*: the concept of an entire building which regulates the use of its related art forms, transforming it into an integrated representation of the underlying structural system. Through the presence of art forms a building becomes a work of art, no longer just the material functions of statics and space creation but a visible expression of a systematic design based on a carefully planned vision.

This new analysis of Greek temple architecture was not simply an exercise in the acquisition of archaeological knowledge; it was revolutionary in its consistently tectonic approach to a topic in which considerations of the correct measurement of the orders and handling of the orders had hitherto prevailed.[6] Bötticher used his tectonics as the conceptual basis of the development of a new style, characteristic of its century and expressive of its age. The new style would accommodate the new building types it required, and make use of the properties of a new material: steel. In a speech commemorating his predecessor at the Berlin Bauakademie, Karl Friedrich Schinkel, Bötticher outlined how the use of steel brought with it a new mechanical and static system of forces in building, and how the new work forms entailed a new range of art forms, a new language of ornaments, which would be derived from the two main styles of nineteenth-century western architecture – the Greek and the gothic.

For the majority of Bötticher's students this aspect of his work was one they could understand and use. Many adepts adopted his distinction between art forms and work forms, but ignored the latter's abstract, conceptual character, identifying them rather with the actual steel or stone framework of a building. Work forms became identified with the visible and tangible structure and art forms with ornament, and his theory was turned into an unnecessarily complicated statement about the rationale and use of organic ornament as well as the use of modern materials.[7] The hermeneutic aspect of his work was largely ignored, but his inquiry into the significance of Greek temple architecture and the way he set about answering it were symptomatic of the loss of the self-evidence of the classical past as the source for contemporary design.

Until the 1750s, when Vitruvianism still reigned supreme as the paradigm for the art of building, the answer to questions about the meaning of architectural forms linked elements of a building – its circular or cruciform shape, the use of caryatids or of the doric order – to elements outside architecture: to religion, myth or ideology. One could say it was an iconological answer, linking architectonic elements to the beliefs and social practices of its surrounding world. In eighteenth-century French theory this iconological system of meanings was enlarged by the new aesthetics of *caractère*, which saw buildings as expressive systems representing their functions and the social status of their inhabitants and users.[8] With Bötticher's tectonics this idea changed fundamentally. In a move that established the autonomy of architecture as an entirely self-referential art and reduced the range of its possible meaning to the expression of the static and mechanic forces at work in a building, Bötticher broke with the Vitruvian tradition of endowing a building with meaning by connecting it with social, religious or political practices.[9] This had the advantage of finding a solution to that vexed issue in German aesthetics: how to argue convincingly that architecture, in spite of its obvious practical utility, may still be called an art in the way that completely non-useful arts such as ballet or music can be called arts, in whose enjoyment the human spirit is temporarily liberated from all material considerations. By showing that building becomes an art through the use of ornament that expresses the inner nature of architecture, and at the same time transcends its materiality because the use of *Kunstformen* is evidence of free artistic design and planning of the building as a whole, Bötticher was able to reconcile the practical demand that a building be solid and functional with the aesthetic demand that it also be beautiful and an expression of the freedom of the human spirit.

There was a price to be paid for this emancipation of architecture: its range of meaning had become very restricted. Architecture possessed meaning because it has the potential to represent. To explain his concept of art forms Bötticher paraphrased Aristotle's *Poetics*. Where the original proposes that all art is representation, imitating significant human action, Bötticher wrote that art forms are imitation of appearance.[10] Meaningful human action in all its variety and scope, and all its implications of intentionality, human agency and – since Aristotle is here discussing tragedy – emotional entanglement, have been reduced to the formalism of imitating the forms of the world that surrounds us.

## Semper

For Gottfried Semper history and design were also closely linked, and like Bötticher he thought that architectural meaning is representational.[11] The range of what architecture represents, however, is for him much broader: it is not limited to the tectonic forces at work in a building, but includes the entire range of human crafts and their historical development. The key notion in Semper's understanding of architecture is style, which is no longer a historical concept referring to the formal vocabularies of the past but, as in Bötticher, a hermeneutic concept based on an interpretation of architecture as primarily representational. Style is 'the correspondence of an artistic appearance with the history of its origins, with all the conditions and circumstances of its becoming'. As in Bötticher, tracing the development of the appearance of a building is a way of understanding why it looks the way it does. For both authors meaning in architecture is both self-referential, in that the appearance of a building explains its meaning, and a representation of the development that led to that appearance.[12] *Der Stil* is both a study in tracing the development and principles of the decorative arts and architecture (even though Part III, devoted to architecture, was never published), and a formulation of their meaning in terms of the laws governing that development. Just as *Tektonik der Hellenen*, *Der Stil* was intended as an intervention in the contemporary crisis in architectural design, as its subtitle – *Eine praktische Aesthetik* – suggests.[13]

Like Bötticher – and practically all architectural theorists of the nineteenth century, Semper sought to develop a new 'theory of invention' or practical aesthetics which would both tell the architect how to design in such a way that his buildings would have style, and unveil the meaning of artefacts from the past with reference to the earliest developments of architecture. What distinguished him from his predecessors was his decision not to stop at the origins of Greek architecture, but to extend his inquiries to the arts of pre-classical and non-western societies. As has often been pointed out, the arrival of a large shipment of the Assyrian treasure of King Ashurnasirpal II's palace in the British Museum in 1848 and the organisation of the Great Exhibition held in London in 1851 were turning points in the development of Semper's thought about the arts. Instead of analysing a key monument of classical architecture – such as Laugier's hypothetical primitive hut or an actual building such as the Pantheon or the ruins at Paestum – Semper concentrated on the crafts that were for him the true origin of building, and their metaphorical meaning in primitive society. Architecture originated in the ritual representation of binding, joining and weaving. As a consequence the meaning of architecture is not located in particular buildings or in ideas associated represented by them, but in human action in its social and cultural context. As Semper put it, 'In a most general way, what is the material and subject matter of all artistic endeavour? … I believe it is man in all his relations and connections to the world.'[14] As a consequence, architecture should not be considered primarily in terms of stylistic development, but as an artistic representation of human life and action. Thus Semper's second design for the Dresden Opera is a statement in stone of this view, as the building itself represents both the development of architecture from primitive crafts, and a representation of the origins of human society; it sets itself up as a religious ritual, similar to drama in ancient Greece which united an architectural setting with the 'representation of significant human action'.

As Mari Hvattum has recently shown very clearly, Semper's analysis of artefacts is in fact conducted almost exclusively in formal terms.[15] Whereas one is led to expect a reading of primitive artefacts in terms of their religious and social meaning, Semper instead gives us a very restricted analysis of the way the object is made and how these crafts are represented in its design. Hvattum argues that this 'immanentisation', this viewing of the symbolism of artefacts in exclusively aesthetic terms, can be understood against the background of Semper's project of developing a comparative science of artistic invention, which could only be feasible if the objects under investigation are completely severed from their external context. Through this reduction of an artefact's meaning to the immanently aesthetic, the search for the meaning of works of art is reduced to a matter of scientific method; the question of its meaning or relevance is transformed into a question of how to acquire legitimate knowledge about the way it was produced.[16]

Both Bötticher and Semper adopted a strategy to find the meaning of architecture by reconstructing the history of architecture, by restricting the range of meaning to the purely architectonical, and by considering architecture as a representation of the acts that give buildings their meaning. Bötticher emphasised the mechanical forces revealed by the building's art forms, Semper the primitive crafts of binding, joining and weaving. Despite the different outcomes of their analyses, their conceptual frameworks owe much to the Aristotelian view of a work of art as the representation of significant human action. In opting for the formal autonomy of architectural meaning they both replace Aristotle's concept of significant human action – the entire range of meaning that lies outside of architecture considered as a craft or tectonic performance – with the strictly tectonic or craft procedures that are at work in a building or which contributed to its construction. They thus contributed to the development of a formalism in design theory and art history that would guarantee the aesthetic status of architecture as an autonomous art, and would become the basis for the sophisticated formal analysis of Wölfflin and Schmarsow. Their theories would, however, contribute very little towards understanding why classical architecture provides such fascination for every generation of architects and viewers.

Bötticher and Semper sought to provide an account of architectural meaning that is at the same time both formal and imitative. Considered from a formalist's perspective this gives great economy and elegance to their accounts: architecture represents itself, and therein resides its meaning. But this account is also depressingly tautological, and not very successful in delivering what they both set out to do: to understand classical architecture in all its richness of meanings. Seen in this light, it is not surprising that Riegl and his followers misrepresented Semper as a materialist who reduced the meaning of architecture to an expression of materials, techniques and functions.[17]

# Warburg

To put Bötticher and Semper's attempts at understanding the significance of classical architecture into a clearer perspective, it is helpful to compare their work with that of an art historian who was obsessed by the same question but who came to quite different conclusions. By the time Aby Warburg (1866–1929) became active as an art historian, classical architecture was no longer perceived as the paradigm for present-day design. It was no longer studied with the same urgency that had driven Bötticher and Semper to it. Instead, it had become the domain of archaeology and academic architectural history. The majority of its practitioners in Germany were either adherents of the documentary school of art history began by Rumohr, or formalists such as Wölfflin, interested in tracing the chronological developments of building types, styles or motifs.[18] The underlying question – what is the present-day significance of classical architecture – was hardly addressed. Hence one way of describing the advent of modernity in the study of classical architecture is to say that its significance was no longer seen in terms of design. Its history had run its course, and any new attempt at a teleological analysis that stayed within the bounds of the purely architectural, as Bötticher and Semper had done, would have been declared irrelevant and based on a fallacy.

Warburg's practice of art history was not to legitimate decisions in artistic design, or to prepare for a renaissance of classical art in his own time, or to study classical forms within the study of aesthetics. He wanted to understand the intentions and motivations of the artists, and of historians who try to make sense of them. Like Jacob Burckhardt he felt that the history of art would become 'the study of what goes on inside the beholder'.[19] Unlike Burckhardt, however, he no longer believed in the unbroken continuum of history; what interested him were the cracks, ruptures and fissures. Though Warburg wrote little specifically about architecture, his analysis of the meaning of classical art was important for a culture which no longer took the significance of that art for the present day for granted.

Warburg's strategy in understanding the significance of antique renewal for renaissance artists was twofold: the search for literary analogies for what is depicted, and the reconstruction of genealogies for particular visual motifs such as rippling hair or floating dresses. The hypothesis underlying this was that renaissance artists turned to antiquity whenever, as Warburg put it in his dissertation on Botticelli, 'life was to be embodied in outward motion'.[20] By comparing works with similar content, Warburg showed how renaissance artists could take the antique as a model, what sorts of problems might be solved by using antique models, and how these models might be used. This approach to history is in many ways the opposite of Bötticher and Semper's methods. It is based on assumptions about the artists' intentions, it is problem-oriented, it looks to non-visual sources such as poems or descriptions, it seeks to understand a work of art by reconstructing its context, and it does not assume an inherent teleology in the development of art.

One of the few cases where Warburg discussed architecture is in the talk he gave in 1923 on images from the Pueblo Indian region, based on his travels to the American south-west in 1896. Like Semper he discussed the artefacts of non-western cultures; like Bötticher he analysed religious art and practice, drawing pioneering parallels between the rites of the Indians and the frenzied rituals of Greek maenads. His aim was to convince his audience that the rituals of the Indians were not childish play, but 'the primary pagan mode of answering the largest and most pressing questions of the Why of things.'[21] Indian artefacts and rituals are read as attempts to deal with the incomprehensibility of life; in enacting the ritual the Indian attempts to become part of the hidden order of things: 'The masked dance is danced causality'.[22] Only gradually is this bonding in ritual between man and nature replaced by symbols: the Pueblo Indians actually clutch living snakes, but in the Old Testament a brazen serpent is adored, and in antiquity Asclepius becomes the serpent-deity, the healer.

Like Semper, Warburg sees a continuum between 'primitive' artefacts and ritual and modern works of art, but unlike Semper his primary aim is not the formal analysis of the primitive techniques that were used in their making. He has little to say about the forms and appearance of the objects he discusses, but a lot about their ritual and religious meaning. In Semper's analyses the context disappears behind an analysis that sees representation of the formal patterns produced by primitive crafts as the meaning. In Warburg's interpretations the formal and aesthetic aspects of a work of art disappear behind the reconstruction of their contextual meaning. Bötticher relegates the ritual aspect of Greek temple architecture to the artefacts and sculptures that he considers as decoration.

It is in Warburg's essay on non-western art that the importance of his approach for architectural history lies. Classical art and architecture are no longer for him the silent greatness and noble simplicity of Winckelmann. It is no longer a stylistic model, but the remains of man's primitive past. By considering antique art in terms of anthropology and the history of religion, he redeemed it from its isolated position as a timeless aesthetic norm that had lost its relevance. He went far beyond Bötticher and Warburg's Aristotelian and hermeneutic approach to architectural history, based on the premise that antique architecture could still speak in a meaningful way to nineteenth-century architects. For Bötticher and Warburg architecture belonged to the realm of the mimetic arts, engaged in representing significant human action. They believed that history should focus on the ways in which architecture could achieve such mimesis, and be written as the gradual unfolding of a meaningful plot. For Warburg classical art became part of modern man's prehistory, comparable to the remains of non-western primitive cultures, or to other human artefacts, and no longer endowed with a special aesthetic status. Writing the history of antiquity's revivals was no longer the reconstruction of the unfolding of its artistic potential, but an attempt to understand what went on in the minds of the artists. By placing it in its historical context, Warburg showed how classical art could still be significant for modern man, because it tells us about our buried prehistory rather than about aesthetic norms. In breaking with the nineteenth-century aesthetic tradition Warburg paved the way for current practices in visual culture. If modernity is the acknowledgement that the past is no longer an unproblematic and accessible model for the present, Warburg is one of the first of art historians to use this realisation to redefine the discipline. His transformation of art history into a history of visual culture is still of great importance to architectural history.

# Notes

1 Schinkel (1979:149).
2 On Bötticher see Herrmann (1992).
3 Bötticher (1874:1–5 and 25).
4 Bötticher (1874:20, 35–36 and 24). The quotation is taken from p. xv of the 1844 edition.
5 On these conceptualisations see Payne (1999:228–35).
6 On the archaeological tradition in the nineteenth century see Salmon (2000).
7 See Van der Woud (2001:52–3) for the way the Dutch reception of Bötticher's theory exemplifies this reduction of his ideas to a design theory.
8 See Robin Middleton's introduction to Mézières (1992:21–31) and Boffrand (2003).
9 Bötticher (1874:28). 'Wenn auch die Bildung der Sculptur und Malerei, gleicher Weise auf der Nachahmung von daseienden Vorbildern in der Natur beruhen, schreiten sie doch, mit der zunehmend sich schärfenden Fähigkeit im Erkennen und Wiedergeben des individuell Charakteristischen dieser Vorbilder, nach Seite des Begriffes wie der Form in einer Weise vorwärts, dass man zuletzt sagen kann, es sei ihr Gedanke aus dem bloss andeutenden Symbole des Anfänglichen, zur höchsten Möchligkeit realer Verwirklichung geführt werden. Solcher Wandlung gegenüber stehen die tektonischen Formen im umgekehrten Verhältnisse. Gerade jenes Streben nach Realität führt zur Abschwächung ihres eigenthümlichen Wesens: denn sie dürfen nicht über eine gewisse Grenze bloss andeutender Symbolik hinausgehen, ohne nicht diese aufzugeben.'
10 Bötticher (1874:34). 'Wie gut die Alten sich der Bildungsweise und des Verhältnisses dieser Gestaltungen bewust gewesen sind, ergiebt sich aus einer bekannten Stelle in der Poetik des Aristoteles. Hier wird von den Gebilden der Kunst gesagt, sie seien eine Nachahmung von Erscheinungen ... und eine Zusammensetzung derselben nach einer bestimmten Nothwendigkeit ...' Cf Aristotle, Poetics 1448a: 'The objects of this mimesis are people doing things.' See Russel and Winterbottom (1972:92), and their introductory remarks on pp. 85–7.
11 On the close links between historical analysis and what Semper sometimes called 'eine Kunsttopik oder Kunsterfindungslehre', using like Bötticher a strikingly rhetorical way of putting it, see for instance his late essay 'Ueber Baustile', based on a lecture held in 1869 and reprinted in Semper (1884:397). 'Wem dieses Ziel der möglichen Begründung einer Art Kunsttopik oder Kunsterfindungslehre auf derartigen Untersuchungen über den Ursprung, die Umwandlung und die Bedeutung traditionellen Typen der Kunst zu hoch gesteckt erscheinen sollte ... '.
12 G. Semper, 'Ueber Baustile', (Semper 1884:402). 'Stil ist die Uebereinstimmung einer Kunsterscheinung mit ihrer Entstehungsgeschichte, mit allen Vorbedingungen und Umständen ihres Werdens.' English translation in Mallgrave and Herrmann (1989).
13 Semper (1860:vi) ' ... die bei dem Prozess des Werdens und Entstehens von Kunsterscheinungen hervortretende Gesetzlichkeit und Ordnung im Einzelnen aufzusuchen, aus dem Gefundenen allgemeine Prinzipien, die Grundzüge einer empirischen Kunstlehre, abzuleiten.'
14 Semper (1884:403). English translation in Mallgrave and Herrmann (1989:269). See also Mallgrave (1996:277–92).
15 See Hvattum (2001).
16 In one of his early lectures, 'Ueber den Ursprung einiger Architekturstile', held in London in 1853, Semper does set out to explain the origins of Chinese and Assyrian architecture in terms of their political context, but ends up giving a very traditional analysis in terms of ground plans, building types and the handling of spaces which is heavily indebted to Sir William Chambers's book on Chinese architecture; he does not discuss early Chinese building in terms

of the primitive crafts at all; see Semper (1884:351–68). For a similar pattern of thought see 'Ueber Baustile', reprinted in Semper (1884:420), where a link is forged between Greek democracy and temple architecture, again not in terms of the original crafts but in terms of the location of the temple in the city.

17 See Mallgrave (1996:355–82).

18 See Podro (1982:1–17) for an overview of these two traditions in German nineteenth-century art history.

19 Warburg (1999:7) and Sitt (1994:227–43).

20 Warburg (1999:108).

21 Warburg, 'Images from the Region of the Pueblo Indians of North America', in Preziosi (1998:202).

22 Preziosi (1998:203).

# Chapter 5

—

# Projecting modern culture

## 'Aesthetic fundamentalism' and modern architecture

*Gabriele Bryant*

'Where once art was silent, politics and philosophy began; where now the politician and the philosopher reach their limit, this is the point of departure for the artist again ...'

Richard Wagner, *Art and Revolution*

'In the aesthetic differentiation ... art becomes its own point of view and establishes its own claim to power.'

Hans-Georg Gadamer, *Truth and Method*

'The multifariousness of the modernisation process constantly produces partial modernities, disjunctions, different mixtures of tradition and modernity, and conflicts stemming from that',[1] writes the late German historian Thomas Nipperdey. The debate about the relationship between social, political, economic, cultural and artistic modernity is still far from resolved.[2] In the case of Germany, for example, its 'incomplete modernisation' has been the subject of discussion amongst historians for some time, and Matthew Jefferies, reflecting on the relationship between politics and culture, writes that 'Without a degree of "modernisation", "modernism" would have been unthinkable, yet the relationship between the two was always problematic ... '[3] This problematic relationship is the theme of this chapter.

The inherently oppositional nature of much of what we have come to refer to as 'modernist art' in relation to its own time and culture has been thoroughly analysed within the areas of painting and literature.[4] Whilst critics speak of the 'literary and artistic subversion of rationalist modernity'[5] in modernist art, the modern movement in architecture is still frequently celebrated as the very embodiment of the modern 'spirit of the age'. However, the ambivalent nature of this 'modern spirit' – itself an essentially modern construct – needs to be analysed further.[6]

Within architectural history the simplistic assumption of a break between nineteenth and early twentieth century architecture in terms of their relative modernity, as well as the teleological structure of the canonical accounts of architectural modernism, have increasingly come under attack.[7] Hypostatising the modern movement in architecture as the culmination, if not the epitome, of modernity appears increasingly one-dimensional, a simplistic and misleading ideological construct. As this construct has frequently provided the starting point for a critique of the triumphs and failures of architectural modernism, it must be challenged if we are to expand our insights into artistic modernism and its complex relationship with modernity. This chapter explores the theoretical background for a different way of thinking about conflicting modernities, and suggests how this different way of thinking might alter our understanding of the development of the modern movement in architecture and its culturally reformist agenda.

It has been convincingly argued by Koselleck and others that a funda-mental aspect of modernity, in contrast to groups or societies since the middle ages that defined themselves as 'modern' as opposed to the 'antiqui',[8] is that with a developing historical consciousness being modern implies not just a distinction from the past, but also an open attitude towards the future. History itself – and with it humanity and culture – become a project.[9] This idea of the creation of modern culture as a project, based on what the philosopher Hans Blumenberg has called modernist self-assertion,[10] implies a belief in the creatability of the future, a Nietzschean faith in the plasticity of the world.[11] This is nowhere more pronounced than in modernist movements which aim to transform and recreate life through art.[12]

The term 'aesthetic fundamentalism' has been introduced by Stefan Breuer to characterise groups, movements and artistic cults which are highly critical of modernity and its effects on the individual and society, and which aim to overcome these ills through the creation of an aesthetic countermodel.[13] Where fundamentalism has been defined in sociology as a 'rebellion against the rationalist tendencies of the Western world as a whole, and at the same time against its deepest institutional foundations',[14] aesthetic fundamentalism is related to religious fundamentalism in that it aims to work against social disinte-gration by turning to other spiritual values. Aesthetic fundamentalism aims to provide deeper meaning not by returning to older systems of belief, but through the creation of new  values in and through art. It has its roots in the literary debates of romanticism and the philosophy of German idealism, which exerted a profound influence on art and architectural theory from 1800 onwards. During the nineteenth century it reached new theoretical heights in the aesthetic conceptions of Richard Wagner and Friedrich Nietzsche, and came fully to the fore in the early twentieth century in the 'aesthetic opposition' of the years before and immediately after World War I.

Responding to what Max Weber famously saw as the 'disenchantment of the world' in modernity, romantic *Kunstreligion* aimed at the poetic reenchantment of the world. But romanticism does not just provide a contrast to the enlightenment project; in many ways it is its mirror image: *Kunstreligion* and scientism go hand in hand.[15] Aesthetic fundamentalism, too, remains entangled in the assumptions implicit in the problems it aims to overcome. The essence of modernity has been discussed by Zygmunt Bauman as the exclusion of ambivalence:

> The typical modern practice, the substance of modern politics, of modern life, is the endeavour to eliminate ambivalence, an endeavour to clearly define everything – and to suppress or eliminate anything that resists such clear definition … intolerance is thus the natural inclination of modern practice … the dysfunctionality of modern Culture is its functionality.[16]

It remains to be seen whether aesthetic fundamentalism does indeed offer an alternative, a corrective principle to modern dysfunctionality, or whether it leads instead to an expansion of the frame for the modern subject's total claim of order in its world.

An analysis of the ideology of aesthetic fundamentalism reveals an important movement within the history of German modernism which can be characterised in seemingly paradoxical terms as antimodern modernism, a tendency within artistic modernism which, while aesthetically and stylistically innovative, is based on a rejection of the culture of modernity. It describes an art which is of its time by being in retreat from, and if not to say in dialectical opposition to it.[17] Antimodern modernism is the result of an antagonistic model of history which was first developed in German idealism and which presupposes the modern projective mode of thinking – the way in which, with the beginning of the *Neuzeit*, history itself had been turned into a project.[18] It also assumes the separation of art from its historical context through an 'aesthetic differentiation' where, in Hans Georg Gadamer's characterisation, 'the work of art loses its world … to ultimately establish its own claim to power.'[19]

With the conceptual shift of art into an atemporal ideality, the idealist paradox of non-synchronicity of history and art was established. Thus arose the possibility of art's – and the genius-artist's – radical dissociation from the immediate historical context to assume the status of 'transcendental doctor' (Novalis). While the proponents of agendas for an artistic 'cure' of modern society frequently make use of the concept of the modern *Zeitgeist*, art in aesthetic fundamentalism is conceived as its antidote, reflecting the rejection of the given and the projection of the new as 'Selbsterzeugung der avisierten Zukunft' (self-generation of the anticipated future – Reinhart Koselleck). The revolutionary myth of a reversal of history[20] is incorporated into the new conception of art as the catalyst of historical change and the creation of a new culture *(Kulturschöpfung)*.

The growing significance of the aesthetic sphere as the arena of religious and sociopolitical change can be traced back to German idealist aesthetics. In the shadow of the French Revolution, German philosophers and poets conceived of an 'aesthetic revolution', in which the newly autonomous aesthetic sphere was to provide an aesthetic utopia as a countermodel to bad reality. The revolutionisation of man and society was to be achieved dialectically through the experience of art. The idealist–romantic agenda of an aesthetic revolution culminates nearly a century later in Friedrich Nietzsche's philosophy of art, in which man and his world are conceived as a 'self-created work of art',[21] and its echoes can still be heard in one of the most famous propositions of modern architectural discourse, Le Corbusier's 'Architecture or Revolution. Revolution can be avoided',[22] which can also be read as 'Revolution through architecture, architecture *as* revolution'.

The aesthetic revolution establishes the notion of art as the *Vorahmung* (prefiguration) of life,[23] of art as an autonomous realm, an aesthetic utopia, which is to serve as a dialectical countermodel to reality rather than a reflection of it, a *Vorschein* (anticipation) of an ideal state. In idealist *Vorschein*-aesthetics, art becomes the platform from which a reformation of life can be achieved. This process constitutes a fundamental change in the status of art, in which 'Art becomes its own point of view and constitutes its own claim to power'.[24] Following Gadamer's critique of the results of aesthetic differentiation, Helmut Kuhn in *Ontogenese der Kunst* discusses the idealist postulation of aesthetic autonomy, art's declaration of independence as the conceptual precondition for modern aesthetic activism. 'Autonomy, as understood by idealist aesthetics, does not mean the isolation of Art, but it formulates a total claim.' The ambivalence of aesthetic autonomy appears as 'exclusive autonomy, which begins by pushing Art into a perhaps splendid, but in any case infertile, isolation, only to then tip over into its instrumentalisation'.[25] This reversal of the idealist agenda of autonomy indeed constitutes an aesthetic revolution with far-reaching consequences for modern art theory and practice.

If the belief in the perfectibility of man and the constructibility of the world are to be the hallmarks of modern utopian thinking,[26] the aesthetic fictionality of the ideal becomes the central creed in aesthetic modernism. Jauß describes the avant-garde's aim to transgress the boundaries between aesthetics and politics 'in order to achieve in an aesthetic revolution the new beginning of history under the directorship of art'.[27] When the utopia of the aesthetic becomes charged with *Geschichtsphilosophie*[28] (or *Geschichtsphilosophie* charged with aesthetics) the activist phase of the aesthetic revolution begins, as art itself becomes the historical subject, an agent of historical change.

The connection between *Kulturpessimismus*, antimodernist polemics and the idea of aesthetic utopias as counterworlds to sociopolitical reality have to be emphasised in aesthetic fundamentalism. As the aesthetic realm is conceived as a kind of *hortus conclusus* by eighteenth century aestheticians, art assumes the role of incubator of the ideal new, the catalyst for turning *Masse* into *Volk*, civilisation into culture, society (*Gesellschaft*) into community *(Gemeinschaft)*.[29] Where *Kunstanschauung* is fused with *Weltanschauung*, it is a small step to an *Aufhebung*[30] of total art in an aestheticised life. This idea found its clearest exponent in Friedrich Nietzsche, and affected many architects and theorists in the early twentieth century, including Walter Gropius, who wrote 'Maybe the artist is called upon to live a work of art rather than create one.'[31]

The idealisation of aesthetic states in past cultures, and the dream of creating these ideal states anew in and through art, frequently go hand in hand. The dream of an aesthetic state first appeared in the writings of Friedrich Schiller, at the end of *Briefe über die ästhetische Erziehung des Menschen* (Letters concerning the Aesthetic Education of Man, 1795) as the 'Reich des schönen Scheins' (realm of beautiful illusion). The idealist concept of an aesthetic state is linked to Schiller's belief in the transformation of political and social reality through self-cultivation, *Bildung* through aesthetic culture.[32] It is the primacy of political intentionality in Schiller's aesthetic ideas that must be emphasised.[33] As Jürgen Habermas has argued, Schiller's *Briefe* constitute the first programmatic exposition of an aesthetic critique of modernity;[34] the aim was not 'Ästhetisierung der Lebensverhältnisse' (aestheticisation of the conditions of life), but 'Revolutionierung der Verständigungsverhältnisse' (revolutionisation of the conditions of discourse).[35] And Paul de Man has insisted in this context that 'the aesthetic … is primarily a social and political model', warning against attempts 'to relativise and soften the idea of an aesthetic state'.[36] The aestheticisation of the old utopia of an ideal state and its role in the aesthetic–political agenda of idealist dialectics – and the modern avantgarde – remains the object of controversy.[37] The aesthetic–political goals of Weimar aesthetic humanism[38] radicalised by the experience of revolution, experienced their *Aufhebung* in the philosophy of Hegel, and culminated in the aesthetic worldview of Friedrich Nietzsche a century later.

In the modern idea of the aesthetic state, a shift takes place from the utopian idea of the Greek *polis* as crucible for the creation of aesthetic–moral beings to the attempted (re)creation of an aesthetic state and its inhabitants by the artist as a virtuoso of *Lebenskunst*. Stephen Collins has convincingly argued for the conception of the modern notion of the state: 'Fundamental for the whole reconceptualisation of the idea of society was the faith that the common good as well as order is a human creation.'[39] This reaches its conclusion in the idea of an aesthetic state. The quest for an artistically constituted new order and the emergent utopia of the new man are two sides of the same coin – a modern faith in the total creatability of man and society.

Examples of aesthetic fundamentalism abound in nineteenth century architectural theory, and one could argue that it lies at the very core of the aesthetic concept of romantic historicism. In the work of the Prussian architect Karl Friedrich Schinkel architecture is elevated into the role of *Kulturträger*, protagonist in the creation of a new culture. But although Schinkel was the most prominent advocate of an idealist architectural theory in the Schillerian vein, he was by no means the only architect of the time in whose declarations we find an echo of the aesthetic concerns of German idealism. The idea of ennobling man, of achieving a transformation of humanity through the art of building, was a central concern in German architectural theory around 1800. While the idealist movement of architecture as art has frequently been criticised for aiming at a mere aestheticisation of building and thus capitulating in the face of growing social change, the moral elevation of society through art, the Schillerian dream of achieving a good society via aesthetic education, was seen as the social duty of the idealist architect.

The 'vorteilhafte Würkung [sic] auf die Veredelung des Menschen', the beneficial effect of fine architecture for the ethical formation of man, had been emphasised by Johann Georg Sulzer in his *Allgemeine Theorie der Schönen Künste* (General Theory of the Fine Arts, 1771–4), and was expanded by Gottlieb Huth, culminating in his essay *Von der Wirkung der Baukunst auf die Veredelung des Menschen* (Concerning the Effect of the Art of Building on Ennobling Man, 1794). This transfer of theories of *Dichtkunst* (poetology) into architecture also marks a general shift from a preoccupation with the 'character' of particular architectural elements towards a new concern with *Wirkungsästhetik* in architectural theory.[40] The idea of architecture effecting individual and, by extension, sociocultural change had an important impact on art and architectural theory for generations to come, and in the early twentieth century the tradition of aesthetic fundamentalism can be traced as a major motif in the emergent modern movement in architecture. Its critical discussion aims at further dispelling what Colin Rowe has referred to as the 'still pervasive dogma of modern architecture's immaculate conception'.[41]

One of the most significant founding documents of architectural modernism is Walter Gropius's Bauhaus Manifesto of 1919, in which the quest for a new social order through a new art is clearly voiced. The context of its conception in the wake of the 1918/19 revolutionary uprisings in Berlin,[42] and the formation of the *Arbeitsrat für Kunst*, have been analysed in detail by Marcel Fanciscono and Iain Boyd Whyte.[43] The example of the early Bauhaus and the ideological debates surrounding serve as a test case for my argument that the role of aesthetic fundamentalism in modern architecture, from the turn-of-the-century *Jugendstil* and *Lebensreform* movements to German expressionism and the programme for social reform in the *Neues Bauen*, warrants further examination.

Many of the themes that dominate the debate about the role of art and architecture in post-war Germany, most crucially the Nietzschean idea of the world as work of art, are familiar from the *Jugendstil* movement. The artist's faith in the role of architecture as the foundation of a new social order was transmitted to the next generation, which was influenced by Nietzsche's 'artists' metaphysics'.[44] At the turn of the century Peter Behrens proclaims that '[artistic] man shall be the creator of Culture',[45] and this Nietzschean topos became commonplace in expressionist circles in the next decade.[46] The equation of art and the formation of man is clearly expressed by Adolf Behne: 'Each art is ultimately *Menschenbildnerei* [the formation of man]. Architecture is just its strongest and most apparent manifestation.'[47] In German expressionism the modern artist was firmly established in the role of *Menschenbildner* and *Lebensgestalter*, with architecture now presented as the heir to the Wagnerian notion of the artwork of the future as the means to 'bring salvation from this most unfortunate time'.[48]

Schiller had developed his ideal for an aesthetic education of man in the enlightenment tradition, in accordance with the utopian conception of a *eupsychia*.[49] In Nietzsche's radical vision, however, the new man and new life as such are conceived as works of art, and this led in turn to the elevation of the idea of the artist as sociopolitical leader. Where the creation of an 'art of the artworks'[50] was rejected for the sake of *Lebensgestaltung*, the aggrandisement of the artist into a cultural and political leader remains but a small step. This artist does not just issue a call for new social order, but is driven by the faith in the possibility of creating this new social order in and through art. The pathos of the expressionist dictum, where the artist is called upon 'to remember his high, magnificent, priestlike, divine calling and seek to raise the treasure that lies in the depths of men's souls'[51] (the echoes of romantic *Kunstreligion* and Zarathustrian discipleship are impossible to miss in Bruno Taut's words here), is replaced by a greater sobriety of discourse. But the agenda of an artistic reformation, a redemption of man and society, remains unchallenged.

In 1918, as a founder of the *Arbeitsrat für Kunst*, Bruno Taut called for the erection of a 'Cathedral of Socialism'. Invoking the romantic notion of an *Ars una*, he spelt out his vision of the great building as a symbol of and catalyst for a spiritual revolution, the foundation of a better future:

> Art! – that is a great thing, when it exists. Today this Art does not
> exist. The fragmented tendencies can only find their way back to a

single unity under the wings of a new architecture ... The immediate vehicle of the spiritual [*geistig*] forces and moulder of the sensibilities of the general public, which today are slumbering and tomorrow will awake, is architecture [*der große Bau*]. Only a completed spiritual revolution will create this architecture. But this revolution, this architecture, will not come of their own accord. Both must be *willed* – the workers of today must prepare the way for the new architecture. Their work on the future must be supported.[52]

The aporia of an artistic programme which is to be both symbolic of and instrumental in achieving an aspired-to state of spiritual and sociopolitical unity – and the 'work on the future' is to be understood here as no mere metaphor, as the creation of the 'project future' becomes the ultimate task of the artist in aesthetic chiliasm – runs through many manifestoes of aesthetic fundamentalism in German modernism. The proposed solution usually sees the artist as a popular medium who can divine the underlying spiritual essence of his time, and give expression to it through his *Gestaltung*, thus advancing as the Nietzschean leader, the artistic *Vorahmer*, of culture. Where the artists of the life-reform movement had been characterised as possessing an almost religious faith in the role of the artist as *Kulturbringer*, in the socio-aesthetic model of the disciples of Nietzsche's Zarathustra the architect emerges as *Weltbaumeister*.[53] As the architect–director of the 'great work' appears as the metapolitical leader of the new man and his new society in Bruno Taut's vision of 'socialism in the unpolitical, suprapolitical sense',[54] the expressionist 'socialism of the artist' proves to be an expansion of the romantic cult of the genius as *Kulturschöpfer*.

The quest for an artistic–spiritual revolution explored by Taut in the wake of the November Revolution of 1918 bears more than a faint trace of the response amongst German intellectuals to the French Revolution, which led to the notion of an aesthetic revolution. It also echoes Richard Wagner's mid-nineteenth century assertion in *Die Kunst und die Revolution* (Art and Revolution) that 'Where once art was silent, politics and philosophy began; where now the politician and the philosopher reach their limit, this is the point of departure for the artist again ... '.[55] And it is taken up by Taut's colleague in the *Arbeitsrat*, Walter Gropius, who wrote in *Baukunst im freien Volksstaat* (Architecture in a Free Republic),

We need a new communal spirituality of the entire people. Government alone cannot provide it. ... Not even political, only complete spiritual revolution can makes us 'free' ... But how do the people achieve this unity of the spirit which alone engenders the natural rhythm of wholeness? A great, all-encompassing art assumes the spiritual unity of its time. It needs the most intense relationship with

its environment, with living people … Today's generation has to
start from scratch, rejuvenate itself, create a new humanism and a
universal form of life for its people … Then the people will once
again cooperate in the great artistic work of time … . From ancient
times, the architect has been called upon to conduct this orchestra.
Architect, that means: leader of art. Only he is able to raise himself up
again to be such a leader of art, to be its first servant, its superhuman
guardian and organiser of its undivided totality … He will have to
surround himself with spritually like-minded coworkers in close
personal contact – just as the masterbuilders of the Gothic cathedrals
in their medieval guilds. And in forming the new, living-and-working
cooperatives for all artists, he will be building a cathedral of freedom
for the future, not hindered, but borne along by the people as a
whole.[56]

From the late eighteenth to the early twentieth, the quest for an
aesthetic utopia as *the tabula rasa* on which to start remodeling society from
scratch, the idea of the artist as leader and spiritual medium in the revolution of
humanity, runs as a leitmotif through German art and thought. The legacy of
Schiller's programme for the aesthetic education of man, and the idea of an
aesthetic revolution in Nietzsche's 'artists' metaphysics', have shaped the early
twentieth century programme of art as precursor of a new social order.

The relationship between the individual and society in expressionist
aesthetic activism has frequently been misunderstood. Though regularly called
into question in recent scholarship, the emphasis of a break between
*Expressionismus* and *Neue Sachlichkeit* (new objectivity) still dominates much of
the discussion of the history of modern architecture. An understanding of the
inseparability of the cult building – the cathedral or crystal dome of Expressionism – from its environment, and of the dialectical structure of this conception for a reformation of society through art, is essential to understanding the
continuity between the seemingly narcissistic crystalline temple-buildings of
expressionism and the ideology of the *Neues Bauen*. The legacy of the expressionist cult building as aesthetic utopia lives on in the modern belief in the spirituality of the world of geometric abstraction. As the critic Adolf Behne was one
of the first to admit, the move towards a new objectivity in art and architecture
shared, rather than rejecting, the romantic–idealist metaphysical constructs of
expressionist art theory: 'There is no question but that the functionalists, even
the most *sachlich* ones, could more readily be classified as romantics than as
rationalists … Their attitude inclines toward philosophy and has a metaphysical basis', he wrote in 1923 in *Der moderne Zweckbau*.[57] In the early twentieth
century quest for abstraction as a path to the absolute, with its faith in the
symbolic power of geometry, *Bauen* and *Gestaltung* are invoked as

quasi-metaphysical constructs.[58] Where *Bauen* is elevated into *Gestaltung der Wirklichkeit* (formation of reality) by Mies van der Rohe and others, it is ultimately equated with Nietzschean man's 'original metaphysical activity'.

In the ideology of the *Neues Bauen* the term 'new' suggests the influence of the expressionist concept of the new man, while also implying a rupture with the past times and with the idea of a world of original creation.[59] The idea of the creation of an autonomous world, what the nineteenth century aesthetician Conrad Fiedler called 'the production of reality', becomes the essence of modern *Formgestaltung*.[60] The basis of this new *Gestaltung* is, in Fiedler's words, 'the creation of *Gestalten*, which only come into existence in this way', and is thus not mimetic, but a creation parallel with (and thus equivalent to) nature.

The shift concerning the foundation of such a new world from art to other realms of *Gestaltung* during the formative years of the modern movement in architecture marks a far-reaching expansion of idealist art theory. In the concept of 'form-creative technology', a transfer of nineteenth century idealist aesthetic categories is made from art to technology, the impact of which can be seen within architectural modernism. The paradigm of a *Vorahmung der Natur* that emerged in the aesthetic revolution is now transferred from art to a faith in the all-encompassing nature of *Gestaltung* as a seminal force, not just in the creation of a new aesthetic, but as a starting point for cultural renewal.

Acknowledging that the phenomenon of modernity seems to elude every attempt to establish a unified set of criteria or devise a systematic theory, Lachtermann proposes that the idea of construction lies at its core. He claims 'that the "idea" giving significant shape to the "constellation" of themes ingredient in modernity is the "idea" of construction or, more broadly, the "idea" of the *mind* as essentially the power of making … '.[61] The crucial point here for my argument is the link between this idea of construction and modern aesthetics. The rationalist position of constructive reason, though often rejected, is in fact complemented in the project of modernity by the idea of an aesthetic constitution of the world in the project of modernity. Dalibor Vesely has described art, understood aesthetically, as an instrumental mode of representation, in contrast with a symbolic mode.[62] The inherently problematic nature of this approximation of the realm of aesthetics to instrumental rationality has been analysed by Cornelia Klinger in an exemplary investigation of the modern legacy of romanticism:

> Thus the attempt to create meaning in the medium of the aesthetic does not lead to a new symbolic order that transcends modernity, but instead (in the more harmless case) to the creation of an illusion (*Scheingebilde*) or (in the more dangerous case) to an extension of economic–technical rationalism, that is, an application of its 'rules for the manipulation of matter' onto those areas which had hitherto been outside its sphere of influence. The realms of *Innerlichkeit,* the

psyche, soul and Culture ... are appropriated in the attempted creation of unified whole and thus in a fusion with instrumental reason subjected to manipulation.[63]

It is this fusion of seemingly opposed forces which emerges as crucial for an analysis of the legacy of aesthetic fundamentalism in modern art and thought. It means that such formulations as the Bauhaus motto of 1923 – 'Art and Technology: A New Unity!' – appears as less of a conceptual break with earlier tendencies than it might at first appear. In the first third of the twentieth century, as artists and intellectuals of diverse ideological backgrounds and orientations sought a 'connection with other spheres of values – especially politics and technology',[64] aesthetic and technological fundamentalism joined forces in the ideology of the modern movement.

The powerful fusion of aesthetic and technological fundamentalism had come fully to the fore in architecture by the 1920s and 30s, constituting an essential part of the ideological legacy of the modern movement. As we expand our understanding of architectural modernism to embrace its multifariousness, we may also have less difficulty in accepting that its various manifestations reach across the whole political spectrum, and that the dream of the creation of a new man and a new culture is not confined to a particular political or artistic movement. The promise of individual and social reform, of historical redemption through the great work, seems irresistible to artists, intellectuals and politicians, from the avantgarde to the forces of 'reactionary modernism'.[65] The idea of a *Vorahmung der Natur*, that is, *Vorahmung der Welt* in art as it was established in the aesthetic revolution, emerges in artistic modernism as the paradigm of the quest for a new social order. We may yet come to recognise that the tradition of aesthetic fundamentalism provides the conceptual basis for the most diverse movements in modern art and architecture.

# Notes

1   Nipperdey (1988:404).
2   As critics of a uniform view of the modernisation process have pointed out, the dynamics of modernisation did not consist of 'the substitution of one set of attributes for another ... but rather in their mutual interpenetration and transformation.' (Tipps 1973:217).
3   Jefferies (1995:5).
4   See for example Silvio Vietta, 'Zweideutigkeit der Moderne: Nietzsches Kulturkritik, Expressionismus und literarische Moderne', in Anz (1994:9–20).
5   Michael Stark, 'Ungeist der Utopie?', in Anz (1994:152).
6   As Anthony Cascardi has argued, what we are dealing with is 'an antinomic configuration ... the result of an attempt to make theory the master of history ... ' which, traced back to the seventeenth century and the 'foundational project' of Descartes, 'is at once strongly revisionist

and staunchly rationalist, [yielding] a series of paradoxes and antinomies which reflect the instability of the synthesis from which it is forged.' (Cascardi 1987:207-8).

7   See for example Schwarzer (1995).

8   For a *begriffsgeschichtliche* survey see Hans Ulrich Gumbrecht, 'Modern, Modernität, Moderne', in Brunner (1978:vol.IV, 93–131), and also Fritz Martini, 'Modern, die Moderne', in Kohlschmidt (1958:vol.2, 391ff).

9   See Horst Thome, 'Modernität und Bewußtseinswandel', in Mix (2000:15–27).

10   Blumenberg (1983).

11   It is generally understood that Nietzsche's is not a *Werkästhetik*, but a *Schaffensästhetik* (the emphasis is not on the artwork as object, but on the act of creation), with *Schaffen* not restricted to the act of creating artworks, but expanded into the principle of creative *Lebensgestaltung* (formation of life). With this emphasis in mind it cannot be surprising that it is precisely those artists who rejected the 'art of the artworks' who aimed at a transition of art into life, the term *Lebensreform* should be taken quite literally here – that is, the artists who conceived of themselves as *Lebensgestalter*, enthusiastically endorsing Nietzsche's 'artists' metaphysics' in the early twentieth century.

12   For an extensive survey and critical discussion of the life-reform movement, including the Nietzschean impact on it, see Buchholz (2001).

13   Breuer (1995).

14   Talcott Parsons, *Soziologische Theorie*, quoted in Breuer (1995:2).

15   Klaus Heinrich, 'Der Untergang von Religion in Kunst und Wissenschaft', in Zinser (1986:263–94).

16   Bauman (1995:20–22).

17   It has to be emphasised that 'anti-modern' is not to be equated simply with 'reactionary' or, even worse, be automatically interpreted, especially in the German context, as pre- or proto-fascist. Rather, 'anti-modern modernism' reaches across the ideological spectrum, from the forces of reaction to the avant-garde, which has been defined as a new position standing in opposition to its immediate past as well as its own sociocultural context. See Weightman (1973).

18   In his discussion of the semantic background of the notion of *Neuzeit* ('new time'; modernity) in German, Reinhart Koselleck mentions the related concepts of *Neuthum* and, particularly significant in our context, *Neuwelt*. Koselleck, "Neuzeit': Zur Semantik moderner Bewegungs-begriffe', in Koselleck (1977:264–99).

19   Gadamer (1986b:88).

20   For a discussion see Reinhart Koselleck, 'Historische Kriterien des neuzeitlichen Revolutions-begriffs', in Koselleck (1989:67–86).

21   Nietzsche refers to 'die Welt als ein sich selbst gebärendes Kunstwerk' (Nietzsche 1988: Vol. 12, 119), and 'uns selber machen aus allen Elementen eine Form gestalten, das ist die Aufgabe!' (Nietzsche 1988: Vol. 9, 361).

22   Le Corbusier (1946a:289).

23   The term *Vorahmung*, which is introduced in this context by the philosopher Hans Blumenberg, is the opposite of *Nachahmung,* the German term for 'imitation'. *Vorahmung* thus implies a reversal of the mimetic process, a Copernican turn in the conception of art, in which art is now not to imitate, but to anticipate or prefigure life (Blumenberg 1993:93). For an art historical analysis of this shift of paradigm, see also Hofmann (1970).

24   Gadamer (1986b:88).

25   Helmut Kuhn, 'Die Ontogenese der Kunst', in Henrich and Iser (1982:81–131).

26   Ferdinand Seibt, 'Utopie als Funktion abendländischen Denkens', in Vosskamp (1985: Vol. 1, 259).

27   Hans Robert Jauss, 'Mythen des Anfangs: Eine geheime Sehnsucht der Aufklärung', in Jauss (1989:65).

28   For an in-depth analysis of this development, see Szondi (1991).

29   See the seminal study by Stern (1961).

30   The Hegelian term *Aufhebung* implies the elimination, elevation and preservation of an idea, the continued existence of a spiritual entity on a different level or in a different manifestation.

31   Walter Gropius, quoted in Prange (1991:176).

32   Chytry (1989:3).

33   Criticism has been levelled against it from exponents of different ideological persuasions, mostly arguing that Schiller's conception led to an exodus from the realm of politics, and that it anticipated the modern 'aestheticisation of *Lebenswelt*'.

34   It is striking how Schiller's critique of the fragmentation and alienation of man in modernity in the *Briefe* anticipates some of Karl Marx's radical analysis.

35   Jürgen Habermas, 'Exkurs zu Schillers Briefen über die ästhetische Erziehung des Menschen', in Habermas (1985b:62).

36   Man (1984:264). For a wider discussion of this theme see Eagleton (1990).

37   For a discussion see Ch. 5, 'Die Wendung zur Gemeinschaft – von der Gemeinschaft als Familie zum Staat als "großes Ich"', in Klinger (1995:171–93).

38   For this theme in modern German thought see Gleissner (1988).

39   Collins (1989:32).

40   *Wirkungsästhetik* places the emphasis on the physical or mental/psychological effect of art on the observer/listener/user. For a discussion see Klaus Jan Philipp, 'Von der Wirkung der Baukunst auf die Veredlung des Menschen: Anmerkungen zur deutschen Architekturtheorie um 1800', in Nerdinger (1990:43–7).

41   Rowe and Koetter (1978:30).

42   Weinstein (1990).

43   Franciscono (1971); (Whyte 1982).

44   See Kostka and Wohlfarth (1999).

45   Behrens (1901:22).

46   See Hillebrand (1978).

47   Quoted in Pehnt (1998:29).

48   This theme is discussed in more detail in my essay 'Timely Untimeliness: Architectural modernism and the idea of the *Gesamtkunstwerk*'. Chapter 10, part 2 of this book.

49   See Manuel and Manuel (1979).

50   Nietzsche (1988, Vol. 2:453).

51   Bruno Taut, quoted in Prange (1991).

52   Taut (1918:16–19).

53   The title of Bruno Taut's 'architectural drama' *Der Weltbaumeister* of 1919 is indicative of his meta-aesthetic intentionality. For a discussion see Whyte (1982) and Prange (1991).

54   Taut (1919b:59).

55   Epigraph at the beginning of Wagner (1983).

56   Quoted in Long (1995:198–9).

57   Behne (1996:122–3).

58   For an analysis of the notion of *Bauen* see Neumeyer (1986:133–4).

59   See Rosemarie Haag Bletter's introduction to Behne (1996).

60   For a discussion of the modern notion of *Formgestaltung* and its philosophical background, see the introduction by Detlef Mertins to Behrendt (2000).

61   Lachtermann (1989:4).

62   Vesely (1985).

63   Klinger (1995:219).

64   Emmerich and Wege (1995:7).

65   Herf (1984).

# Chapter 6

# Modernity and the question of representation

*Dalibor Vesely*

An overwhelming number of people in modern society think of architecture and the city as 'a given to be endured, an art to be designed, a madness separate from reality, or as a fragment that cannot endure.'[1] They occasionally respect or admire it, but more often they tend to 'flee it, condemn it, ignore it, try to live in it, or just use it to create their own fragments. Everyone emotionally or intellectually, politically or economically grabs his fragment, which is partially real and creates a total reality with it. The splintered identities, the competing ideologies, the fractured parties and the glaring, cluttered advertising of competing businesses assault the person and the society from a thousand sides'.[2]

Architects themselves are obsessed with the differences that separate them and give their work a dimension of novelty and originality. This leaves behind the common references and goals which contribute to the long-term cultural relevance of their work. The exclusive emphasis on difference and originality leads not only to a problematic merit of the results, but also to their separation from the common world which we all, in one way or another, share.

There is a temptation, quite understandable, to describe the world which we all share as the given or real world. However, to use the term 'real' becomes extremely difficult in a situation dominated by competing ideologies and opinions where even 'virtual reality is just another reality', and where the 'fact that it is computer generated with no physical existence makes it no less real'.[3]

In everyday parlance there is a tendency to save the meaning of 'real' by associating it with the domain of practice, in our case with the practice of the office or the building process. This kind of practice is usually considered to be radically different from the non-reality or lesser reality of a project, from the deeper understanding and foundation of the design problem, or from a clearly defined vision. It does not need much imagination to see that there is a certain truth in such a definition, but that in most cases it is rather misleading. The character of practice is not always practical; in fact it is more often theoretical.

To see this elementary truth it is sufficient to look more closely at the nature of a typical brief or programme, the design criteria and the conditions of its execution. If we take as a preliminary criterion of reality the horizon of our everyday common-sense world, a whole book would have to be written in order to explain how the process of design and building is related to this horizon. The content of this book would have to be devoted almost entirely to the different aspects of representation and to its history. We may already sense that representation is not limited to the visible physiognomy of buildings and spaces, but that it is related more to the situational structure and meaning of architecture. It is in this relation that the nature and the degree of the reality of architecture can be established. However, before we can ask what is the reality, the structure and the meaning of architecture, in other words what architecture is and what it represents, we have to understand the role of representation in the creation and experience of architecture.

## The nature and limits of representation

The problem of representation is closely linked with the process of making (*poiesis*) and with creative imitation (*mimesis*). Each project, however small or unimportant, begins with a programme or vision of the anticipated result. The formation of a programme or a vision always takes place in the space of the experience and knowledge which is available to us. The result can be seen as the actualisation of one of an infinite number of possibilities. The formation of the programme can be modified or improved through words or drawings because the field of possibilities is potentially present and available in them. It is under such conditions that the actual result becomes a representation of all related latent possibilities which bring into focus their typical characteristics and enhance their presence. This happens each time we succeed in understanding and grasping what is essential about a performance space, concert hall, or particular urban space. In contrast to conventional understanding, 'representation does not imply that something merely stands in for something else as if it were a replacement or substitute that enjoys a less authentic, more indirect kind of existence. On the contrary what is represented is itself present in the only way available to it'.[4]

This brings us to the point where representation practically coincides with the essential nature of making, and in particular with the making of our world. In the original Greek sense, making as *poiesis* is a bringing into being of something which did not previously exist.[5] The bringing into being is a creative step in which the open field of creative possibilities is transformed into a representation articulated by gesture, word and image. This rather limited mode of possible representation is the only way that we can, due to our finite abilities, come to terms with the inexhaustible richness of reality. Because we have no other access to reality, certainly not a direct one, the unity of representation and what is represented is for us the only possible criterion of the reality of being.[6]

The problem of representation as we know it from European history can be followed through a development in which the primary unity of representation gradually became a question of continuity between representation and what is represented. The articulation and preservation of continuity took place in a framework which was until relatively recently dominated by cosmological thinking. It was only in the second half of the eighteenth century that the cosmological paradigm was replaced by a historical one, characterised by a new definition of the origins of representation, the concept of the primitive hut, the formation of new typologies, and the beginning of historicism which culminated in the cultural relativism of the early twentieth century. It is virtually impossible to say what reference framework characterises the twentieth century. Despite the *Gesamtkunstwerk* legacy of art nouveau, the expressionists' unifying vision of the cathedral of the future, the surrealists' dream of the reconciliation of all opposites, the world of the twentieth century remained fragmented and torn apart by deep conflicts. The situation was not changed or improved by concerted efforts to establish an international framework for creative co-operation. The international constructivist movement united progressive artists, international surrealism, and the call for an international centre for modern architecture. The intentions behind this effort were formulated in many different ways. The following is typical: 'From all over the world come voices calling for a union of progressive artists. A lively exchange of ideas between artists of different countries has now become necessary ... the long dreary spiritual isolation must now end. Art needs the unification of those who create ... Art must become international or it will perish'.[7]

The failure of these attempts has much to do with a deep dilemma in the nature of the avant-garde, the dilemma between the need for participation and the desire for individual freedom and emancipation. The possibilities of genuine participation were compromised by the naive belief that the main forces of unification, objectivity and universality would come from technology. The quasi-religious status given to art in the nineteenth century was transferred in this century to technology. After World War I it was generally assumed that 'from amidst the hardest struggles an architectural style will arise which bears the stamp of the new age; for above everything that has happened stands the historical meaning of the new facts, ensuing from the victories of technology over matter and the power of nature. Every style is enforced on an age like fate; it is the manifestation of the era's metaphysical significance, a mysterious imperative'.[8]

The elevation of technology into a universal metaphysical foundation of a new era of culture was the last step in a process in which it seemed possible to reduce all that is worth knowing about the making of architecture into a transparent productive knowledge. It did not seem to occur to those who enthusiastically believed in such a possibility that technology itself has no particular content, that it is only a method of invention and production, and that it cannot therefore be a source of order of any kind. Order is always constituted in the communicative space of a particular culture as a whole. When culture itself is reduced to its most elementary characteristics and is represented in a manner compatible with technical thinking, then and only then it is possible to believe that 'technology is far more than a method', that 'it is a world in itself'.[9] Under such conditions some members of the avant-garde believed that architecture 'should only stand in contact with the most significant elements of civilisation. Only a relationship that touches on the innermost nature of the epoch is authentic'.[10]

Mies van der Rohe, whose later work represents probably the most interesting interpretation of the relationship between architecture and technology, was convinced that technology reveals its nature most explicitly in construction and in large-scale structures, but that as well as its nature technology also reveals something else. He describes this enigmatic else as 'something that has a meaning and a powerful form, so powerful in fact that it is not easy to name it' (Fig. 6.1).[11] To clarify the enigma, Mies asks if the application of technology is still technology, or rather architecture. 'Some people are convinced', he writes, 'that architecture will be outmoded and replaced by technology. Such a conviction is not based on clear thinking. The opposite happens. Wherever technology reaches its real fulfilment it transcends into architecture'.[12] This conclusion becomes clearer once we realise that 'technological fulfilment' is an idea which goes back, via Semper, to Goethe and Schinkel, where it is seen as an idea of material transformation, revealing the poetic function of architecture.[13] In the process of material transformation the inner logic of a building and its material realisation manifest themselves as an ideal material form. This corresponds with Mies' own conclusion: 'Architecture depends on its time. It is the crystallisation of its inner structure, the slow unfolding of its form. That is the reason why technology and architecture are so closely related'.[14]

The primary conditions for the new relationship between architecture and technology were established in the seventeenth century, in a development which resulted in the ambiguity between traditional symbolic and new instrumental representation. It was in this period, in the late seventeenth and early eighteenth century, that the long history of architectural thinking, which was always closely associated with the mathematical representation of its principles, was thoroughly influenced by new developments in natural science. In a relatively short time a point was reached where the former interpretation of traditional thinking and the new instrumental thinking became practically

6.1
**Lake Shore Drive,
Mies van der Rohe**

indistinguishable. Some of the symptoms of this new situation are the foundation of engineering schools, which had already begun to compete with traditional architectural education during the eighteenth century, the formation of
modern aesthetics as a new formal appreciation of art, and the general formalisation of culture. Other symptoms, less visible and less obvious, are the dimin-
ished relevance of tradition, apparent most clearly in the problematic nature of
modern classicisms; the growing arbitrariness of architectural decisions; and the
discontinuity between the means and the content of representation.

It is interesting to realise how long the ambiguity of the symbolic and
the instrumental representation was preserved in our cultural memory. This is

apparent in all the main architectural movements of this century from constructivism, Bauhaus and De Stijl to French purism. In all of them the formal representation of reality was no longer differentiated from the mathematical representation of technical knowledge. Mies van der Rohe himself made this clear when he wrote 'our real hope is that technology and architecture grow together, that some day the one be the expression of the other. Only then will we have an architecture worthy of its name. Architecture as a true symbol of our time'.[15]

This hope did not last very long. It soon became clear that it was not architecture but technology that had become the symbol of our time. The fact that architecture was particularly open to technical interpretation has much to do with the general technisation of everyday reality and with the already accomplished level of organisation and formalisation of typical situations, particularly of those related to work, administration and domestic life. The level of achieved formalisation is reflected in the history of architectural typologies and standardisation, where the original purpose of situations based on religious, cultural or broader everyday meaning was reduced to types and standards which proved to be technically and economically successful. These are the rules in a historical context where technical perfection and economic efficiency are considered to be 'the most significant elements of civilisation and the innermost nature of the epoch'.[16]

The technisation of everyday life was also, no doubt, strongly influenced by the possibilities of representation developed in great diversity and on a large scale in the domains of architecture, urbanism and landscape design. I am thinking here not only of the representational power of perspective, descriptive geometry, topology and surveying, but also of the power of these techniques to transcend the unity of representation and establish a new horizon of autonomy. This brings us to the essence of what is manifested as a difference between the participatory and emancipatory nature of representation. It is well known that we experience the surrounding world in its plenitude and in its given state as pure otherness. We have already seen that there is no such thing as original or direct experience of the given reality, only a mediated one, and that the most important mediating role is played by representation and its unity. Only the unity of representation can bring us closer to the depth and the plenitude of phenomenal reality, which would remain otherwise inaccessible. A line of poetry or a single painting can tell us about the hidden meaning and beauty of the landscape in the same way as a light in a sacred space tells us about the intelligibility of the sky and the divine.

We may conclude that the primary purpose of representation is its mediating role, which can be described as participatory because it enhances our ability to participate in the phenomenal reality. However, the process of representation can also move in the opposite direction towards the emancipation of the results and, as a consequence, towards their separation from the original communicative context. This is a tendency which we know well from the attempts of the avant-garde movements to create a new language of expression and representation, a language which is fully emancipated from history and tradition and which can support the autonomy of a particular avant-garde position.

The most radical manifestation of emancipatory representation can be seen in some recent tendencies which, despite the time distance, still share the main intentions of the earlier avant-gardes.[17] The technical homogenisation of whole areas of modern life make it much easier to live in the illusion that even the most abstract architectural solutions, based on a narrow technical criteria, may be adequate and appropriate. Human adaptability is one of the important factors which contributes to the cultivation of this illusion. However, much more important seems to be the overwhelming and persuasive power of emancipated representation in which only the level of reality which can be expressed in technical language is addressed. It is quite astonishing how many different forms and masks this language can adopt. Yet, behind all the masks, there is a common set of characteristics which we can find not only in the areas normally associated with production and technology, but also within the field of creative activities and disciplines.

## From creativity to productive mentality

The difference between creativity and production coincides to a great extent with our earlier distinction between participatory and emancipatory representation. Creativity is always situated in a particular communicative context from which it grows and in which the creative results participate. This circular process is not only the essence of creativity, but also the essential moment in the disclosure and constitution of the human world. Production, on the other hand, though it may grow from the same context, separates itself from it and establishes its own operation in the autonomous domain of reality. What makes this separation possible is this know-how in technical knowledge and the autonomy of the formal structures embodied in emancipated representation. In real life, the distinction between creativity and production is never entirely clear and absolute. There is always an initial element of inventiveness of production in each creative act, and a certain creativity in any production, at least in its initial stage. In terms of attitudes and goals, however, the distinction between them remains strong and clear. What is produced, unlike that which is created, is not in a communicative relation with its cultural setting; its purpose and meaning are established entirely in accordance with its internal logic. There are many structures and buildings – industrial plants, supermarkets, schools, hospitals – and also many works of art which are produced in the same way as any other industrial product. A typical example is a product designed for a precise purpose and for a particular place, people or culture. In his vision of the new art, which was supposed to be universal, Theo van Doesburg describes it in purely productive terms: 'The work of art must be entirely conceived and formed by the mind before its execution. It must receive nothing from nature's given forms or from sensuality or from sentimentality. We wish to exclude lyricism, dramaticism, symbolism, etc. In painting a pictorial element has no other element than itself. The construction of the picture, as well as its elements, must be simple and visually controllable. Technique must be mechanical, that is exact, anti-impressionistic'.[18]

The productive attitude to art and architecture, which profoundly influenced the nature of creativity during this century, became particularly conspicuous in its last decades. One of the main characteristics of the productive attitude is the tendency to accelerate the development of its own intrinsic productive possibilities. This characteristic is directly linked with the nature of emancipated representation, in which reality is translated and reduced into an image structured more by our inventive possibilities and visions than by the given conditions of reality itself. To invent or produce under such conditions is like moving at high speed through thin air. It is perhaps not surprising that in the fragmented culture of this century it proved to be easier to produce than to create.

There are many documents which can help us to see deeper into the intricate nature of production, but few are more rewarding and enlightening than the drawings of D. Libeskind, who described them as 'deconstructive constructions' (Fig. 6.2).[19] They are conscious explorations of 'the relation between the intuition of geometric structure as it manifests itself in a pre-objective sphere of experience and the possibility of formalisation which tries to overtake it in the objective realm'.[20] The drawings give us a unique insight into the constructive possibilities on the boundary of actual and imaginary space, an insight into the representative power of our imagination challenged by the conceptual power of invention. The transition from actual to imaginary space, from the geometrical representation of actual spatial relationships to their formal equivalents, is in essence a transition from the space of real possibilities to the space of possible realities. In this process, which illustrates the emergence of the autonomy of geometrical representation, the original continuity of meaning is replaced by the transformational meaning of the process itself. The open-ended and enigmatic nature of the results is the price which must be paid for the gained productive freedom. This seemed to be the rule and demand of the current situation. What are the reasons for this? 'Contemporary formal systems present themselves as riddles – unknown instruments for which usage is yet to be found. Today we seldom start with particular conditions which we raise to a general view; rather we descend from a general system to a particular problem. However what is significant in this tendency, where the relation

6.2
**Daniel Libeskind,
'deconstructive
construction'**

between the abstract and the concrete is reversed, is the claim which disengages the nature of drawing as though the "reduction" of drawing were an amplification of the mechanisms of knowledge'.[21]

The tendency to extend, and where possible to move beyond, the limits of visual representation is one of the main characteristics of the contemporary avant-garde in its effort to transcend the confines of traditional culture and the human condition. It is perhaps not surprising that geometry and mathematical thinking play a dominating role in such an effort. Mathematics has always been the major instrument of transcendence, because it generates its own development regardless of whether or not the accomplished results can be directly reconciled with the world of phenomena. The extension of mathematical thinking into a broader sphere of culture brings architecture itself close to mathematics and, as a result, into the stream of productive thinking.

Because architects are usually not much concerned about the sources and the nature of knowledge received from other fields, and tend to see such knowledge either with an uncritical aura or as a pragmatic tool, they contribute to a rather paradoxical situation. In the case of mathematics, great effort was invested during the twentieth century to better understand its logical foundations and applicability, and to achieve a more comprehensive vision of the relationship between mathematical representation and reality. In all these studies and investigations the recurring questions were the ontological structure of the conditions and possibilities of formalisation, the structure of formal systems, and the continuity of meaning in mathematical operations. It is strange that architects, who encounter practically the same problems in their own work, do not seem to be much concerned about the nature and implication of these structures. The words of a leading mathematician, speaking about his own field, nevertheless apply to architecture as well: 'The abstract is not the first. It is by a perpetual return to its intuitive origins and to the reality of its problems, by a close fidelity to the imperatives of this hidden life which traverses theories like fertilising sap, that mathematical thought reconquers, through the inevitable snares of a necessary abstraction. This original concrete, which is always present, at the core of its movement, and which manifests in most characteristic fashion its permanent activity in the highest moments of creation ... To detach itself from these roots, would in reality be to condemn itself to asphyxia, to enclose itself in a kind of mortal solitude which would result in the emptiness of a system void of all content.'[22]

The danger of emptiness haunted modern architecture from its very beginning. However, it is important to realise that the emptiness was caused not only by the buildings, but also by the absence of articulated public culture. In a situation where the continuity of shared meaning was broken into fragments of understanding, it is difficult to expect that ambitious abstract structures and their implied meaning will be understood as was intended by their authors. When Mies van der Rohe speaks about the spiritual meaning of construction, or Michel Seuphor about 'architecture which by the technical and physical methods peculiar to the age, reflects in its particular organisation the magnificent order of the universe'[23], it no longer sounds convincing.

We may feel, quite rightly, that there is a deep gap in communication, not only between people or people and buildings, but also between the different areas of culture. The presence of this gap, it seems to me, is illustrated by the amount of verbal explanations and commentaries that accompany visual art. Their purpose is no doubt to bridge the gap between the personal introverted meaning of the work and its public reception. This also illustrates in a small way a much larger problem – the gap between the achievements of modern science and technology, including their deep influence on contemporary society, and the communicative nature of the phenomenal world. This is, I believe, the main source of our current difficulties in meaningful communication, reflected most clearly in the impossibility of reconciling the abstract, conceptual representations of our world and the particular conditions and aspirations of our life.

There is a tendency to believe that the emancipation of technological possibilities and powers affects reality as a whole in an equal manner, and that this leads to the emancipation of human life and existence. This would be true only if it was possible to reduce life and nature to transparent knowledge, but as we know this is inconceivable. Whole areas of nature and life are beyond our capacity of rational comprehension, and yet it is these areas that exert the greatest influence on the nature of our world. This becomes increasingly apparent in the growing knowledge accumulated by current anthropology, human ecology and environmental medicine, and is illustrated very well in the following statement: 'The evolutionary development of all living organisms, including man, took place under the influence of cosmic forces that have not changed appreciable for very long periods of time. As a result, most physiological processes are still geared to these forces, they exhibit cycles that have daily, seasonal and other periodicities clearly linked to the periodicities of cosmos. As far as can be judged at the present time, the major biological periodicities derive from the daily rotation of the earth, its annual rotation around the sun and the monthly rotation of the moon around the earth.'[24] This is only a brief description of the conditions under which the regularity of certain vital processes of our lives were constituted and under which they became eventually the source of other regularities and movements that structured the more articulated layers of our life and culture. The fact that the articulation of cultural life is directly linked with conditions that remain relatively unchanged, while at the same time the path of culture open to technological transformation undergoes a radical change, creates a tension and ultimately a deep void in the very heart of culture.

The vision of modern society undergoing a steady technological transformation *en bloc* is very misleading. There is a great difference between the levels of reality which can be directly manipulated and those levels that resist such manipulation. In the case of dwellings for instance, the development of new construction techniques, materials and services is taking place on a different level and with a different rate of change than the development of thinking about the nature and purpose of the dwelling, which is rooted in tradition, customs, habits, and the relative stability of primary human needs.

How can the differences between natures and rates of development be reconciled? The typical answer refers to technology and to the necessity of

6.3
**Volkshaus,
Hans Scharoun**

adaptation to its imperatives. Just how one-sided and problematic such an answer is is demonstrated by the complex history of adaptation, which goes back at least to the end of the eighteenth century, when the monopoly of a disengaged emancipated rationality was for the first time seriously challenged by romanticism and its influence on later generations.[25] We have to remember and acknowledge that romanticism was not only a reaction to the enlightenment, an artistic movement or an impossible dream, but also a science, philosophy and general attitude to culture as a whole.[26] In the dialectical development of modern culture during the last two centuries, the romantic tradition in different forms and under different names was the main source of the continuity of humanistic culture, of creativity and the sense of wholeness. It is mostly through its more recent manifestations in expressionism and surrealism, but also in certain aspects of constructivism and high tech, that the romantic tradition exerts its influence on contemporary architecture. This may not be immediately apparent; it is easier to see the influence of romantic culture where it is most explicit. It is difficult to find a better example than the work of Hans Scharoun. His whole life was devoted to a thoughtful and highly personal interpretation of culture, which under the relatively narrow term expressionism represented a rich, long-term contribution to philosophy, literature, theatre and the visual arts. In the expressionist epoch, most of German culture was dominated by the desire to transcend the fragmentary experience, to attain a vision of the whole and by the attempt to achieve a union with the inward reality of the world.

# The inwardness of modern culture

The phenomenon of inwardness is the main characteristic not only of expressionism but also of the twentieth century as a whole. It is the result of a long-term transformation of European culture which made it possible to believe that our life can be represented in its entirety in terms of scientific, technical rationality, leaving behind all that cannot be subordinated to this vision – mainly the domain of personal experience, praxis and the natural world. The emancipation of scientific rationality led to the formation of a culture with its own criteria of intelligibility, and to a new sense of wholeness based on the continuity of the humanistic tradition accessible through personal and introverted experience. In the field of architecture the typical mode of embodiment of this culture can be found in the romantic notion of genius. In the creativity of genius the traditional complexity of culture is reduced to a single, creative gesture and to a direct communication with the assumed creative powers of nature. In his 1925 lecture at the Breslau Academy, Scharoun said 'The creator creates intuitively in accordance with the impulse which does not correspond only to his temperament but also to the time to which he belongs and with which he is, to a great extent, one. And if we want to explain this impulse, then we must understand the real tasks of our time. The law which drives and leads an architect can be perhaps grasped only metaphysically'.[27]

The law which drives and leads an architect is very closely linked with the mystery of architectural form (*Gestalt*) to which Scharoun explicitly refers. 'The great mystery in the creative work is undoubtedly gestalt, gestalt in the sense of organic and multiple form'.[28] The mystery of form has much to do with the question of authenticity, which for Scharoun was synonymous with the organicity of design, measured by the correspondence between the functional form (*Leistungsform*) and the essential form (*Wesenhaften Gestalt*). The functional form is a result of an investigation (*Gestaltfindung*), in which the appropriate solution is determined by the given purpose, material and construction. Together with Hugo Häring, with whom he shared many ideas, Scharoun believed that the functional or organic form, as he sometimes calls it, is a result of an anonymous process, in which the intrinsic laws of nature or human life determine the design. Despite the importance of functional investigation, the goal of each project was the essential form which was supposed to reconcile the formal solution with the spiritual principles of the epoch. However, the presumed anonymity and objectivity of the process is a deceptive illusion. The ways in which design is determined by the laws of nature and of human life are conceivable only as an interpretation in which the role of the architect, his experience, imagination and intentions are decisive. This is even more obvious in the search for the essential form, which in the absence and negation of all precedents requires not only a great deal of experience and knowledge but also a high level of inventiveness.

Under such conditions the task is not only to invent a particular building from one's own cultural resources, but also to invent a culture which would make the building meaningful. The result is a cycle which seals the introverted nature of the creative process and opens the door to potential arbitrariness and relativism. It is very difficult to imagine how a culture articulated in an introverted dialogue can substitute the richness and wisdom of a culture which was cultivated and shared for many centuries. This is a dilemma which is clearly apparent in the discrepancy between Scharoun's buildings and his stated intentions. In the Berlin Philharmonie, for instance, the main hall was no doubt deeply influenced by the history of music auditoria, and yet Scharoun describes the process of its making as a direct dialogue with the nature of music and with the nature of space seen as a landscape. 'The construction', he writes, 'follows the pattern of a landscape with the auditorium seen as a valley and there at its bottom is the orchestra surrounded by a sprawling vineyard climbing the sides of its neighbouring hills. The ceiling, resembling a tent, encounters the landscape like a skyscape'.[29] The indeterminate, changing perceptual structure of the whole is held together by the constructive imagination of the architect and the musical experience of the audience. It is interesting to see how early Scharoun anticipated the close link between his own imagination and public experience. In one of his drawings for the Glass Chain, he illustrates the place and the role of the artist among the people, his ability to embody and represent their will and elevate it to the higher level of spiritual existence (Fig. 6.3).[30]

It is a sign of the avant-garde mentality that the architect sees himself as a sole agent, fully responsible for everything related to creativity. This illusion culminates in the belief that the world is essentially his own world. Everything created under such conditions is bound to be unique, and yet it very often claims universal validity. This paradox can be sustained only by a self-centred culture, prepared to share the paradox as a norm. However, this does not resolve the real problem of the relation between the universality and particularity of design. We

6.4
**Berlin Philharmonie,
Hans Scharoun**

can see this not only in the architecture of Scharoun, but also in the work of his antipode, Mies. The universality of Mies' structure, it is conventionally believed, represents not only the universal but also the specific aspects of the programme and of the broader context of culture. In fact, the deeper content is present only enigmatically and is accessible only through very cryptic personal interpretations. No amount of wishful interpretation can bridge the gap between the promise of meaning and its fulfilment. In the end Mies' buildings remain what they are, cultivated material structures which can at best be appreciated aesthetically. The talk about Mies' classicism, his own arguments about the expression of the essence of the modern epoch through technology, are only parallel and rather empty intellectual constructions. It is due to these constructions that the emancipated and isolated reality of Miesian structures is situated in a broader sphere of meaning. Such meaning may be available to the author and to those who are persuaded by the thrust of the argument, but to those who are not initiated or have their own critical understanding the argument must appear hermetic and illusory. It is quite astonishing to what extent the twentieth century avant-gardes succeeded in fabricating their positions, their promises of new meaning, coherence and wholeness through publicity, exhibitions, manifestos and utopian projects, rather than through the convincing quality of buildings, to say nothing of cities.[31] In a sense, the career of Mies shows similar characteristics.[32]

The critical role which the media, the secondary and derivative mode of representation, play in the making of modern architecture, illustrates how tenuous the link between architecture and its cultural context became. In the Miesian way of thinking, the universality of the solutions is, contrary to the intentions of its author, only a form of universality. In the work of Scharoun the most important thing is the process of creation starting from, and cultivating, the particular. 'We know', he writes in the last years of his life, 'that all our attempts are only a modest beginning in detail'.[33]

In the development from the particular and from the detail there is always a certain anticipation of the result in the form of an idea or conceptual image. However, the aversion towards the *a priori* presence of all universality leaves Scharoun's work isolated from the broader meaning of the common culture. In this sense it is complementary to the work of Mies.

# The grey zone and the discontinuity of contemporary culture

It is a curious historical coincidence that Scharoun's Philharmonie and Mies' National Gallery, the two most typical representations of the polarity in modern architectural thinking, share the same space on the Kulturforum in Berlin. The grey zone which separates them can be understood both literally and metaphorically. The space of the forum in its contemporary state is a sad memento of twentieth century inability to create a genuine public space. This is reflected in the broader and deeper metaphorical meaning of the grey zone which shows the true depth of the gap between the universality of modern culture, represented by modern science and technology, and the domain of introverted culture, represented mostly by arts, humanities and personal experience. The depth of the gap was already apparent in the contrast between Mies' conviction that 'the individual is losing significance' and that 'his destiny is no longer what interests us', and Scharoun's doubts about the role of rational knowledge and structured creative process. 'Do we reach pure creativity through reflection, through knowledge? – No – man is the centre'.[34] In one sense the grey zone is a metaphor of a deep discontinuity in modern culture; in another it is a metaphor of the problematic attempts to resolve the discontinuity from a single, relatively narrow, position. The typical example is a loose and arbitrary connection established between a highly personal experience and ideas of universal validity.

6.5
**Kulturforum, Berlin**

In the history of modern architecture attempts to resolve the problem of cultural discontinuity ended in the formation and consolidation of several distinct positions. The most obvious, which we have already discussed, were formed around the belief in the universal role of technology and around personal expressive epiphany. Among the others we might mention belief in the restorative power of the vernacular tradition, classicism and, more recently, the historicising improvisations of postmodernism and conceptual deconstruction.

The arbitrary nature of the relationship between the sphere of experience and the sphere of concepts or ideas is the most problematic characteristic of the grey zone. This ambiguous relationship is a source of an unprecedented freedom to produce, but also of an overwhelming relativism, loss of meaning, narrowing down of the sphere of reference and, as a result, a general cultural malaise.[35] The nature of the malaise can easily be illustrated by the dilemma facing most contemporary architects. It is generally assumed that true creative architecture should be free of historical and other unnecessary cultural references in order to be as original and unique as possible, yet at the same time it is expected that the result should be understood, appreciated and accepted by everybody. The originality of design achieved in the atmosphere of arbitrariness and relativity is manifested primarily in the visibility of the result. Visibility always presumes, even in its most abstract form, some form of continuity with the natural world. This is its main virtue. It is for the same reason that visibility can be pushed to its limits and serve as a transition to the derivative quasi-visibility in the conceptual domain. This is particularly relevant for understanding the fragile nature of visibility in works structured under the strong influence of technical thinking – today considered to be the main source of originality. In many of these works questions of visibility do not usually precede, but follow, the diagrammatic stage of the project, and often remain residual.

The residual nature of the primary visibility in modern buildings was anticipated by Mies when he wrote 'The visible is only the final step of a historical form, its fulfilment. Its true fulfilment. Then it breaks off and a new world arises ... Not everything that happens takes place in full view. The decisive battles of the spirit are waged on invisible battlefields.'[36] The invisible battlefields are the domains of conceptual thinking, calculations and diagrammatic imagination. The extent to which contemporary architectural projects are conceived on this level can be illustrated by many examples, mostly from the area of architecture inspired by pure structural possibilities.

The fragile nature of visibility can be extended to other areas of our experience. What we experience in front of an incomprehensible building or structure escapes explicit understanding, but is reflected in our tacit response. This was recognised by the commentators on the virtue of constructivism, particularly in the discussion of the problem of beauty. 'The beauty of the machine is the rational value of an irrational product. Irrationality is the essence of the inexplicable beauty of the machine. It is for that reason that machines can be not only an example of modern, logically functioning mind, but also of a nervous modern sensibility. There is nothing more nervous than a vibrating dynamo.'[37]

This understanding of the nature of beauty illustrates the transformation of modern sensibility, in which the richness of a fully articulated world revealed in the works of art and in buildings is reduced to a personal aesthetic experience based on elementary sensations. In the closed world of aesthetic experience it is virtually impossible to differentiate between the nature of reception and the nature of production or creation. 'The neo-Avant Garde moves today within a more or less non-binding pluralism of artistic means and stylistic schools while no longer able to enlist the force of an enlightening originality released in the violation of established norms, in the shock of the forbidden and frivolous, in irrepressible subjectivity'.[38]

The difficulty of enlisting the force of originality pushes contemporary avant-garde deeper into a more radical form of self-centredness and self-referentiality. The result is a higher level of autonomy and separation from everyday reality, accompanied by a desperate search for new sources of originality in the inventiveness of current technology and in the domain of private fantasies. In this situation it is no longer clear what the difference is between the product of imagination and imaginary reality. In the production of imaginary solutions the dialogue with phenomenal reality is replaced by a monologue of conceptual imagination which often relies on the quasi-visibility of geometry as its scaffold. Under such conditions 'the illusion of seeing is therefore much less the presentation of an illusory object than the spread and so to speak running wild of a visual power which has lost any sensory counterpart'. This loss is the main characteristic of a situation which leads to hallucinations, 'because through the phenomenal body we are in constant relationship with an environment into which that body is projected and because when divorced from its actual environment, the body remains able to summon up, by means of its own settings, the pseudo-presence of that environment'.[39]

This almost sounds like a description of some recent projects with a strong affinity towards virtual reality, generally acknowledged as a consciously structured and controlled hallucinatory world. Contexts in which hallucinations can take place are limited to certain spaces and media and cannot be identified with the reality of the whole. There are structures in our culture which resist hallucinations. Merleau-Ponty is more specific when he writes 'What protects us against delirium or hallucinations are not our critical powers but the structure of our space'.[40] The structure of our space has its source in the depth of culture, and coincides with the overall coherence of our cultural world. Because our existence is always spatial, the nature of lived phenomenal space determines the topography, orientation, meaning and sanity of our existence.[41]

# Notes

1    Stackhouse (1972:3).
2    Stackhouse (1972:3).
3    Dreyfus (1987:41–55).
4    Gadamer (1986a:35).
5    Gadamer (1986a:92–115); Aristotle (1982:1448b2–1450a8); Grassi (1962:118).
6    Gombrich (1960:203–12). On a deeper level there is a close affinity with the representation and the articulation of the world in the Heideggerian tradition (Heidegger 1992:91–149; Gadamer 1975:397–449).
7    Bann (1974:59).
8    Platz (1975:161).
9    Neumeyer (1991:324).
10   Neumeyer (1991:332).
11   Neumeyer (1991:324).
12   Neumeyer (1991:324).
13   Bauer (1963:138–47); Horn-Oncken (1967:9–29).
14   Neumeyer (1991:324).
15   Neumeyer (1991:324).
16   Neumeyer (1991:332).
17   The history of the problem is outlined in Poggioli (1968), Bot (1968) and Osborne (1979).
18   Doesberg, 'Art Concrete', April 1931, number 1, p.1, trans. in Osborne (1979:128).
19   Libeskind (1980).
20   Libeskind (1980:22).
21   Libeskind (1980:20).

22  Ladriere, J., 'Mathematiques et Formalisme' in *Revue de Questions Scientifiques*, Louvain, 20 October 20 1955, 538–73, trans. in Kockelmans (1970:481–3).

23  Seuphor (1971:38).

24  Dubos (1965:42).

25  Gusdorf (1976); Abrams (1953); Behler, Horisch and Schoning (1987).

26  The attitude of the romantic generation to science is exemplified in Goethe's view of Newtonian science, see Nisbet (1972). For a general history and assessment of romantic science see Lepenies (1978).

27  'Intuitiv gestaltet der Schaffende nach einem Impuls, der nicht nur seinem eigenen Temperamente entschpricht, sondern die Zeit, der er dient, zum guten Teile eigen ist. Und soll dieser Impuls erkennbar und erklarbar gemacht werden, so bedarf es dazu der Haranziehung der der Zeit eigenen realen Aufgaben. Das Gesetz, das den Architekten treibt und leitet, ist vielleicht nur metaphysisch a priori zu erfassen.' Janofske (1984:22)

28  Janofske (1984:17).

29  Hans Scharoun, quoted in Blundell-Jones (1995:178).

30  Janofske (1984:35).

31  Honey (1986); Tegethoff (1989:28–94); Colomina (1994:193–223).

32  Tegethoff (1989:43).

33  Scharoun (1967:157).

34  Janofske (1984:137–8). This statement stands in sharp contrast with Mies' position, who claims that 'the individual is losing significance. His destiny is no longer what interests us.' Mies van der Rohe 'Architektur und Zeit' in *Der Querschnitt*, Berlin 1924, quoted in Johnson (1947:186).

35  Ricoeur (1965:271–87); Taylor (1989); Vattimo (1988).

36  Neumeyer (1991:326).

37  Teige (1966:129–44).

38  Habermas (1985a:23).

39  Merleau-Ponty (1986:340).

40  Merleau-Ponty (1986:291).

41  The subject of this essay is further developed in the forthcoming book *Architecture in the Age of Divided Representation*, Cambridge, Mass., MIT Press, 2004.

Part Two

# Architecture

**Chapter 7**

# 'How is it that there is no modern style of architecture?'
## 'Greek' Thomson versus Gilbert Scott

*Gavin Stamp*

'With us architecture has all but ceased to be a living art … ' Alexander 'Greek' Thomson complained in 1871 in a lecture which addressed the central architectural problem of the nineteenth century, *How is it that there is no modern style of architecture?* 'The present age, so rich in achievement in other departments, is seen making the most ridiculous efforts to insinuate its overgrown person backwards into empty shells of dead ages, which lie scattered about on the old tide marks of civilisation, rather than secrete a shell for itself according to the ordinary course of nature.'

Thomson's presidential address to the Glasgow Institute of Architects has often been selectively quoted, particularly by those anxious to present its author as a sort of proto-modern, who rejected the tyranny of historical precedent and groped towards a non-historical functional architecture through the encircling fog of eclecticism. That, of course, is to fall into the trap of interpreting the past by the preoccupations of the present, for while Thomson (1817–75) was undoubtedly a most original and thoughtful designer, anxious to exploit the possibilities of new materials like cast iron and plate glass, he was far from being a pioneer of the modern movement. However 'modern' his buildings may seem in their abstraction and innovation, Thomson's visual imagination was firmly rooted in antiquity.

Nevertheless, 'in both thinking and building [Thomson] was at least half a century ahead of his contemporaries,' the *Architects' Journal* could claim when the Royal Institute of British Architects held its conference in Glasgow in 1964, giving him almost the same mythic status as the second winner of the Alexander Thomson Travelling Studentship, Charles Rennie Mackintosh. 'He stands head and shoulders above his contemporaries, a lone genius who not only designed with singular imagination but also built with splendid skill.'[1] A few years later Nikolaus Pevsner gave a less hyperbolic assessment of Thomson's unique approach to architecture, but insisted that 'he was a rationalist first and foremost in his domestic and commercial buildings, and in his three grandiose churches he was overpoweringly original … For the purposes of office buildings Thomson accepted iron as naturally as the non-Grecian Glasgow and London architects of the fifties. So he was not a historicist, strictly speaking … '[2]

Pevsner was discussing Thomson as a writer on architecture, and in the context of one of the most celebrated exchanges in the Victorian 'battle of the styles', the Glasgow architect's public attack in 1866 on the gothic revival in general and the selection of the design for the new buildings for Glasgow University by George Gilbert Scott (1811–78) in particular. This remarkable, angry lecture by Thomson – 'a greater architect than Scott' – has been interpreted not just as classic versus gothic and a Scottish architect versus a well-established English outsider, but as a forward-looking architect opposed to an historicist designer (to use Pevsner's misuse of that word). But while many of Thomson's criticisms of Scott's gothic hit home, and he maintained that the gothic revival was a 'retrograde movement' and that 'a great part of the work of the last two or three centuries has consisted in rectifying the errors of the mediaeval system,' in truth the Glaswegian was no more or less progressive than his London rival.[3] Both Thomson and Scott combined cast iron structures with masonry, and both employed large sheets of plate glass. Indeed, the two architects had much in common; the argument was about style – not so much as a matter of taste and historical associations (although these mattered) but about which – the Greek or the gothic, the lintel or the arch – was the more sound and practical to be the basis of development of a distinctive modern architecture for their time. Both knew that a new style could never be invented; it must evolve.

The central question Thomson posed in his characteristically elegant and powerful 1871 oration – 'how is it that there is no modern style of architecture?' – was, as he himself said, often asked in his time, both in Britain and abroad. With their detailed knowledge of world history and of other and earlier civilisations, the Victorians were acutely conscious that their own time had produced no recognisable style like the styles of the past, a language of architecture that could be categorised and labelled, its grammar and vocabulary codified – a distinct Victorian style that could join the Egyptian, the Greek and the gothic. Why was it that a civilisation that clearly justified belief in progress, that was conquering the world with steam railways and iron ships, with the electric telegraph, scientific medicine and parliamentary government, could not manage to

7.1
**Alexander 'Greek' Thomson
(1817–75), ambrotype,
late 1850s**

7.2
**Sir George Gilbert Scott
(1811–78), carte-de-visite
portrait, 1860s**

7.3
The Choragic Monument of
Thrasyllus, Athens, engraving
from Stuart and Revett, *The
Antiquities of Athens*, Vol. II,
1789

7.4
Acton Burnell Church,
Shropshire, east elevation,
woodcut from Thomas
Rickman, *An Attempt to
Discriminate the Styles of
Architecture in England*, 5th
edition, ed. J. H. Parker, 1848

produce a distinctive manner of building as had earlier and more primitive societies?

'We need an indigenous style of our own,' asserted Thomas Harris in his polemical pamphlet of 1860, *Victorian Architecture: A Few Words to Show that a National Architecture adapted to the ... Nineteenth Century is Attainable.* 'This is an age of new creations; steam power and electric communications, neither the off-shoot of any former period, but entirely new revolutionising influences. So must it be in Architecture if it is to express these changes.'[4] Similarly that confident goth, William Burges, could write the following year that

> when we remember that the distinguishing characteristics of the Englishmen of the nineteenth century are our immense railway and engineering works, our line-of-battle ships, our good and strong machinery – or, to go to other points, our free constitution, our unfettered press, and our trial by jury, it will naturally suggest itself, whether any style of architecture can be more appropriate to such a people than that which ... is characterized by boldness, breadth, strength, sternness, and virility,

and then conclude that the best style for Victorian Britain was, of all things, early French gothic.[5] A decade later, however, Thomson was less optimistic when writing about the present in a similar vein. 'If we have no architectural style, it is not for lack of material, for we know nearly all that has ever been done,' he lamented. 'It is not for lack of wealth, for our undertakings are most extensive, and exhibit a lavish expenditure of money. It is not for want of intellectual talent, for we have excelled all former ages in the number and grandeur of our discoveries. How is it, then, there is no modern style in architecture?'[6]

Thomson's ostensible solution to the problem was 'to abandon with all convenient expedition the whole mass of accumulated human traditions under which we have been, as it were, smothered, and take earnestly to the study of the Divine laws, and by-the-by we shall find it more difficult to keep running in the old rut than hitherto we have found it difficult to get out of it. Let us once fairly comprehend the living law, and we will at once and for ever get freed from the bondage of dead forms.'[7] This may seem a radical and ruthless answer, but it was born of the awareness that the tragedy of his time was that architects simply knew too much. Architects were victims of the sophisticated antiquarian and historical understanding of the age; creative innocence and spontaneity were undermined by the inescapable historical sense. In this Thomson was simply echoing the worries of his earlier adversary, George Gilbert Scott, who in discussing 'The Architecture of the Future' in 1857 had concluded that

The peculiar characteristic of the present day, as compared with all former periods, is this, – that we are acquainted with the history of art. We know better whence each nation of antiquity derived its arts better than they ever knew themselves, and can trace out with precision the progressions of which those who were their prime movers were almost unconscious. What, for instance, did the Greek know of his joint debt to Egypt and Assyria for the elements from which he developed his noble architecture? … It is reserved to us, alone of all the generations of the human race, to know perfectly our own standing-point, and to look back upon the entire history of what has gone before us, tracing out all the changes in the arts of the past as clearly as if every scene in its long drama were re-enacted before our eyes. This is amazingly interesting to us as a matter of amusement and erudition, but I fear is a hindrance rather than a help to us as artists.[8]

Thomson's solution to this profound dilemma was not to adopt the gothic in all its apparent flexibility, as Scott did, or to make a desperate attempt to design in a conspicuously new style, as was attempted in vain by such naïve extrovert architects as 'Victorian' Harris (who in fact designed in a wild, eclectic gothic). As Ruskin had realised, 'if you are not content with a Palladio, you will not be content with a Paxton, and pray you get rid of the idea of there being any necessity for the invention of a new style.'[9] Thomson concurred. He was too wise to wish to forget the past, to throw out the study of history as so many architectural schools in Britain and America did in the 1960s; he wanted not revolution but evolution, development of an existing style that was suitable to modern needs. And that style was, of necessity, historic. 'Greek' Thomson knew well that a new style could never be invented by one man, but must be based on existing traditions, for he went on to emphasise that 'these old forms are not to be despised; far otherwise. They are there to teach us what has been already discovered – to place us upon an elevated starting-point for yet higher attainments – to connect our sympathies with the men whose thoughts they represent, and with the Creator whose laws they reveal to us.'[10] And the old forms he would have modern architects learn from were Greek.

Nor – for all his ability to combine the Egyptian and the Greek and to spice up the result with a dash of the Indian – was Thomson's solution to be as eclectic as possible, choosing useful or attractive features from a wide variety of styles. By 1871 the great moral crusade of the gothic revival was failing, and, having run successively through so many phases of gothic – both English and continental – architects were now experimenting with yet more precedents: Elizabethan and Jacobean, Dutch gables and leaded-light windows, to produce those styles loosely categorised as Old English and Queen Anne in a subtle and creative but nonetheless doomed attempt to escape from the essential architectural problem of the century.

Thomson's alternative answer was to return to fundamentals, and for him that meant the pure, trabeated language of the Greeks – a language which, as he argued persuasively, could be developed both to accommodate contemporary conditions and to incorporate modern materials. As Henry-Russell Hitchcock put it, 'He does not *continue* the Greek, he *returns* to the Greek, finding in it in the '50s a personal vehicle quite as the leading English architects of the time were finding in continental mediaeval work their personal vehicles.'[11] Thomson was surely thinking of the Greek when he said that 'there are in some of the more highly developed styles features which are as well near perfection as we can well conceive.' So the past was not to be rejected – far from it. The answer was to adhere to the eternal, divine laws of architecture, which were always there to be followed if only architects would see and learn. As Thomson put it in his Haldane Lectures three years later, discussing the work of his predecessors, 'the promoters of the Greek revival … failed; not because of the scantiness of the material, but because they could not see through the material into the laws upon which that architecture rested. They failed to master their style, and so became its slaves.'[12]

In the belief that an architecture of the future must be based on fundamentals, on the development of one pure constructional style, Thomson was not so far from Scott. The English gothicist may not have gone on and on about architecture being a response to divine or eternal laws, but he did believe that the gothic revival could generate a comprehensive, practical modern architecture by developing one style, that which was generally recognised at the time as the best period of medieval English architecture, 'second pointed' or geometrical decorated. It was, he argued, 'the duty of those who guided that revival to see that its course should not be wildly eclectic, but that we should select once and for all, the very best and most complete phase in the old style, and taking that as our agreed *point de depart*, should make it so thoroughly our own, that we should develope [*sic*] upon it as a natural and legitimate nucleus, shaping it freely from time to time to suit our altered and ever altering wants, requirements and facilities … '[13]

In his influential *Remarks on Secular & Domestic Architecture, Present & Future*, first published in 1857, Scott had no doubt that an architecture of the future would be based on the gothic, but incorporating elements of the classical and other styles as well as new materials and technologies. 'As I started on the supposition that the developments of the future are to be founded on our own pointed architecture as a nucleus, I of course assume that the style will be essentially and mainly of that family;' he wrote in the concluding chapter, but I at the same time hold that it will be a perfectly new phase, differing more from any form that pointed architecture has hitherto assumed than any of those varieties (whether chronological or national) from one another. It will differ, not only through expressing the ideas of another age, but also from its comprehensiveness. This must, in fact, be its leading characteristic. As before stated, the remarkable feature of our age, so far as it bears upon this subject, is our knowledge of all arts which have existed. This, in my opinion, absolutely precludes us from generating a perfectly new art, spontaneously growing as a plant from its seed, as has been the case in former periods. Knowing all which has gone before, we can scarcely fail to cull from among them such ideas as seem to have value for ourselves. Some of these ideas may possibly appear incongruous with the nucleus on which we are working, but this will not often occur; and a living art has marvellous facility for the admission of ideas apparently incongruous, and of harmonizing them with itself. Especially, however, will all that is good in the various phases of pointed architecture be united under one head.[14]

Thomson took pains to point out the impracticalities and illogicalities of modern gothic, while Scott emphasised that the classical tradition was both alien and inflexible. Both architects, however, insisted that their chosen styles could absorb the practical requirements of the modern age, and both revealed in their work a similarly open attitude towards using the new building technologies of their time: cast iron and plate glass. Not only were both architects masters at designing elegant and appropriate decorative ironwork; they used iron extensively in their structures. Thomson's commercial warehouses were iron-framed behind their masonry walls, and he used iron columns in both the Queen's Park and the St Vincent Street churches. In using new materials in places of worship, religious differences were significant. 'Instead of being crowded with stone piers,' complained the Scots Presbyterian, the church interior 'should be as open as possible. But the mediaevalists never give us such forms. Of course iron pillars and lath and plaster arches are not to be thought of.'[15]

It is true that Scott, as an Anglican and as a disciple of Welby Pugin, used 'correct' ecclesiological plans, and would not allow a base material like iron in a sacred building (unlike nonconformists in England and Presbyterians in Scotland), but he was very happy to use metal in his secular buildings. In the Midland Grand Hotel at St Pancras and in Glasgow University, staircases are visibly supported by decorated iron beams, and in his 1857 book he argued that 'It is self-evident that this triumph of modern metallic construction opens out a perfectly new field for architectural development ... The fact is that all

7.5
**A. Thomson, St Vincent Street
Church, Glasgow, interior
(1857–9)**

7.6
**G.G. Scott, St George's Church,
Doncaster, interior (1853-8)**

these iron constructions are, if anything, *more* suited to Gothic than classic architecture … '[16] Thomson, on the other hand, seems to have considered that iron as a structural material was not so much the potential generator of a new architecture but was complimentary to his chosen architectural language: 'the simple unsophisticated stone lintel,' he liked to observe, 'contains every element of strength which is to be found in the most ingeniously contrived girder.'[17]

When it came to plate glass, Thomson and Scott were both happy to use very large sheets of the material in a manner that had not been technically possible before the 1840s. In the clerestories of his churches Thomson fixed large panes of rolled glass direct to the masonry, without timber frames or glazing bars, while his commercial warehouses and domestic buildings are conspicuous for the use of large sliding sash windows containing single sheets of plate glass. 'Another inconvenient peculiarity of Gothic windows is that they are narrow, or subdivided into narrow compartments,' he argued,

> thus preventing us from getting the full advantage of that useful and very beautiful material, plate glass, which has been all but universally adopted, to the exclusion of the Gothic lattice, composed of small fragments of bad glass set in a network of lead, which has the properties, ascribed by the mediaevalists to the classic porticoes, of keeping out light and letting in the rain. As our clients insist upon having wide windows with large sheets of plate glass, and object to steep roofs, it is quite impossible to make good Gothic houses; and consequently a kind of nondescript, almost unworthy of the name of architecture, has come to be the prevailing style.[18]

This was unfair. Scott was perfectly willing to use large sheets of plate glass in his country houses, like Kelham Hall and Walton Hall, often in windows with straight lintels rather than arched heads. 'We must have large windows, plate glass in large sheets, sash windows if we like, and every convenience of our day. These clearly demand a new expansion of the style … ' he argued in 1857.[19]

Two decades later Scott was dismayed by the advent of the eclectic semi-classical Queen Anne style for domestic architecture, a manner of building which had evolved in reaction to the perceived modernity and undomestic vulgarity of much High Victorian gothic, but which has been seen as somehow more progressive by many historians. He complained that

> the Queen Anne-ites … freely adopted lead lights, iron casements, and all kinds of old fashions which a Gothic architect would have hardly dared to employ, so much so, indeed, that a so-called 'Queen

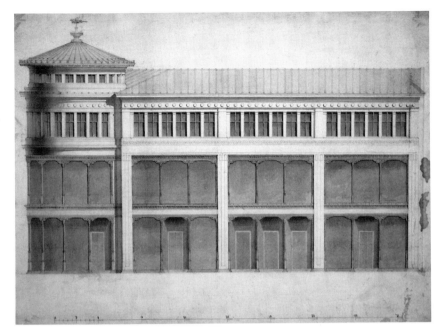

7.7
**A. Thomson, warehouse in Howard Street, Glasgow, unexecuted design c.1851, elevation**

7.8
**G.G. Scott, Hunterian Museum, University of Glasgow (1867–70)**

7.9
A. Thomson, Holmwood
House, Cathcart (1857–8)

7.10
G.G. Scott, Kelham Hall,
Newark-on-Trent (1858–62)

Anne' house is now more a revival of the past than a modern Gothic house ... The aim of the Queen Anne architects now seems to be to show that nothing can be too old-fashioned for their style.[20]

As it happened, Thomson was a friend of one of the leading Queen Anne-ites, the expatriate Glaswegian J.J. Stevenson, but it is not recorded what he thought of this development.

Thomson had given a number of lectures to the Glasgow Architectural Society over the years, most famously his bold attack on the designs for the new university buildings on Gilmorehill. His 1871 presidential address to the Glasgow Institute of Architects seems to anticipate some of the other themes developed in the Haldane Lectures on the history of architecture a few years later, as in the insistence that, *pace* Ruskin and many of the goths, architecture does not bear the least resemblance to anything in nature, that it is peculiarly and exclusively a human work. As Thomson put it in 1874,

Whence then come Music and Architecture? There is nothing in Nature like either; for, although they may have been slow of growth, the fact is before us that they are something that by man or through his agency has been added to the word of God, and that, not presumptuously or sinfully, as some would tell us, but by destiny and duty; for, being made in the image of God, man was made partaken of the divine nature so far as to become a fellow-worker with God – in however humble a sense, a co-creator.[21]

This was certainly contrary to the understanding of Scott, who as a goth had maintained that 'All works of pure decoration, whether in sculpture or painting, must be remodelled by reference to Nature, – an implicit and unconditional falling back upon which, not as our copy, but as our guide and starting-point, must be the great, all-pervading characteristic of the future style.'[22]

Above all, there is the message here, as in the Haldane Lectures, that architecture must respond to laws, and that these laws – an expression of physical and mathematical truths apprehended by all civilisations – are eternal and thus an inescapable aspect of creation; that is, of the Divine. In this religious view of architecture, he might perhaps be compared with A.W.N. Pugin, with his insistence that gothic was the only true Christian (Catholic) style, while the Greek was to be discarded a product of a mere 'pagan' civilisation (a notion Thomson, of course, impatiently rejected), except that the Presbyterian Scot's understanding of divine law was deeper and more inclusive, less sectarian and partisan. His attitude might also be compared with that of Scott's distinguished first pupil, the church architect G.F. Bodley who, as Michael Hall has argued, seems to been entirely unconcerned with ideas of modernity and wished to seek perfection in an English late gothic manner which was beyond time, beyond development: 'In heaven there are no revivals.'[23]

But for Thomson the gothic was irrational, an expression of a short and superseded phase of human history and of Christian development: 'I hold that it never reached the highest place amongst architectural styles, and that, from its nature, it never can – that the spirit and circumstances which produced it do not now exist; and condemn as not only vain, but mischievous, all attempts to apply it to modern purposes.'[24] In contrast, he maintained that the essence of the trabeated language of classicism had far deeper roots.

> Long before man came to need it, long before the foundation of the world, at the very beginning, in the councils of eternity, the laws which regulate this art were framed, and … it cannot be supposed that they have been drifting down the stream of time, unheeded by their Author. Emanating from such a source they cannot be trifled with blamelessly. I am inclined to think that they cannot be perverted with impunity.[25]

At a time of confusion and doubt, Thomson – with his almost mystical interpretation of architecture – committed himself not to any one style, or to any one religion, but to universal, perennial truths.

Few others shared his interpretation of these truths. For Gilbert Scott, who had nailed his colours to the gothic cause, 'the arch must ever in future claim precedence over the beam and lintel,' although he also wrote that

> I think we may, in the first place, lay down for our architecture of the future, that it must unite in itself the two great normal principles of construction – the lintel and the arch. It is impossible that either of these can ever again be relinquished; each must in future be adopted as convenience dictates. The buildings of the middle ages admitted both, but it is for us more systematically to unite them.[26]

Thomson, however, was tenacious in his commitment to trabeation, and was pleased to observe that it was a principle followed by modern engineers as with the Britannia Bridge over the Menai Strait of 1846–50: 'In this way Stephenson laid a lintel over the opening formed by the sea between Caernarvon and the island of Anglesea which is considerably greater than any opening ever spanned by an arch. Now, you will observe that the simple, unsophisticated stone lintel contains in its structure all the scientific appliances of strutting and straining used in the great tubular bridge. In short,' he concluded, repeating his favourite debating point, 'Stonehenge is more scientifically constructed than York Minster.'[27]

In the end, perhaps, Thomson had the last laugh over Scott, for over a century later most new structures are of reinforced concrete or steel, and are trabeated rather than arcuated (Thomson – like Schinkel – can, of course, be seen as a precursor of the rectilinear abstraction of Mies van der Rohe). Even so, as J. Mordaunt Crook argued in discussing Thomson's lecture, 'by spurning all use of the arch, he chose to fight his battles with one hand tied behind his back. Like Pugin in reverse, his religious convictions … cut him off from whole sections of architectural experience.' Nevertheless, 'within the parameters of his chosen medium, 'Greek' Thomson was unbeatable. He knew the power of line, mass and gravitational expression. He understood, above all, the communicating role of metaphor.'[28]

What is particularly notable about his 1871 lecture text is that although he referred to those two great neoclassical sculptors of an earlier generation, Canova and Flaxman, 'Greek' Thomson did not once mention the word 'Greek'.[29] Possibly by 1871 Thomson was relaxing his rigid adherence to the virtues of one style. The 1870s was a decade of change, of doubt and confusion, and at the age of fifty-four Thomson had his best work behind him. Perhaps he was beginning to become disillusioned, recognising that there is more than one road to salvation, and that to achieve the progress he desired a more broad-minded approach to style was necessary. He admitted that 'no two minds are exactly alike, and as all our work should be done "on soul and conscience", it is better that everything should be cast into the crucible,' so this lecture might be interpreted as a qualified change of mind. Surely not: he was then building his Egyptian Halls in Union Street, his last great commercial structure, which the following year *The Architect* considered the 'most successful effort' in 'Mr Thomson's well-known "Egyptian-Greek" style – a style which he has made his own, and in which he has no rival.'[30] David Page has recently argued that there is nothing Egyptian about this extraordinary building at all; its facade is pure Greek and a presentation of 'idealised sources of his architectural beliefs'.[31]

Following the architect's death in 1875, the Glasgow-born designer J. Moyr Smith extravagantly praised Thomson's Egyptian Halls in his musings on 'The Style of the Future', in which he advocated a creative use of precedent – whether trabeated or arcuated. 'If a man has the head-power,' he wrote,

> he can use a style and adapt it to himself; if not, he adapts himself to his style. From materials supplied by a far less promising and far less tractable style than the English, Mr Thomson, of Glasgow, was able to produce perfect specimens of civic and domestic architecture, which were at the same time perfect specimens of advanced Greek; which is rather extraordinary, as everybody thought that Greek was

perfected a couple of thousand years ago. Mr Thomson's life and practice, it is true, were different from that of many of our architects; he was acquainted more or less with all styles, and selected Greek as the basis of future work; he mastered the style; was thoroughly imbued with the Greek feeling, and, gathering riches of fancy from sources unknown to, or overlooked by, the later Greeks, the style advanced in both flexibleness and fulness of fancy under his hands. And to carry out this work consistently and persistently, he resolutely refused all work in which he had not full power to use his own style; but his steady progress must have amply repaid him for the sacrifices he made, and the consciousness of reviving and carrying out a style until it reached that splendid culmination in the Union-street Warehouse, was surely a reward greater than has been vouchsafed to any other architect of this century.[32]

A belief in the virtues of a trabeated architecture – Greek architecture – is surely implicit in every reference to divine and eternal laws in Thomson's 1871 lecture. Three years later, in his Haldane Lectures, Thomson would return to a spirited defence of the transcendent superiority of the Greek. Perhaps the tactful, literally un-Grecian tone of the 1871 lecture may simply be explained by the fact that it was a presidential address to a professional institute which, by its nature, must needs be a broad church, including goths as well as classicists amongst its membership.

Even if it was a tactful performance, this lecture still serves to emphasise what an isolated and independent figure Thomson was towards the end of his career, heroically maintaining the virtue of developing a single style when all around him – even in Glasgow – were collapsing into eclecticism. Such tenacious adherance to an unfashionable ideal may have made Thomson seem old-fashioned, but we might well argue today that he was in fact ahead of his time – not so much in anticipating modern architecture (whatever that may have been, or be), but in looking forward to the revival of the authority of classicism in the early twentieth century. The eclecticism of the late Victorian decades, culminating in the feverish experimentation of *art nouveau*, was succeeded by a profound reaction – 'The Morning After,' as Goodhart-Rendel put it [33] – when the grand manner returned to fashion, when the great Burnet of Glasgow was invited south to enlarge the British Museum (no London architect seemed up to the job), and the new architectural schools loosely conformed to the model of the École des Beaux-Arts and taught a system based upon understanding the orders.

In Glasgow it meant the appointment of Eugène Bourdon as Professor of Architecture in 1904, the countering of the waning Mackintosh influence, a growing taste for the neo-Grèc and American classicism, and a revival of interest in the architecture of 'Greek' Thomson.[34] Thomson represented modernity; Mackintosh and the Arts and Crafts Movement, with their roots in the gothic revival, were seen as old-fashioned. It is surely no accident that the Edwardian belief in discipline and authority, combined with an exploration of the possibilities of abstraction, was accompanied by the publication of the first serious articles on Thomson in the London journals since his death. Reginald Blomfield could write, in the *Architectural Review* in 1904, that 'Thomson of Glasgow was possibly the most original thinker in architecture of the nineteenth century.'[35] Then there were the articles by Lionel Budden in the *Builder* in 1910 and by Trystan Edwards in the *Architects and Builders Journal* in 1914, the fateful year that also saw tribute paid to the greatest of Glaswegians by Albert Richardson in his revisionist study of *Monumental Classic Architecture in Great Britain and Ireland*. At the end of this magnificent book he argued that

> Thomson's predilection for abstract form in its enthralling mystery and dramatic intensity was the outcome of an original mind. His work in this respect stands alone, and while it reveals no sympathy for the broader and more academic rendering of the antique, as exemplified by the works of Professor Cockerell, Elmes or Playfair, within its own sphere it is unique.[36]

Posterity has been rather less kind to the reputation of Thomson's rival, George Gilbert Scott, though more merciful to his legacy of buildings. Even before his death in 1878 Sir Gilbert was pained to see the dissolution of the gothic crusade to which he had been so committed and his type of modern gothic rejected as too churchy and vulgar. His son, the brilliant but flawed George Gilbert Scott junior, took up with the Queen Anne style and became deeply pessimistic about any possibility of generating a truly creative modern architecture. In words that echoed both the conclusions of his father and the thoughts of Thomson as well as expressing the doubts of his Late Victorian generation, the younger Scott concluded that

> Those ages which have been the most fruitful in great works of architecture have been, as a rule, singularly ignorant of architectural history ... Nothing is more striking at the present day than the absence of true creative power in architectural art ... We have produced no national style, nor do we seem likely at present to do so. We have broken the tradition which maintained the continuity of art history, and made each successive style the natural outcome of its predecessor. Everywhere we meet with reproductions of ancient styles, attempted revivals of lost traditions, nowhere with any

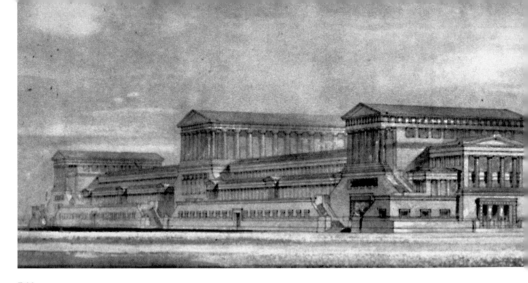

7.11
**A. Thomson, Natural History Museum, London, unexecuted competition design, 1864, perspective**

genuine power of creating new forms of beauty united to new requirements. Indeed, it is difficult to see how, when a tradition is broken up, or has exhausted itself, a new and genuine architecture is to be originated. We must look for this among the unknown possibilities of the future. But for the present we may well console ourselves for the deadness of the creative power in the vigour of the critical faculty. Our age is, in matters of art, eminently antiquarian, and in the minute acquaintance with the history of past styles which we possess, we may find some amends for the want of one of our own.[37]

As for Sir Gilbert Scott, in the twentieth century he was seen as representing the worst aspects of High Victorian architecture as well as being an architect whose large office produced too much and who restored far too many churches and cathedrals. By the 1920s the Victorian gothic was generally considered self-evidently ridiculous and hideous, and in his book on *The Gothic Revival*, published in 1928, Kenneth Clark could write of Scott that 'he believed that he built very good Gothic, we that he built very bad.'[38] Thomson's work, in contrast, was still regarded as modern and worthy of study – at least in Glasgow. When the young John Summerson as an architectural student first visited Glasgow in 1926, all he drew in his sketchbook were buildings by Thomson; 'Thomson was Glasgow's architect hero,' he later recalled.[39] Scott was, perhaps, too successful and too prolific to be a hero, and only in recent decades has his industry, his intelligence and his skill as a restorer, as well as his ability as a designer, begun to command respect and admiration.

It was the most important new building commission in mid-Victorian Glasgow which brought these two very different architects – one Scottish, one English – together, if in conflict. Given the city's history and the peculiar and discreditable circumstances of the 1860s in which local architects were denied

7.12

**G.G. Scott, University of Glasgow, approved design 1866, perspective**

any chance of designing the new University buildings, no doubt it would be far better and more appropriate if powerful Doric porticoes and sublime Thomsonian colonnades stretching towards infinity now commanded Gilmorehill, rather than the over-assertive gothic tower, arbitrary tourelles and crow-stepped gables designed by Gilbert Scott. While there can be no doubt that Scott did far too much and produced much dross, the best buildings of both men are impressive and intelligent answers to contemporary conditions, while their writings reveal two fine minds extolling the virtues of their chosen styles and trying to come to terms with that insuperable problem of finding a non-derivative, universal, characteristic architecture for their complex, confident and yet uncertain age. Although Scott once wrote that, after he first encountered Pugin's writings, 'modernism had passed away from me and every aspiration of my heart had become mediaeval,' and despite the fact that, as far as Thomson was concerned, 'In artistic forms the Greeks aimed at ideal perfection; and so far as we can comprehend the matter, they attained it,' both these great Victorians might nevertheless be seen as striving for something new, something modern.[40] But that modern was not necessarily the modern of the early twentieth century.

Perhaps the whole idea of 'modernity' was and is a snare and delusion. It is for many historians, who seem less concerned with the context and intrinsic merits of a building than with aspects of it which might seem to point towards a particular future. The architectures of Thomson and Scott, different as they were, were both rooted in a deep and subtle understanding of the past, whether ancient Greek or medieval. As intelligent public-spirited practitioners, both were uncomfortable about the inescapable importunities of history in an age which seemed in many respects so progressive and enlightened. Yet they also understood the truth that a new style for the Victorian age could not just be invented. Instead, both Alexander 'Greek' Thomson and George Gilbert Scott chose to express themselves in an historic style which they loved, and which they made into personal language, developing and adapting it to meet the challenges of their own time. No more should be asked of any architect.

# Notes

This chapter is a revised and enlarged version of an article first published in *The Alexander Thomson Society Newsletter* 21, May 1998, and then in *Mac Journal* 4, 1999.

1    Beazley and Lambert (1964:1015).
2    Pevsner (1972:183).
3    Thomson, 'An inquiry into the appropriateness of the Gothic style for the proposed buildings for the University of Glasgow, with some remarks on Mr Scott's plans' (1866), in Stamp (1999b:74 and 65).
4    Quoted in Harbron (1942:63).
5    *Builder*, 15 June 1861, p.403, quoted in J. Mordaunt Crook, 'Early French Gothic', in Macready and Thompson (1985:50).
6    Thomson (1871) in Stamp (1999b:102).
7    Thomson (1871) in Stamp (1999b:102).
8    Scott (1857:263–4).
9    Ruskin, 'The Influence of the Imagination in Architecture', address to the Architectural Association, 1857, in Cook and Wedderburn (1905:349).
10   Thomson (1871) in Stamp (1999b:103). By the time he published *The Three Periods of English Architecture* in 1894, Thomas Harris had also concluded 'That all endeavours to *invent* a new style, as such, must be abortive. It must grow out of *something*; it is therefore submitted that all question of a new style must be subordinated to the consideration of *a new construction* which will prove to be the "something" required. That is, iron.' Quoted by Harbron (1942:66).
11   H.-R. Hitchcock to Graham C. Law, 13 March 1950, quoted in Stamp (1999a:18).

12  Thomson, *Art and Architecture: A Course of Four Lectures ... delivered at the Glasgow School of Art and Haldane Academy, 1874*, Manchester 1874: Lecture III, p.8, in Stamp (1999b:147).

13  Scott (1879) and new edition Stamp ed., Stamford 1995, p.208 (passage written in 1864).

14  Scott (1857:271–2)

15  Thomson, 'On the Unsuitableness of Gothic Architecture to Modern Circumstances' (1864), in Stamp (1999b:58).

16  Scott (1857:109,111).

17  Thomson (1866) in Stamp (1999b:67).

18  Thomson (1864) in Stamp (1999b:58).

19  Scott (1857:213).

20  Scott (1879:376); this passage was written in 1878.

21  Thomson, Haldane Lecture I (1874), in Stamp (1999b:123).

22  Scott (1857:268).

23  Hall (2000:94).

24  Thomson (1866:384).

25  Thomson (1871) in Stamp (1999b:101–2).

26  Scott (1857:267, 266).

27  Thomson, Haldane Lecture IV (1874), in Stamp (1999:169–70).

28  Crook (1987:196-197).

29  Earlier, for the dining room at Holmwood House, Thomson had designed a frieze made up of enlargements of Flaxman's illustrations of Homer's *Iliad*.

30  *The Architect*, 13 July 1872

31  Letter to the author, 14 May 1998, and see Page and Stamp (2001:6–10).

32  Smith (1875:425), reprinted in *Ornamental Interiors, Ancient and Modern*, London 1887. Oddly enough, in the decorative work of John Moyr Smith and other artists within the aesthetic movement 'advanced Greek' and progressive gothic were fused in a distinct style: see Stapleton (1996). Indeed, *pace* Moyr Smith, Thomson himself sometimes designed in gothic, and not just at the beginning of his career: see Stamp (2003).

33  Goodhart-Rendel (1953).

34  Stamp (1992) and Stamp (2004).

35  Blomfield (1904:194).

36  Richardson (1914:104).

37  Scott (1881:1–2), and Stamp (2002).

38  Clark (1950:182). In the second edition, in his 'Letter to Michael Sadleir', Clark described the unthinkingly hostile 'feeling towards nineteenth-century architecture which prevailed in 1927.'

39  Summerson (1994:3).

40  Scott (1879:373) Thomson, Haldane Lecture III (1874), in Stamp (1999b:158).

# Chapter 8

# 'A complete and universal collection'

## Gottfried Semper and the Great Exhibition

*Mari Hvattum*

In the spring of 1851 London witnessed an event described by Queen Victoria as 'The most beautiful and imposing and touching spectacle ever seen'.[1] *The Great Exhibition of the Works of Industry of All Nations* opened in May and closed six months later, having been visited by almost one fifth of Britain's population. The Great Exhibition presented an encyclopaedic overview of human ingenuity in the midst of the unfolding industrial revolution – machines and weapons, art and architecture, possible and impossible products of industry including

8.1
**Crystal Palace, 1851, view along the main nave**

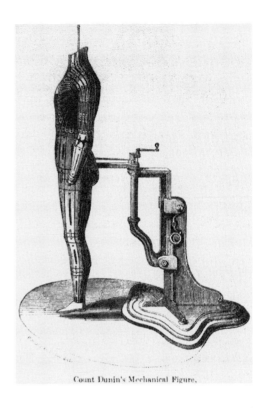

8.2
**'Expanding figure of a man composed of 7,000 working parts'**

Count Dunin's Mechanical Figure.

'philosophical instruments' such as the 'expanding figure of a man composed of 7,000 working parts'. Artefacts were assembled from past and present, far and near, on an unprecedented scale, comprising more than 100,000 objects from all corners of the world.

Among the incredulous audience witnessing the Hyde Park extravaganza was a young and ambitious German architect, Gottfried Semper. Temporarily stranded in London as a political refugee, Semper found opportunity both for employment and contemplation at the Great Exhibition. Organising several of the national exhibits, he had intimate knowledge of the display and ample time to reflect on its significance.[2] His verdict was mixed. He lamented the vacuous historicism of the mass-produced goods, their borrowed styles and *faux* materials. Yet he admired the comparative principle upon which the exhibition was based, seeing its comprehensive overview of human ingenuity as a possible key to contemporary design. This hope would become a predominant aspiration in Semper's theoretical work, pursuing as he did a 'method of inventing' for modern architecture. The following chapter investigates this aspiration as it came to expression in the Great Exhibition and in texts inspired by the event, shedding new light on the inherent modernity of nineteenth-century historicism and encouraging reconsideration of the often misconstrued relationship between historicism and modernism.

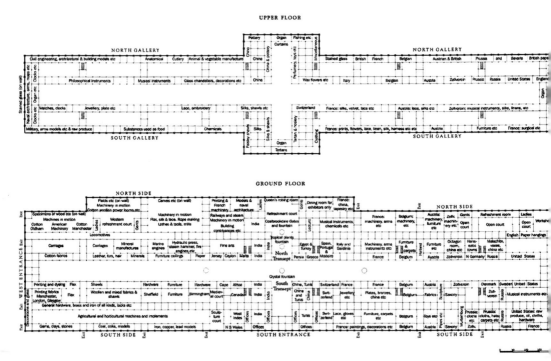

# The Crystal Palace: dreams of the conquest of history

For Semper and his contemporaries the Great Exhibition was an epoch-making event. It was a manifestation of the Victorian era, summing up its dedication to progress, industrialisation and market economy. It was a symbol of a new, liberal world, intended to encourage manufacture and to 'improve the application of art and design to industry'.[3] Prince Albert, patron of the exhibition, defined its two principles as 'the unity of mankind' and the 'division of labour'. Through these principles mankind was to fulfil its 'great and sacred mission': to use its God-given reason to discover the laws by which the Almighty governs His creation. By making these laws his standard of action, man was to 'conquer nature to his use'. The purpose of the exhibition, according to the Prince, was to 'give a true test and a living picture of the point at which the whole of mankind had arrived in this great task, and a new starting point from which all nations will be able to direct their further exertions'.[4]

8.3

**Plan of the Crystal Palace, 1851**

Although the Great Exhibition was dedicated to the future – to the progress of art and industry in the modern world – its strongest assertion was about the past. Underlying its progressive optimism was a new notion of human history, a notion that had emerged only in the late eighteenth century and come to full articulation in the nineteenth. It was this novelty that led Thomas Hardy to describe the event as 'a precipice in Time', and the year 1851 as 'an extraordinary chronological frontier'. Here for the first time, Hardy wrote, 'we had presented to us a sudden bringing of ancient and modern into absolute contact', as in a 'geological fault'.[5] The exhibition seemed to make the whole history of civilisation accessible simultaneously, furnishing a complete and comparative overview of human enterprise past and present. It fuelled a dream nurtured since the enlightenment: the possibility of gaining a complete insight into the laws of historical change. In this sense the significance of the exhibition exceeded even Prince Albert's ambitious expectations. While the prince saw it as a demonstration of man's control over nature by means of technology, his contemporaries went further, hoping to reveal not only the laws of nature but those of history as well. The exhibition embodied one of the most far-reaching ambitions of nineteenth-century modernity: the scientific mastery of culture and history.

# The axis of progress

A look at the way the exhibition was laid out will reveal some of its underlying ambitions. Entering the Crystal Palace through the south transept, the visitor encountered a twofold starting point; the beginning of the exhibition itself and the 'point zero' of human civilisation.[6] From south to north along the central transept of the palace he would progress through various degrees of 'primitiveness' – from Tunis, through China and the Middle East, to India, Turkey and Egypt. The developmental line continued in the main nave, where the visitor was led, east to west, from the new world, through the European continent, and finally to the conceived culmination of the exhibition and culture alike – the industrial products of the British Empire. The Great Exhibition was a carefully choreographed exercise in progressive historiography.

The axis of progress in the Crystal Palace described the various states of civilisation reached by contemporary peoples, yet underneath this synchronic comparison lurked a diachronic agenda. The argument was simple. Cultures remote in space could be regarded analogous to those distant in time. This comparative principle was not the invention of the exhibition organisers; it was in fact the principle upon which the emerging disciplines of anthropology and ethnology rested. These disciplines sought to overcome the scarcity of historical evidence by means of comparison. The study of early man, it was implied, could be substituted with a study of primitive man.[7] Our historically remote ancestors are analogous with our geographically remote contemporaries.[8] Within the matrix of progress, the old and the primitive became commensurable. The exhibition illustrated this idea in an extraordinarily clear way. Here, as one visitor observed, different stages of cultural evolution were revealed 'in their simultaneous aspect, like the most distant object revealed at the same moment by a flash of lightning in a dusky night'.[9] The exhibition provided a tour not only through the world but through world history, furnishing, as Semper later put it, a 'longitudinal section' of human culture and history.[10]

## 'A complete and universal collection'

Semper's remark is taken from a text written in the immediate aftermath of the Great Exhibition. *Practical Art in Metals and Hard Materials: Its Technology, History and Styles* was a work commissioned by Henry Cole to give theoretical support to the reform of British art education, and duly earned Semper a professorship in Cole's new Department of Practical Art.[11] Yet the piece was hardly a success. Incomplete, with an unwieldy title and an idiosyncratic language, *Practical Art in Metals* was not even deemed worthy of publication and remains today a pale pencil manuscript in the Victoria and Albert Museum Library.[12] One aspect of the text has proven to be of enduring interest, though, namely Semper's introductory outline of an ideal museum. This 'complete and universal collection' was, according to the author, to provide nothing less than

> the longitudinal Section – the transverse Section and the plan of the entire Science of Culture; it must show how things were done in all times; how they are done at present in all the Countries of the Earth; and why they were done in one or the other Way, according to circumstances; it must give the history, the ethnography and the philosophy of Culture.[13]

Semper outlined the organisation of such a collection in considerable detail, the structure of which anticipated the organisation of the new Victoria and Albert Museum. The universal collection would form a great comparative matrix in which the artefacts were arranged not according to chronology or aesthetic value (the two most common criteria for museum classification at the time), but according to the four primordial techniques of making, and their corresponding 'elements'.[14] [Fig 8.4] The section comprising textile art, for instance, would begin with the simplest wickerwork, expand to more refined textile products, and culminate in the metamorphosed motif of *Bekleidung* in its different guises. Similarly, the other elements of architecture would be traced

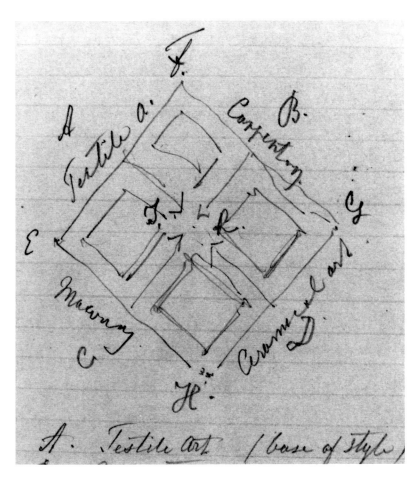

8.4
**Gottfried Semper,
diagram for an
ideal museum,
from *Practical Art
in Metals and Hard
Materials: Its
Technology, History
and Styles*, 1852**

from their simplest origins to their most sublimated expressions, and presented in their development through time and place. In this way Semper hoped to establish 'a good Comparative System of Arrangement', 'a sort of Index to the History of Culture' that would enable 'the Student to see the things in their mutual relations, to observe their mutual affinities and Dissimilarities, and to find out the Laws and Premises, upon which all these mutual positive and negative relations depend.'[15]

Semper's fictitious collection carries more than just museological interest. Patterned on the grand comparative display he had encountered in the Crystal Palace, the 'complete and universal collection' was meant to provide a comprehensive overview of human making. It would be not simply another museum but a complete encyclopaedia of human culture: a system of axioms by means of which a science of art could be established. Semper's collection, then, was not created so much out of an interest in the past as an ambition for the future. If its comparative display could unravel the secrets of human creativity,

Semper hinted, it would also give us the key to today's and tomorrow's artistic efforts: 'Supposing that such a sound basis were given, it would be possible to remedy the imperfections and irregularities of national education by Instruction, and to bring about an harmonious, intellectual, moral and practical Development of a Nation.'[16] This attitude is characteristic of nineteenth century historicism and testifies to its inherent modernity: the dream of controlling tomorrow by extracting the laws of yesterday.

# The unity of epochs

Nowhere did this new attitude to past and present come to a more forceful expression than in the Great Exhibition. If the exhibition as a whole was emphasising progress and evolution, drawing a 'longitudinal section' through human culture, the individual exhibits had their own significance as 'transversal' sections through particular nations and epochs. One interesting example is the medieval court. Designed by A.W.N. Pugin, the famous English campaigner for the gothic revival, the court was more than a display of a historical style. For

8.5

**A.W.N. Pugin, The Medieval Court, Crystal Palace, 1851**

Pugin, gothic art was a vehicle for spiritual and social reform. His dream, as a recent critic points out, was 'the dream of a generation which thought it could redeem the evils of industrialism by reliving the art of the Middle Ages'.[17] The middle ages, Pugin argued, possessed everything the present lacked: spiritual unity, aesthetic congruence, and most importantly, harmony between its *Zeitgeist* and its artistic style.

Pugin's medieval court is symptomatic of a new historical imagination that had come to dominate the nineteenth century. Up to the eighteenth century the past had been largely regarded as one continuous historical 'space', defined by the biblical notion of *genesis* and *eschaton*. The emerging historicism of the eighteenth and nineteenth centuries had broken this tradition, construing the past as a series of commensurable but separate 'spaces'. Thinkers as diverse as Winckelmann, Herder, Saint-Simon and Comte had seen history as a succession of distinct epochs, each marked by a particular character or style. While the present was in a sorry state of confusion, the past – and particularly the middle ages – displayed a coherence pervading every expression of life, marked, as it were, by a common fingerprint. This epochal coherence could be described as organic: a purposeful interaction of parts in an autonomous and harmonious whole.[18]

With the emergence of this new epochal consciousness the study of art and craft took on a new urgency. No longer seen as an ahistorical manifestation of eternal aesthetic norms, art could now be seen as a document of civilisation, an 'index' and 'encyclopaedia', as Semper wrote, of human culture. This was the underlying assertion of the Great Exhibition, and was even more explicit in the 'new' Crystal Palace as it was rebuilt at Sydenham in 1854. Here the medieval court was supplemented by courts illustrating everything from the ancient Egyptians to the renaissance, together forming a complete world history. The visitor could wander through successive styles, each indicating a different epoch, seeing before his eyes the gradual unfolding of civilisation.

Art history was not the only field to be affected by the novel notion of epochs and their evolution. Nineteenth century historicism seemed to promise the possibility not only of a science of art, but of human history and culture in general. If history could be seen as the successive formation and disintegration of organic systems, then the historian had a new instrument at his disposal. As the new discoveries in biology and anatomy had successfully proven, all organic systems develop according to laws. Behind the seemingly arbitrary flux of history, it was concluded, similar laws must be in action. By a careful study of individual epochs and their constitution, and by an extensive comparison with other epochs, these laws might be revealed. This comparative methodology was what informed Semper's dream of an ideal museum. By means of this method, he argued, it would be possible to 'find again those connections between things, and of transforming into an organic system of comparison what was before only

8.6
**Steam machine in the shape of an Egyptian temple**

an exterior and more or less arbitrary system of co-ordination and of exterior order'.[19] The 'complete and universal collection', far from simply displaying the past, contained the key to the present and future. The comparative method was for Semper the means to expand the laws of history into a 'method of inventing'; a procedure to reveal and realise the true 'style of our time'.

# The crisis of the present

Before we look more closely at Semper's 'method of inventing', we should ask why the pursuit of such a method had become such an obsession for architects and theorists of the nineteenth century. Once again, the Great Exhibition is a useful example to study. While intended to demonstrate the superiority of the industrialised west, the exhibition seemed to some of its critics to prove the exact opposite. To be sure the breathtaking progress of modern technology was displayed, but also exposed with regrettable clarity were the discrepancies and confusion that such industrialisation brought with it. For Semper, as for many of his contemporaries, the grotesque eclecticism and 'unnatural' designs seen among the exhibits – steam machines with the appearance of Egyptian temples and wheelchairs looking like the royal throne – revealed the degree to which western culture had become estranged from a true artistic sensibility.[20] [Fig 8.6] In a time when stone could be cut like cheese and papier-mâché made to look like marble, the natural appropriateness of material to purpose had been irrevocably lost. 'The present conditions are dangerous for the industrial arts, decidedly fatal for the traditional higher arts', Semper concluded gloomily.[21] The decay in western art was made even more apparent by the presence of artefacts from non-industrial nations, and Owen Jones was only one of many who eulogised over the compelling simplicity of 'primitive' craft, 'harmonised as by a natural instinct' with 'no superfluous and useless ornament which an accident may remove'. The authenticity and appropriateness of these products, Jones argued, were in stark contrast to the excesses of western industrial art.[22] In the midst of such a crisis the idea of a 'method of inventing' which could guide the contemporary artist in his pursuit of truth and beauty was undeniably attractive.

This severe verdict on the present was not unrelated to the new notion of history. If history is envisioned as a succession of coherent epochs, each appearing as an organic whole in which all expressions of life adhere to a dominant *Zeitgeist,* then the discrepancies of the present must seem all the more conspicuous. The sense of crisis expressed by nineteenth century thinkers may well have been a response to a time of unprecedented social, intellectual and material upheaval, but this disorder must have seemed particularly acute against the image of a past construed as coherent and organic. The cry for a unified and unifying 'style of our time' was a result not only of a chaotic present, but also of a notion of the past that had by now become normative. The construction of the past as a succession of organic epochs justified the demand that the present should also unite in one organic unity, in one coherent epoch, and in one true style. To contribute towards such unity was the ambition of Semper's theory of architecture.

# The method of inventing

The Crystal Palace's display of world history as a comparative field seemed to provide a promising new way of interpreting the past. Semper was greatly inspired by the event, and proclaimed that the 'comparative method applied to the study of the history of art is the only way to achieve a true knowledge and appreciation of these important moments of the monumental style.'[23] The comparative method was not, however, only a historiographical tool. In the practical mind of the architect it provided both a key to the interpretation of the past and to the prediction of the future. This was the aspiration of Semper's *Practical Aesthetics*. It was a theory that should provide a logical method of invention, a vehicle for historical interpretation and a basis for educational reform. In short, it was to provide a total method for the interpretation, production and transmission of architecture and art. By a careful study and comparison of historical epochs and their corresponding 'styles', the laws of both history and art could be found, and these laws could in turn be applied to the present. They could guide the search for a true style and thereby solve the most burning question of nineteenth century architectural discourse: 'In which style should we build?'[24] In opposition to many of his contemporaries, Semper rejected the idea of style as a matter of choice or invention. Style is rather the result, he insisted, of certain forces in society working upon ancient motifs of art.[25] Only by discovering the laws governing these forces could a true style of our time be revealed. 'The shackles would fall by themselves if the urge that drives the present became more generally aware of its aim. Here is victory and freedom!'[26]

Semper's 'universal collection' was to allow human culture in all its aspects to be captured and displayed in the simultaneity of the comparative matrix. By means of this matrix, which would grant 'a clear insight over [art's] whole province', the laws of artistic making were to be revealed, and a practical aesthetic formulated.[27] What is extraordinary about Semper's ambition is not so much its breadth as its depth. Universal collections and general histories were favourite pursuits in nineteenth-century scholarship. Semper's 'ideal and universal collection', however, was not only supposed to display everything but to explain it, capturing the full meaning and manifold of human creativity in one universal overview. Its significance was to be guaranteed not by the particular meaning of the artefacts displayed, but by the methodological arrangement itself, displaying human culture and history as an immanent system whose laws are available for explanation and prediction. Within the laboratory of the comparative matrix, the riddles of art and history were to be solved once and for all.

We can now start to appreciate the significance of the 'complete and universal collection' in Semper's thinking. This all-comprising assemblage was to furnish the overview of human history, unravelling 'why [things] were done in one or the other Way, according to circumstances'.[28] It was to reveal the very constitution of an epoch and explain its corresponding style. Art is at all times based on the same elementary motifs, Semper explained, but circumstances of time and place mean that these motifs are modified differently in different epochs. If the key to these modifications could be found, then a style of our time would reveal itself. By providing access to the laws of history, the 'complete and universal collection' was to grant control of both past and present, promising a potentially unlimited knowledge of the past and the predictability of the future. Lurking underneath Semper's 'method of inventing' is a dream of what Reinhart Koselleck has called the 'makeability of history': if you know the laws of history you may design your own future.[29] History has become a malleable material for the crafting of progress.

# Towards a theory of architecture

This brief glimpse into the 'magical glass' of the Crystal Palace and into Semper's theory of style is not meant as a complete picture of either. What it does is sketch out some themes occupying the historical and architectural imagination of the nineteenth century. A theme which emerges conspicuously is that of a theory of architecture and culture. The kind of theory envisioned by Semper is quite novel, and intimately linked with the unfolding modernity of the nineteenth century. While the eighteenth century still operated within a Vitruvian tradition, seeing architectural reflection as an interpretation of an eternally valid canon, the nineteenth century came to understand theory in a productive sense, as 'simple but powerful generalisations about the world and how it operates, that enable us to predict accurately future operations'.[30] Theory has become the instrument for introspection into the foundations of human society, requiring, as it were, a transparency of history and culture. It presupposes a world in which all historical, spiritual and practical factors are present and defined, so that the result of their interaction can be calculated and serve as a paradigm for a correct style.[31] Within this double framework of positivism and historicism, architecture could be understood as a branch of problem-solving, and its success could be guaranteed by a careful compliance to method.

Semper never completed his project for a 'method of inventing'. Its precondition – the complete overview of all aspect of history and culture alike – was ultimately a futile pursuit. Semper himself must have had a creeping suspicion of this futility, warning that the ideal collection 'will perhaps never be practicable, nor would it yet be desirable to try it … '.[32] Full introspection is not granted to man, situated as he is within a particular horizon. Yet the dilemma in Semper's thinking should not escape us. He was trying to balance continuity and innovation, the need to maintain tradition and at the same time find genuine expressions for contemporary culture. He attempted to find a domain beyond caprice and determinism, an attempt both more ingenious and more problematic than most of his interpreters have granted. It is this attempt that makes him so relevant to our own attempts to understand the uncertain ground of modernity.

# Notes

1   Quoted in Sparling (1982:v).
2   There are conflicting opinions as to what exhibits Semper was involved with. Mallgrave lists the Swedish, Danish, Canadian and Turkish exhibits as Semper's design (Mallgrave 1995:197). Germann and others include the Egyptian exhibit among Semper's commissions (Germann 1976:224). For a description of the exhibits and details of Semper's involvement, see Lankheit (1976:23–47).

3  Quoted in Beaver (1993:54).

4  Prince Albert's speech at the Lord Mayor's banquet in honour of the Great Exhibition, quoted in Commission of the Great Exhibition (1851a:xi).

5  Hardy (1979:286).

6  Stocking (1987:1–6).

7  See for instance the writings of Gustav Klemm, an early anthropologist to whom Semper was greatly indebted. For their relation see Mallgrave (1985). For an account of British anthropology and its reliance on the comparative method, see E.E. Evans-Pritchard; 'The comparative method in social anthropology', (Evans-Pritchard 1964:13)

8  The French missionary Lafiteau had already promoted the use of a comparative method in the early eighteenth century, comparing American Indians with the ancient Greeks.

9  Whewell (1852:3).

10  Semper (1852:§8).

11  Mallgrave (1995:208–18).

12  The manuscript exists as a handwritten copy in the Victoria and Albert Museum Library. A second copy, translated by Semper himself into German, is in Vienna. The manuscript was never completed, and apart from an extract in Wingler (ed.), *Gottfried Semper, Wissenschaft, Industrie und Kunst und andere Schriften über Architektur, Kunsthandwerk und Kunstunterricht* (Wingler 1966) it has not to my knowledge been published. I have not corrected Semper's English grammar or punctuation, nor have I altered the somewhat erratic use of capital letters, except when this is absolutely necessary for comprehensibility.

13  Semper (1852:§8).

14  On nineteenth-century criteria for the arrangement of art collections see Jenkins (1992:56–102). Semper's four elements of architecture comprised enclosure, structure, mound and hearth, corresponding to four techniques, namely textile art, tectonics, stereotomy and ceramics. For a presentation of the elements, see Semper's essay 'The four elements of architecture' (1851), in Mallgrave and Herrmann (1989:74–129)

15  Semper (1852: §7–10).

16  Semper (1852: §4).

17  Crook (1981:16).

18  See Eck (1995).

19  Semper (1853: §3).

20  See for instance Semper's essay 'Science, industry, art' where he proclaims: 'England has become estranged from art for the time being through the force of the progressive spirit of the time, and this estrangement is approximately in proportion to how fast she hurries ahead of other civilised nations along a course that, one hopes, they are all going to traverse'. (Mallgrave and Herrmann 1989:157)

21  Semper, 'Science, industry, art' (1852) in Mallgrave and Herrmann (1989:157).

22  Owen Jones in *Journal of Manufacture and Design*, quoted in Sparling (1982:35).

23  Semper, Second Prospectus to *Der Stil* (1859), in Mallgrave and Herrmann (1989:179).

24  Hübsch (1828).

25  Semper (1856-9:241).

26  Semper, 'Science, industry, art' (1852) in Mallgrave and Herrmann (1989:130).

27  Semper (1853:§5).

28  Semper (1852:§8).

29  Koselleck (1985:35).

30  Lang (1987:13).

31  Vesely (1987:24).

32  Semper (1852:§9).

# Chapter 9

# The interior as aesthetic refuge

## Edmond de Goncourt's *La maison d'un artiste*

*Diana Periton*

## Introduction

In a lecture delivered in Paris in 1819, politician and writer Benjamin Constant compared the ancient classical notion of liberty with its modern counterpart. He explained that liberty, and the sense of pleasure and fulfilment it affords, consisted for the ancients in the ability to participate constantly and actively in public collective government. This participation came at the expense of exercising individual choice over affairs of personal interest, since these were closely scrutinised by the same collective government. In contrast, he suggested, the modern ideal of liberty is concerned with the enjoyment of our private existence. In the modern state an individual, absorbed within the multitude, can exert very little public influence. Constant argued that this loss of influence was amply compensated by a vast increase in the possibilities for individual happiness, and ultimately for fulfilment in modern private life.[1]

Constant's assertion that liberty, happiness and fulfilment were to be found in the private rather than the public realm was concomitant with the new value society had begun to place on the notion of home. If in the eighteenth century the aristocratic home could still be understood in its traditional role as a representative civic institution, by the late nineteenth century the ideal of the home for all had become a considerably more private and personally expressive receptacle of meaning. Reflecting this change, Flaubert's *Dictionary of Accepted Ideas* caricatured the nineteenth century home as 'a castle inviolate', while his entry for 'domesticity' instructed that one should 'never fail to speak of it with respect'.[2]

César Daly, France's most prominent architectural journalist between the 1840s and the 1870s, believed that since the private residence could no longer be required to display its owner's ancient noble lineage, it should express instead the character and personality of its occupant.[3] This role for the home was emphasised in the increasingly popular handbooks on etiquette and style. Henri de Noussane's *Le goût dans l'ameublement* of 1896 declared that 'an apartment must have the character of those who inhabit it, it must bear their mark. There is no other way to be truly at home'. Noussane suggests that the salon, or drawing room, the principal room in which guests are received, might even be dedicated to a cult of its owner's personality: 'the main salon is the room for show, the temple where each of us displays to the world the cult which he demands from his devotees'.[4] Paul Bichet agrees that the home should vary according to its owner's identity – 'that of an artist cannot be that of a doctor', but the salon in his *L'Art et le bien-être chez soi* (1890) has a less overtly missionary role. He comments that the room itself is a modern invention, unnecessary in the warmer climates of Greece and Rome, which had the agora, the forum or the baths as settings for debate. The salon emerged in the eighteenth century as the scene for serious but witty and glittering discussion; by the late nineteenth century it had become the place for a quiet and comfortable chat, and should be decorated accordingly: 'For hangings, you have the choice of the richest: plush in traditional fabrics, or, in modern fabrics, silk or velvet; the mistress of the house should also take care to choose fabrics which are kind to her colouring: blue if she is blonde, cherry-red if she is dark'.[5]

If the home was properly to be an expression of personal character, reckless freedom of expression was to be moderated and honed by the exercise of good taste. The proliferating handbooks of advice for the home-dweller emphasise that a tastefully decorated home will be simultaneously conventional and original. They use the term taste as a mode of aesthetic judgement which has two poles. On the one hand, taste allows one to be in touch with that which is right or appropriate. On the other, the tasteful individual will be able to transcend prevailing habits of judgement in order to influence and direct them through the free play of his imagination. It is through the application of taste that the interior of the home becomes more than a collection of mere furnishings, and can be manipulated to arouse our feelings.

These handbooks discuss taste with the conviction that it can be educated by example, not by imposing fixed rules. Kant's late eighteenth century re-evaluation of taste provides a framework in which the interplay of its claims to universality of judgement and to fullness of individual expression can be comprehended. For Kant, taste is a judgement that recognises a perfect harmony which potentially unites imagination and understanding, but which cannot be reached through, or limited by, conceptual knowledge. Things that are beautiful 'dispose the spirit to ideas', but without the vehicle of conceptual thought.[6] It is in this lack of intellectual regulation or definition, this lack of limitation, that the apparent freedom afforded by the aesthetic realm is found. Taste, although rooted in convention, also allows us to rise above that convention, and to act with independence. While it arises through the senses, the aesthetic experience of taste goes beyond either abstract understanding or any particular sense perception, and provides an unfettered free-play of the imagination. Practised within the decorated interior of the home, where the expression of personal character is demanded, it is through exercise of taste that the inhabitant can potentially both create and experience the full possibility of his identity.

In this context, Edmond de Goncourt's *La maison d'un artiste* of 1881, and the home it describes, help to illuminate the value and meaning which had been ascribed to the home at the end of the nineteenth century, and the version of freedom and fulfilment it was meant to embody.[7] The two volumes of *La maison d'un artiste* take the reader in painstaking detail through the house in which Goncourt lived on the outskirts of Paris from 1868 until his death in 1896, a house he shared with his younger brother Jules, who died in 1870 of syphilis. The volumes themselves hardly constitute a handbook of domestic living, but in 1891 the editor of the *Revue des arts décoratifs* ran two articles on the house, proposing it as a perfect example of the modern dwelling.[8] In the preamble Edmond introduces the reader to 53 boulevard Montmorency,[9] a 'nest crammed with more eighteenth century objects than any other in Paris'. Once the front door is open, 'the visitor is welcomed by terra-cotta bas-reliefs, bronzes, and porcelains from that most agreeable period, mixed with objects from the Far East, in that same happy co-existence which they had in the collections of Madame de Pompadour and all the curio-gatherers of the time'.[10] He explains why the collection of objects has become so important:

> Life today is a life of struggle; in every field it demands concentration, effort, sheer hard work. A man, whose existence is no longer out-and-about as in the eighteenth century, who no longer flits around in society from his teenage years until his death, is confined to his hearth ... In this restricted, sedentary, domesticated life, the most ordinary of human beings has no choice but to insist that within the four walls of his home[11] all should be agreeable, pleasant, amusing to the eye; and, naturally, he has found this pleasure in his surroundings through the objects of fine art, or of industrial art, which is more accessible to the general taste. At the same time, these less worldly practices have diminished the role of woman in masculine thought; she is no longer the noble pursuit of our entire existence, the vocation which was formerly the career of the majority of men. The outcome of this modification in habits is as follows: man's interest, having distanced itself from these charming creatures, is

now concentrated largely on pretty inanimate objects, which passion can invest a little with the nature and character of love.

The 'House of an Artist' is Edmond de Goncourt's refuge, a place full of objects which refresh and renew him, and allow him to forget himself in 'the satiation of art'.[12]

## La vie littéraire

Throughout the second half of the nineteenth century the Goncourt brothers wrote with increasing influence as novelists, dramatists and journalists, art critics and historians. In 1887 Edmond began to publish their diaries. These open with an account of the appearance of their first novel, *En 18—*, which coincided with Louis-Napoléon Bonaparte's coup d'état, and thus the start of the Second Empire, in December 1851. Although awkward, disjointed and self-conscious, *En 18—* already demonstrates their enthusiasms – for the femme fatale, for the art of the eighteenth century and the far east, and for the depiction of vignettes of contemporary life. It employs the rapid, acutely rendered dialogue and the close, evocative descriptions which recur, more skilfully handled, in their later work. When they returned to novel writing after a decade spent concentrating on eighteenth century history, they noted their intentions explicitly: 'One of the special features of our novels is that they will be the most historical novels of this era, the novels which will contribute the greatest number of facts and truths to the moral history of this century.'[13]

In their writings the Goncourts concentrated on small but significant details, on anecdotes, on what Edmond referred to as the *document humain* (letters, diaries, objects, physiognomies which might reveal intimate thoughts).[14] Both the past and the present could be represented through these fragments, for 'the present-day novel is based on documents taken from hearsay or from life, just as history is based on written documents. Historians are storytellers of the past, and novelists are storytellers of the present'.[15] They used these details to capture 'billowing humanity in its momentary truth'.[16] For them the work of the artist, literary or otherwise, is to render with accuracy the impression which an object, event or place produces; their descriptions are written not for the sake of the description itself, but as 'a means whereby the reader is transported into a certain atmosphere favourable to the moral emotion which should arise from these things and places'.[17] The artist reveals things, makes them visible, in such a way that it is also apparent how they should be seen, how they should be felt.

The publication of *Germinie Lacerteux* in 1865 first brought the Goncourts as novelists to a wider visibility than their own circle of friends. In the preface they referred to it as a 'true' novel which 'came from the street', a 'clinical study of love' amongst the lower classes, and thus a 'scientific study of contemporary moral history'.[18] This was the novel that attracted Zola to them, and led them to claim to be founders of the naturalist or realist school, even though they denigrated these labels.[19] But the brothers' main concern was that 'artistic writing', in its direct description of the 'human document' should not only present 'that which is base, that which is repugnant, that which stinks'; it should also portray 'that which is noble, that which is agreeable, that which is sweet smelling, … to show all the aspects, the profiles, of persons of refinement and objects of value. But this must be done in a careful and exact … manner'.[20] The focus on depicting the truth of reality might appear to objectify all details equally, to reduce any hierarchy between the things described, but the Goncourts' ultimate interest was in revealing the extraordinariness of particular things and the possible refinement of feeling which might be provoked through description of them. This refinement, this ability to see and to capture directly the beauty and wonder of that which is given, is what enthused them so much in the art of the Far East and that of the eighteenth century. While in the midst of checking the proofs for *Germinie*, they wrote that:

> Chinese art, and particularly Japanese art, those arts which appear to bourgeois eyes to be unrealistic fantasy, are drawn direct from nature. Everything they do is taken from observation. They render what they see: the incredible effects of the sky, the stripes on a mushroom, the transparency of the jellyfish. Their art copies nature …
>
> Basically, there is no paradox in saying that a Japanese album and a Watteau painting are drawn from an intimate study of nature.[21]

The Goncourts sought ways in which the aesthetic impact of the world as it presents itself could be selected and heightened, ideally for its pleasurable effects – this was what the artist, whether author, painter or skilled craftsman, should aspire to achieve. They were well versed in contemporary writings on medical psychology, which 'defined the human being as suspended between stimulus and response; the external world acted directly upon the internal world of the nerves'.[22] They had consulted these works in order to describe more precisely and scientifically the degeneration of their characters when faced with particular external circumstances,[23] but they were no less interested in their own mental state. Neurasthenia was a state of mental hypersensitivity and physical exhaustion identified by medical psychologists as the result of strenuous mental effort coupled with lack of physical action; the Goncourts cultivated such a state as 'the ground of existence for the modern artist. They defined themselves as aristocrats of the spirit precisely because of the extreme refinement of their nerves'.[24]

The concentration on the exact minutiae of things and the feelings provoked by sensing those things, preferably feelings demonstrating a certain refinement, was more important to the Goncourts than any attempt to compose a plausible narrative or coherent scene. Emile Zola, in an introduction to a later edition of *Germinie Lacerteux*, proposes that 'it is not a question of a more or less interesting story, but of a complete lesson of moral and physical anatomy ... The reader feels the sobs rising in his throat, and it happens that this dissection becomes a heart-rending spectacle, full of the highest morality'.[25] The Goncourts' only hierarchy was that of sensation, according to its force and its sensitivity. They criticised their friend Flaubert for attempting, in *Salammbô*, a total reconstruction of a vanished civilisation in which 'feelings are eclipsed by landscapes'.[26] Their concern was to pursue the description of a detail until its strangeness emerged, or to juxtapose fleeting sensations unexpectedly, always in order to heighten the emotions. By focusing on specific details of the appearance and feelings of their characters, they too are left incomplete – they are a series of vivid physiological studies which collect to suggest and to exaggerate a type, rather than an attempt to build up a convincing personality. The Goncourts habitually described states of being rather than actions which relate character to plot.[27] According to their first biographer, Alidor Delzant, who had discussed the writing process with Edmond, they wrote their novels as a variety of tableaux, chosen for their striking effect, then loosely threaded together,[28] rather as the plays, ballets and operas of the nineteenth century were often constructed – they wrote to produce novels in which 'the descriptive matter is laid on in brilliant patches'.[29]

## *La maison d'un artiste*

The seven hundred pages of *La maison d'un artiste* are assembled similarly, as tableaux supplemented and reinforced by transcribed *documents humains*, such as archived inventories of objects, menus from eighteenth century dinners, or historic personal letters. The structure of the books is straightforward: the house is examined sequentially, room by room, item by item, from bottom to top, inside to outside. Through documenting the different objects, their aesthetic qualities, their provenance and the memories they induce, what might simply be a dry inventory itself becomes instead an overwhelming and indigestible succession of vivid pictures, each constructed to provoke the sensations. Edmond frequently moves from a quick, evocative list of the main determining features of a room to detailed descriptions of the specific yet generic qualities of particular

9.1
**53 bvd de
Montmorency,
hallway, June 1886**

pieces. Thus the hallway, paved with red and white marble, has for its wall and
ceiling covering

a contemporary leather, populated by fantastic parrots gilded and
painted on a sea-green background. On this leather, in a studied
disorder – the picturesque mode of the antechamber of an artist's
studio – are all sorts of loud, gaudy objects, shining pieces of
fretworked copper, gilded porcelains, Japanese embroideries, a
plethora of bizarre, unexpected, astonishing things …

Among the specific objects are

this little globe of imperial yellow porcelain, delicately woven; a cage
for a cricket or a buzzing fly, which chinamen like to hang at the head
of the bed; and this faience plaque which shows the branch of a

9.2
**53 bvd de Montmorency, staircase, July 1883**

peach tree in bloom, modelled in open-work in a wooden frame to make a screen, the kind of decoration you would find in the myste-rious, shrine-like corner of a tea-house prostitute's room, a sort of altar piece in front of which she puts a flower in a vase ... and amongst these oriental bibelots, a French marvel, a bas-relief by Clodion!'[30]

In the dining room the tapestry wall-coverings, which originally decorated an eighteenth century garden pavilion dedicated to music, are an exact fit. The panels clothe the walls in

a fantasy landscape, where Boucher's rustic theatre is combined with Lajoue's perspectives of balustraded terraces, with Watteau's enchanted isle in the distance. This conventional landscape is inhab-ited by an adorably illusory creation: beribboned shepherdesses,

144

9.3
**53 bvd de
Montmorency, dining
room, July 1883**

many a Thyrsis white with dust, country spinning girls alluring in lace, huntresses clad in the bright red outfits of Vanloo's hunting parties, and little Bacchus-like peasants riding on goats: this entire world depicted on a white ground, this precious ground which is the soft atmosphere that envelops pretty eighteenth century tapestries; and in this rich milky harmony, by daylight, pink, blue and sulphur yellow are constantly streaked with the brilliance of silk shimmering through wool. Cheerful pictures framed by arabesques entwining on a background of moss green, garlanded with swags of flowers and amaranth-purple lambrequins.[31]

Gently lit by candlelight, these tapestries, together with the small marble statuette of a bathing girl by Falconet, have witnessed some lively dinners. 'Janin, Gautier, Murger, de Beauvoir, and Gavarni who always arrived late, so we put a watch on his plate to reproach him for his lackadaisicalness' have eaten exotic food which 'spoke to the imagination of the stomach', and 'other very spirited people, not at all in the public eye, have been charming and full of verve and gaiety amongst these hangings'.[32] After the first publication of *La maison d'un artiste* Edmond had photographs taken of the interior worlds he had so carefully assembled. In the *Journals* we read that

> Lochard is photographing the corners of my rooms. Looking at the proof-print of my dining room, where I can see the fireplace of the small salon through the open door, I am amazed at the accuracy of the shadows in the recessed rooms at the back of a Peter de Hooch'.[33]

Like his novels, Edmond constructed his rooms with precision so that they would arouse the sensations of those who inhabited them. The first floor was his own private realm, a veritable laboratory of sensation, where he slept (in the bed which had once belonged to the Princesse de Lamballe, chatelaine of Rambouillet) and worked. Next to his study, he labelled two of the rooms the boudoir and the cabinet 'of the Far East'. The cabinet contained his collection of Chinese porcelain,

> this earthly matter fashioned by man into an object of light, of soft shades gleaming like precious stones! I know of nothing else which can be arranged against a wall to bring such enchantment to the eyes of a colorist! The lovely colours invented in this land of the quintessential delicacy of coloration, the detailed research into the infinite graduations of the palette of the universe! Where else would you find an emperor artistic enough to demand, as did the Emperor Chi-Tsong one day, that 'in the future the porcelain for palace use will be blue like the sky that one sees after rain, in the gaps between the clouds?'[34]

These two rooms represented for Edmond a sanctum of aesthetic feeling essential to his own creative fervour. '… when I am preparing to write something,' he confides,

> anything, a piece which doesn't even mention bric-a-brac, in order to get myself going, to let the words of the real writer flow rather than … painfully extracted stylizations … , I need to spend an hour in this cabinet and this boudoir de l'Orient. I have to fill my eyes with the patina of bronzes, the different golds of the lacquers, the iridescence, the bright reflections from the vitrified materials, the jades, the coloured glass, the shimmering silk of the foukousas and the Persian rugs, and it is only after this contemplation of bursts of colour, only

9.4
**53 bvd de
Montmorency,
Cabinet of the Far
East, June 1886**

after this vision which excites me, irritates me, so to speak, that, little by little (and I repeat that all this has nothing to do with the subject of the written composition) I feel the hairs on my neck begin to rise, and very gently a little fever begins to occupy my brain, without which I can write nothing worthwhile. But once I have procured the excitation produced by these glinting gewgaws, … then, in order to write, I have to be in a room with nothing on the walls, a room which should ideally be naked and completely white-washed.[35]

The writing itself must take place in a monkish cell, for only in such austere surroundings can you give life to this 'little world animated by you, spurting forth from your entrails'.[36] The artist must be able to activate his passions, his instincts, but also to master them. He must be both aroused and purged in order to reach that ecstatic state which allows him to grasp the reality of experience and reveal it to others.

Sometimes, though, the excitement caused by contemplation of shiny things simply led Edmond to neurasthenic exhaustion: the *Journal* plaintively describes the 'reordering of my shelves to research the opposition of tones and the contrast of materials, reordering at the end of which tiredness and my dazzled eyes put me in a state of vertigo where it seemed as if the floor would give way beneath me and my bibelots'.[37]

# 53 boulevard de Montmorency

The Goncourt brothers moved to their house in Auteuil after eighteen years spent living together in the centre of Paris. In 1868 Auteuil was still a semi-rural retreat, but the city was rapidly encroaching. The house was built as part of the Villa Montmorency, a private garden suburb laid out in the 1850s, when the domains which had previously constituted Auteuil and Passy were being sold for the development of the new *chemin de fer de ceinture*, Paris' orbital railway line. The railway was buried in a cutting immediately in front of the building, just inside the city's fortifications. Auteuil station is two minutes from the house on foot; from the 1860s this station had frequent and regular services to St Lazare on the right bank and to Montparnasse on the left; at the weekends it was thronged with visitors to the Bois de Boulogne, hard against the ramparts. To the back of the house, at the end of the Goncourt's garden, was the rest of the gated community of the Villa Montmorency, occupied by artists, engineers, businessmen and a surprising number of English families.[38] The conditions of sale for its plots excluded

> industrial establishments, people involved in the sale of wines, liquors, also restaurants, delicatessens, charcuteries and butchers, dairymen or cafés, or those holding public dances, or those practising a noisy trade, or one which spreads unpleasant and unhealthy odours, or women of ill-repute ... such that the said plots can be occupied only for well-to-do residential purposes (*ne pourrons jamais être occupées que bourgeoisement*)'.[39]

In order to buy their suburban idyll they had first to sell the property in Lorraine which had given their great-grandfather the title of de Goncourt, 'that bit of family pride, that great landed estate, that venerable, respected and sacred ground ... that assurance of position and income which only land could furnish in the eyes of an old family ... Finally, after eight months of negotiation, of correspondence, of title searching, we succeeded in getting rid of this chief nuisance in our life'.[40] They exchanged the nobility which carried with it responsibility for the land for a new version which the *Revue des arts décoratifs*, extolling the Goncourts' *habitation moderne*, dubbed the 'aristocracy of the intellect', of those with 'true taste and spontaneity, freed from the conventional frame of old ideas'[41]. The house for which they sold their family seat was a conventional little neo-Gabriel villa of the type seen so often in Paris' outer districts. Inside they upheld a louche but fastidious decorum in the ground floor reception rooms – dining room, small salon, large salon – by installing furnishings originally made for rooms in which the pre-revolutionary nobility had entertained.

9.5
**53 bvd de
Montmorency**

They covered the hall and staircase with an eclectic and brightly coloured mix of
Japanese hangings, drawings from the French school, and exotic bric-à-brac to
suggest the creative minds which occupied the more private rooms above, where
Edmond, as we have seen, arranged his apartment almost as a laboratory of
sensation. Jules died in one of the bedrooms on the top floor, and Edmond left
the room as it was for many years. Eventually in 1884 he asked his friend the
architect Frantz Jourdain to convert three of the attic rooms (Jules' bedroom
included) into his *grenier*, his attic storehouse; here he would hold court
amongst guests invited to his *parlote littéraire le dimanche*,[42] his Sunday literary
gossiping society. Regulars included Jourdain, the writers Emile Zola, Alphonse
Daudet, Joris Karl Huysmans and Guy de Maupassant, poet and dandy Robert de
Montesquiou, and the critic for the *Revue des arts décoratifs*, Gustave Geffroy.

In 1882 Edmond had publicly announced his intention to establish
on his death, and from the proceeds of a sale of his assets, an academy for the
encouragement of writers spurned by the official Académie Française. Three or

four famous names would be joined by six or seven deserving authors, who would receive a small annual stipend. These members would also act as the judges for a yearly Goncourt prize for literature. Some of those invited to the *grenier* on Sundays knew that they were being vetted for membership of this future academy. Edmond received them without getting up from his oversized divan draped in oriental tapestries, where he lounged like a Turkish sultan on his throne. The *grenier* was less formal than the ground floor reception rooms, and less a personal, sensual cabinet of curiosities than his far-eastern rooms on the first floor. Edmond saw it in terms of an artist's studio or a smoking room.[43] In a continuation of the eclecticism of the stairs, it was a place where Japanese embroideries co-existed with a showcase of portraits of his friends accompanying their favourite works, and where an oriental wooden biscuit bowl might sit next to an eighteenth century *pendule* clock. In the reveals of the mansard windows were built-in seats below exhibition cases containing Jules' watercolours, sketches by Japanese artists, and drawings by Watteau and Chardin.

Edmond mentions that from his bedroom window at the back of the house he could see the fort at Issy, to the south of Paris.[44] From the windows of the *grenier*, on the rare occasions when the blinds were opened, it must have been possible to take in the view Horace Walpole saw from the terrace of the gardens that had previously constituted part of the Domain of Montmorency, this 'glorious prospect … over which is extended all Paris with the horizon broken by the towers and domes of Notre Dame, St Sulpice, the Invalides, the Val de Grâce, etc; the whole … goes off in hills decked with villages and country houses that are closed by Meudon and forests on higher hills'.[45] Edmond's closing lines in *La maison d'un artiste* describe him leaning out of his window in the darkness of the night to savour the smell of the trees and listen to the frogs in the Auteuil marsh: 'I experience a sort of enjoyment to feel myself so close to Paris and yet so far away'.[46]

## Conclusion

The Goncourts' home – as described in their diaries, the two volumes Edmond dedicated to it, and the photographs he commissioned – was clearly a place they were proud of. For them a house should be full of beautiful and interesting things; Edmond criticises the house of his closest friend Alphonse Daudet which

> appeared … in a melancholy light on account of the absence of any elegant or artistic refinement, any fanciful or amusing touches … there is not a picture nor a knick-knack nor even a mildly exotic straw hat to be seen. There is nothing, absolutely nothing, that is not utterly ordinary, commonplace, and banal.[47]

The late nineteenth century handbooks on home decorating and etiquette also emphasise the importance of understanding one's abode as an artistic collection. Bichet's *L'Art et le bien-être chez soi*, which was aimed at 'the amateur artist, the man of taste who is arranging his home, the lady of the house who is need of quick advice on the organisation of her rooms, her furniture, her well-being',[48] opens with a chapter dedicated to educating the collector in the history of furniture, hangings and carpets, books, porcelains and medals, so that he or she will have a passing knowledge of the ingredients of the mix which is required. As usual, the Goncourts were scathing about other people's attempts to see the home in this way. 'The collection has entered completely into the habits and the distractions of the French people,' they wrote. 'It is a vulgarisation of the proprietorship of the work of art or of industry, which was reserved in the past for museums, *grands seigneurs*, and artists.'[49] To house a collection properly meant, quite literally, to give it a home – Edmond commented that even the banker Cernuschi, whose superior purchasing power was a constant irritation, could not achieve that.

> I lunched today at Cernuschi's. This rich collector has given his collection a setting that is at once imposing and cold like the Louvre. He has not known how to give it the warm and hospitable atmosphere of a dwelling, how to recreate a little corner of its fatherland. On these bare walls against the brick-coloured background favoured by our museums, these objects from the Far East seem unhappy … '[50]

Providing a proper home for oneself and one's objects meant, for the Goncourts, providing a refuge from the city. Removed from the city's noise and bustle, and from the duties it imposes, one could be fully and freely susceptible to the feelings provoked by the objects and their setting. Again, the handbooks on the home reinforce the Goncourts' assumptions. 'Which of us has never dreamed of a home which belongs just to us, away from the hubbub, in a pleasant suburb or handsome village, surrounded by greenery, at the gates of the city where our affairs imprison us?' asks Noussane in *Le goût dans l'ameublement*.[51] In the imaginary ideal home which dominates an entire chapter of his handbook, Noussane invents two specific retreats, one profane, the other sacred. The *retrait profane* will be filled with a light like that which filters through Murano glass. On a mosaic floor he will pour a thick layer of white sand which he can run through his fingers. An artist or thinker, sitting in this room on the swan-shaped folding chair, on a cushion embroidered with stars, facing his granite copy of the sphinx, would 'escape from the prevailing banality, the suffocation of the existing horizon'.[52] The *retrait sacré* will be a tiny private chapel, containing a white statue of Christ and vases of white lilies, where his soul will be soothed and his energy revived.

Noussane's retreats clearly echo Edmond's far-eastern rooms and monkish ideal study. Between them, these rooms have the effect of titillating the imagination then allowing for action – in Noussane's case it gives him the ability to calm the din his children are making in his imaginary atrium, in Edmond's the ability to write, so that he can confront the world with the truth of his observations. For the Goncourts, it was through their literary art rather than through producing heirs that the continuity of life might be preserved:

> … one of the proud joys of a man of letters – if that man of letters is an artist – is to feel within himself the power to immortalize … Insignificant though he may be, he is conscious of possessing a creative divinity. God creates lives; the man of imagination creates fictional lives which may make a more profound and as it were more living impression on the world's memory.[53]

The Goncourts' particular version of art – capturing the vividness of a sensation evoked or expressed – was their attempt to bring immortality to a fugitive moment.[54]

The house had various roles in procuring this immortality. Through its careful organisation and reorganisation it provided aesthetic moments which were in themselves material to be described and thus preserved. By stimulating the passions it gave feelings in general a greater immediacy, so that they could be more poignantly remembered and rendered more permanent. It was itself art, an 'artistic arrangement … a personal creation'.[55] Beyond its status as a place for the continuity and preservation of sensual stimuli, it served as a sanctuary in which 'work here seems … like work in an enchanted place, and it is hard to leave these things for the streets of Paris'.[56] Despite this lack of inclination, the *Journal* documents days spent dining out, visiting friends, attending plays, café-concerts, operas and cheap dance halls, scouring antique shops and consulting art dealers, for the Goncourts' main laboratory of sensation was the city itself. Even as they condemned it, they were determined to sample its luxuries and to contemplate the 'healthy reek' of its gutters,[57] in order to recycle and embellish these experiences in their writings. Theirs was a tight-lipped hedonism. They lived their lives not for their own sake, but as 'conscientious police agents'[58] discovering data to be eternalised, and Paris, with 'that feverishness which is characteristic of [its] heady life',[59] was overflowing with data. As Edmond grew old the gossip-gathering at the Sunday meetings in the *grenier* became a substitute for experiencing the city in its fullness. This was why it was so important to be 'in the open, at a few steps from the Bois de Boulogne, in light filled rooms filled with art treasures – living at the entrance to Paris, as though retired from the early ardours of the profession, and ready to produce masterpieces'.[60]

In 1819 Benjamin Constant spoke of a modern liberty which would allow all citizens who benefited from it to examine their most inviolable interests, to ennoble their souls and their thoughts. This was the great moral benefit, far beyond mere happiness, of the modern ideal of liberty, potentially available to all, afforded by the emancipation of private life.[61] Constant's ideas were framed in the aftermath of the French Revolution and the failure of Napoléon I's despotic empire. He was attempting to define the conditions in which a new kind of representative democracy, very different from any classical version, might become stable and viable; a complete recasting of the notion of liberty, and the sense of enjoyment and fulfilment that accompanies it seemed essential to this endeavour. By the second half of the nineteenth century the Goncourts, and the writers of handbooks on the home, took for granted this notion that liberty is found primarily in one's private life. The Goncourts were fully aware that they had rejected any active participation in political life, and wrote in the *Journal*

> I do not believe that any admirer of lovely things can be a patriot. My portfolios and my salon are my fatherland. A work of art gives you a distaste for the forum, it creates in you a kind of spiritual and idealistic egotism'.[62]

As artists, they knew that they 'lack[ed] the warm-bloodedness which leads to action; but from that, perhaps, comes the power of observation'.[63] The Goncourts accepted that the most important attributes of public life had been subsumed by the values now attributed to private life, and hoped that it would stay that way. The city itself, which as *polis* had been the setting for its own representation, and through that representation its own continuity, had become in the Paris of the nineteenth century a place of constant and dramatic change. Absorbed into the home, and recycled through the home as art, the city's 'billowing humanity'[64] could be stilled, absolved, and made to endure.

# Notes

All translations are by D. Periton unless otherwise stated.

1   Constant (1980). For a discussion of Constant's views concerning liberty, see Geuss (2001:1–3).
2   Flaubert (1954). Flaubert compiled the dictionary from the 1850s onwards; he seems to have intended it to accompany the novel *Bouvard et Pécuchet*, which was left unfinished when he died in 1880.
3   Daly (1864:14). Daly was the founding editor of France's foremost architectural journal of the nineteenth century, the *Revue générale d'architecture*.
4   Noussane (1896:8,150).
5   Bichet (1890:128, 214–15).

6  For a discussion of Kant's doctrine of taste see Gadamer (1989:42–4).

7  de Goncourt (1881).

8  Victor Champier and Gustave Geffroy, in *Revue des arts décoratifs* 12, 1891–2, discussed in Silverman (1989:216–19). Silverman's thorough study traces the important influence that the Goncourts and their house had on the development of France's particular and official version of art nouveau.

9  It is now number 67, and was last used as a karate club.

10  de Goncourt (1881: Vol. I, 1–2).

11  The word 'home' is written in English in the original.

12  de Goncourt (1881: Vol. I, 2–3)

13  de Goncourt and de Goncourt (1956) 14 janvier 1861, translated in Baldick (1960:17).

14  See preface to de Goncourt (1882:iii).

15  de Goncourt and de Goncourt (1956) 24 octobre 1864, translated in Baldick (1960:17).

16  Preface to 1887 edition (dated 1872) of de Goncourt and de Goncourt (1956).

17  de Goncourt and de Goncourt (1956) 23 juillet 1865, translated in de Goncourt and de Goncourt (1937).

18  de Goncourt and de Goncourt (1864, Preface), summarised and translated in de Goncourt and de Goncourt (1955).

19  See preface to de Goncourt (1882), de Goncourt and de Goncourt (1879), de Goncourt and de Goncourt (1956) 1 juin 1891, etc.

20  Preface to de Goncourt and de Goncourt (1879:viii), translation adapted from de Goncourt and de Goncourt (1966:11).

21  de Goncourt and de Goncourt (1956) 30 septembre 1864, translation based on that given in de Goncourt and de Goncourt (1971).

22  Silverman (1989:36) The Goncourts read the works of Moreau de Tours and the more contemporary Charcot; after Jules' death Edmond became acquainted with Charcot. Silverman notes that A. de Monzie, a colleague of Charcot's at the Salpêtrière hospital in Paris, considered the Goncourts to be part of a new *école psychiatrique* of writers, influenced by Charcot's teachings – see de Monzie (1925:1159–1162).

23  See Baldick (1960:21, 29 and 37) for their use of various medical treatises.

24  Silverman (1989:37).

25  de Goncourt and de Goncourt (1887:xi, Preface by Zola).

26  de Goncourt and de Goncourt (1956) 6 mai 1861, translated in de Goncourt and de Goncourt (1962).

27  See Bourget (1886:166–7).

28  See Baldick (1960:13).

29  See de Goncourt and de Goncourt (1915), introduction by West.

30  de Goncourt (1881: Vol. I, 4–5 and 12).

31  de Goncourt (1881: Vol. I, 14–15).

32  de Goncourt and de Goncourt (1956:19). Jules Janin was a novelist and literary critic, Théophile Gautier a poet, novelist and travel writer, Henry Murger a novelist and playwright, Roger de Beauvoir (aka Eugène Roger de Bully) a writer of melodramas, Gavarni (Sulpice Guillaume Chevalier) a cartoonist and artist – see Galantère's 'Biographical Repertory' in de Goncourt and de Goncourt (1937).

33  de Goncourt and de Goncourt (1956) 2 juillet 1883.

34  de Goncourt (1881: Vol. II, 227–8).

35  de Goncourt (1881: Vol. II, 349).

36  de Goncourt and de Goncourt (1956) 13 juillet 1862, translated in de Goncourt and de Goncourt (1937).

37  de Goncourt and de Goncourt (1956) 20 septembre 1877.

38   See Plum, 'Villa Montmorency', in Farguell and Grandval (1998:138–43).

39   Conditions for sale for the plots of the Villa Montmorency, 1853, quoted by Plum in Farguell and Grandval (1998:139).

40   de Goncourt and de Goncourt (1956) 18 juin, 1868, translated in de Goncourt and de Goncourt (1937).

41   Victor Champier, *Revue des arts décoratifs* 12, 1891–2;400, quoted in Silverman (1989:215).

42   de Goncourt and de Goncourt (1956) 18 novembre 1884.

43   Billy (1954), translated in Billy (1960:299).

44   de Goncourt and de Goncourt (1956) 12 avril 1871.

45   Horace Walpole, letter to William Mason, 10 September 1775, in Lewis (1955: Vol. 25).

46   de Goncourt (1881: Vol. II, 382).

47   de Goncourt and de Goncourt (1956) 8 juillet 1874, translated in de Goncourt and de Goncourt (1962).

48   Bichet (1890:v).

49   de Goncourt and de Goncourt (1956) décembre 1858.

50   *Journal*, 1 juillet 1875, translation based on de Goncourt and de Goncourt (1971). Cernuschi, director of the Banque de France, bequeathed his house and collection to the state as a house-museum in 1896.

51   de Noussane (1896:117).

52   de Noussane (1896:219).

53   de Goncourt and de Goncourt (1956) 8 février 1868, translated in de Goncourt and de Goncourt (1962). See also 10 octobre 1865: 'Everything decays and ends without art. It embalms still lives, and nothing has any immortality unless it has been touched by art.'

54   See Caramaschi (1971:100–1) for a discussion of their attempt to immortalise the fleeting moment through art.

55   Geffroy, 'L'Habitation moderne II: La Maison des Goncourts', in *Revue des arts décoratifs* 12, 1891–2:151, quoted in Silverman (1989:218–19).

56   de Goncourt and de Goncourt (1956) 4 novembre 1875.

57   de Goncourt and de Goncourt (1956) 3 mai 1860, translated in de Goncourt and de Goncourt (1937).

58   de Goncourt and de Goncourt (1956) 22 août 1875.

59   Preface to de Goncourt and de Goncourt (1956), translated in de Goncourt and de Goncourt (1937).

60   de Goncourt and de Goncourt (1887:x, Preface by Zola).

61   Constant (1980:513).

62   Extract from de Goncourt and de Goncourt (1956), quoted in Billy (1954:113–14), translation based on Billy (1960:89).

63   de Goncourt and de Goncourt (1956) 25 novembre 1865.

64   See note 16 above.

# Chapter 10

# Timely untimeliness
## Architectural modernism and the idea of the *Gesamtkunstwerk*

*Gabriele Bryant*

> Mais la nécessité d'un *Gesamtkunstwerk* domine toute l'époque ...
> Jean Cassou, *Les Sources du XXe siècle*

> Every artistic form is the expression of something spiritual, something that relates to the soul ... Thus it is also spiritual things that determine the modern industrial architect.
> Adolf Behne, *Pathetiker und Logiker im Industriebau*

The Hoechst Administration Building (1919–24) in Frankfurt by Peter Behrens is representative of many of the prominent artistic and intellectual concerns of the early inter-war period in Germany. It is a rare example of so-called expressionist architecture which was actually realised, and the building can be read as a programmatic reflection on the critical role of art within the culture of modernity.

Despite being an integral part of one of the major German chemical plants, the Hoechst Building has become a place of pilgrimage for the modern art lover. Though the striking tower–bridge motif dominates the exterior approach and features in the company's logo, it is the cathedral-like interior of the central hall which provides the most spectacular experience of colour and light. The Hoechst Building is regularly cited as a *'Gesamtkunstwerk* of modernity',[1] but this characterisation poses some important questions. What is a *Gesamtkunstwerk*? What is its significance in the modern context? And what role does the idea of the *Gesamtkunstwerk* play in the conception of the Hoechst Building and the oeuvre of its architect?

Before examining in more detail the Hoechst Building and some other architectural examples from the early twentieth century which have been given the label *Gesamtkunstwerk*, let us look at the background of the idea and

**10.1
Hoechst
Administration
Building,
Frankfurt
(1919–24), tower
and bridge**

10.2
**Hoechst Administration Building, cathedral-like interior of the central hall**

its ambivalent place in the history of modernism. As Werner Hofmann has pointed out, 'what is demanded is by no means self-evident. Only with the end of the *Gesamtkunstwerk* can the programmatic "*Gesamtkunstwerk*" appear on the scene. The former appeared, unburdened by reflection, before the arts were separated into autonomous domains, while the latter is the retrospective attempt to reunite the individual arts again.'[2]

Trying to approach the *Gesamtkunstwerk* via definitions, it soon becomes clear that the characterisations found in most dictionaries of art do not reflect the crucial issues as they talk about a unity of different art forms in one work, for example a combination of architecture, sculpture and painting. Such purely formal definitions, however, cannot explain why *Gesamtkunstwerk* is seen by some as a 'precursor of redemption',[3] and by others as a dangerous travesty. Richard Wagner called upon it as a means to 'bring salvation from this most unfortunate time',[4] yet Theodor Adorno a century later referred to 'that suspicious synthesis, the traces of which still frighten in the name of the *Gesamtkunstwerk*.'[5] Jean Cassou wrote that the *Gesamtkunstwerk* was a necessity for twentieth century art that 'dominates the whole period'.[6] Robert Klein called it 'a persistent, or more exactly, a periodic desire, of which both artists and the public dream',[7] and in a recent contribution to the journal *October*, the *Gesamtkunstwerk* is referred to as 'a specter which haunts the theory and practice of the arts throughout our century'.[8] Much more than aesthetic unity must be at stake here. Are we trying to track down a phantom, or a collective dream?[9]

The programme of a *Gesamtkunstwerk* was first explicitly formulated in the mid-nineteenth century by Richard Wagner, although the idea has its theoretical foundations in early romanticism and German idealism. In what has been called an 'aesthetic revolution', art is elevated to a metaphysical level,[10] and religious concepts and political aims brought into the realm of art, as exemplified in the notion of an 'aesthetic state'[11] and the project of a 'new mythology' as the 'most artificial of all artworks'.[12] 'Aesthetic revolution', a term first coined by Friedrich Schlegel,[13] also refers to the redefinition of the relationship between art and nature, or between art and society, which results in the reversal of their roles. Art is no longer to imitate its historical and social context, but rather, society should follow the ideals and rules set out in art. Art serves as *Vorschein* (aesthetic anticipation), and is the catalyst for the creation of something fundamentally new. The transfer of spiritual as well as revolutionary or reformist sociopolitical aspirations into the arena of aesthetics, with autonomous art as the carrier of utopian ideas for the creation of a better society, was to form the conceptual basis for the quest for a modern *Gesamtkunstwerk* in the next two centuries.[14]

Richard Wagner lays out the programme for a *Gesamtkunstwerk* in his essay *The Artwork of the Future*. Writing in exile in Zurich after the failure of the German revolution in 1849, he saw the disintegration of the community of the arts as a symbol of the selfishness of individual members of a greater social whole. He writes of 'the great *Gesamtkunstwerk* that comprises all artistic forms, in order to use all individual forms as a means, to annihilate them in order to arrive at the *Gesamtpurpose*, that is, the absolute, immediate representation of the perfect human nature.'[15] Though it is customary to begin any discussion of the *Gesamtkunstwerk* with a reference to Wagner, the profound influence the idea has had on twentieth century artists stems as much from its discussion and adaptation in successive generations, and particularly from the philosophy of Friedrich Nietzsche. By the early twentieth century the *Gesamtkunstwerk* had become a collective artistic ambition, and, as Vergo has observed, 'the term *Gesamtkunstwerk* was hurled around like a kind of verbal projectile'.[16]

It must be emphasised that the *Gesamtkunstwerk* has always stood for more than an aesthetic programme for the formal reunion of the arts,[17] and the regular resurgence of the *Gesamtkunstwerk* agenda in the context of failed political revolutions over the past two centuries can be no mere coincidence.[18] From Wagner to the present day it has been offered as an answer to social crisis, cultural despair and the modern experience of alienation, claiming to have a major impact on all aspects of life and ultimately to achieve the fusion of life and art and anticipate a new social order. Its aim is not just the elimination of the separation between different artistic forms, but the transcending of the artwork as such, leading to the aesthetic transformation of society, an agenda referred to by Hofmann as 'aesthetic imperialism'[19] and by Bisanz as 'habitational engineering' and 'environmental dramaturgy'.[20] Ambivalence towards modernity lay at the heart of the *Gesamtkunstwerk*, which grew out of the idealist dream of an aesthetic utopia that can offer an alternative to existing reality. The quest for a *Gesamtkunstwerk* was also linked closely to the problem of aesthetic fundamentalism,[21] and it has been put forward variously as 'the ultimate modern myth of the origin'[22] or paradigmatic manifestation of 'anti-modern modernism'.[23] Although the emphasis here is on the German context, important parallels exist with contemporary artists and artistic movements outside Germany.[24]

The German painter, architect and designer Peter Behrens (1868–1940) was an exemplary exponent of the early twentieth century quest for a *Gesamtkunstwerk* in the visual arts. His work over the first four decades of the century marks him as a major contributor to many of the most important artistic movements of the period, categorized as *Jugendstil*, geometric abstraction *sachlichkeit*, neoclassicism and expressionism.[25] He started his career as a painter before turning to the applied arts, becoming a major influence in the theatre as well as within the garden reform movement. Though he never trained formally, he became one of the most powerful architects in Germany, with a large successful office in Berlin. Behrens received some of the most prestigious commissions of his time, from embassy buildings to major factory complexes,

10.3
**Mathildenhöhe in Darmstadt, founded in 1901, Peter Behrens' house**

and exerted an important influence on the younger generation of architects including Walter Gropius and Mies van der Rohe. He is probably best known for his work as an artistic consultant to industry, especially as chief artistic consultant for the Allgemeine Elektrizitäts-Gesellschaft (AEG) in Berlin from 1907 to 1914, in which capacity Julius Posener referred to him as 'Mr Werkbund'.[26]

Most historical accounts of Behrens's oeuvre emphasise his stylistic development, but few examine the thematic continuities in his work.[27] Behrens' concern with the synthesis of the arts remained a *leitmotif* throughout his life. He stated time and again his artistic creed that 'Architecture … among the arts is the foundation, on which the unfolding of the other arts should be fully achieved … The idea of the *Gesamtkunstwerk* must start from architecture',[28] and always saw the *Gesamtkunstwerk* as a project that is predominently ethical rather than aesthetic.

His career as an architect began with his involvement with the artists' colony on the Mathildenhöhe in Darmstadt, founded in 1901, where he designed his own house. The Mathildenhöhe project conceived each artist's house as a focal point of the transformation of life as art, and it was in this context that Behrens wrote that 'Man shall be the creator of Culture … an artist who is his own material, creating something nobler through and out of himself'.[29] 'Each art only partakes in Style,' Behrens continued. 'Style is the symbol of the *Gesamtempfinden* (totality of feeling), the whole concept of life of a time, and it is revealed only in the universe of all arts.'[30] The role of the artist–craftsman as messiah of a new *Lebensgestaltung*, so polemically attacked in Adolf Loos's 'Story of the Poor Rich Man',[31] had become a commonplace by the turn of the century. As well as ensuring a formal unity, the ornamental scheme of the Haus Behrens in Darmstadt gives the house a unifying Nietzschean theme. In a letter to Nietzsche's sister, answering her invitation for a visit to Weimar, Behrens wrote 'I regard myself fortunate that I may bring all my admiration and respect for the wise artist in front of you … I only wish I possessed the power to put my feelings into words.'[32] Critics have coined the term *Zarathustrastil* for Behrens' Darmstadt period, and his villa unmistakably identifies its owner as one called upon to follow in Zarathustra's footsteps.[33]

The iconological programme of Behrens's house on the Mathildenhöhe demonstrates a combination of the *Jugendstil* interpretation of the gothic combined with a reading of Nietzsche.[34] The *Jugendstil* poet Hermann Bahr wrote of his time's need for a *Seelengotik*, a gothic of the soul,[35] that is, the religious transcendence symbolised in the gothic cathedral could be internalised and transformed into the idea of an immanent, secular transcendence of self. Equally, the crystal, which provides the central motif in Behrens's house as well as playing a central role in the colony's opening ceremony, symbolises the transformatory power of art, aspiring to a metamorphosis of the individual for the sake of the whole.[36] The *Jugendstil* disciples of Zarathustra saw themselves as missionaries: their art was to inspire a new secularised spirituality and redemption through modern aesthetics. As Goethe's Faust had proclaimed, and many that followed in his footsteps and became disciples of Nietzsche's Zarathustra were to echo, 'This is the youth's most noble profession. The world, it did not exist until I created it'.[37]

The *Jugendstil* experiment has sometimes been criticised as elitist, ineffectual, and too far removed from the reality it wished to transform. It is frequently argued that instead of achieving a social reformation, reality was effectively excluded.[38] The Wagnerian *Festspielhaus* in Bayreuth, Nietzsche's Zarathustra, and the Darmstadt artists' colony all resided in splendid isolation on their magic mountain, where they formulated their mission for a renewal of society far above the world they intended to ennoble. The *Jugendstil* villa has been described as the swan song of a grandiose bourgeois illusion, 'recommended for very young and very beautiful honeymooners',[39] and Meier-Graefe later commented on the dream 'to revise the history of the last one hundred or three hundred years and bring back the *Gesamtkunstwerk* in the form of the bourgeois house. The attempt … failed even more miserably than Richard Wagner's phantom.'[40]

Behrens' turn away from the excesses of art nouveau ornament came with his call to the *Kunstgewerbeschule* in Düsseldorf in 1903. From this point on his work is characterised by a concern with geometry and proportional schemes, though still pursuing the essential idea of a synthesis of the arts as *Gesamtkunstwerk*. As Meier-Graefe, one of the closest observers of Behrens' development, writes about him at this time 'The unity of the arts can only be achieved through a universality of the principle. This always lies in geometry only …'[41] The idea of achieving a unity of the arts through a common principle, an 'inner transformation of all art', was echoed more than a decade later by Adolf Behne, who called for a new *Gesamtkunstwerk* on the basis of the arts coming together of their own accord,[42] and the idea of geometry as the guarantor of a deeper unity – of the arts as well as of art and life – dominated much of the architectural discourse of the time. In Behrens' case this transformation eventually led to a shift from the aesthetic quarantine of the *Jugendstil* community to the twentieth century's industrial everyday, the 'designer's pact with sobriety … that leaves the idea of a total formalisation intact, the difference being that it is presented no longer in the form of "poetry" but of "prose"'.[43]

Peter Behrens was appointed artistic consultant to the AEG in 1907, where his duties gradually extended to designing everything from letterheads and advertising posters to industrial products and factory complexes. In what Buddensieg has described as 'probably the widest-ranging artistic commission of modern times',[44] Behrens created a new corporate image for Germany's most powerful industrial conglomerate. At the same time he was promoting the idea of a fusion of art and technology in a great synthesis of forces, an idea which became the most important reformulation of the *Gesamtkunstwerk* in the early twentieth century.[45] In a lecture of 1910 entitled *Kunst und Technik* he wrote 'It is in the hands of industry to create culture by bringing together art and technology'.[46]

As a designer for industry, Behrens took the idea of the *Gesamtkunstwerk* from the nineteenth century *Festspielhaus* to the bourgeois villa, and from there to the industrial corporation as cult building. As art and technology became united, the role of bringing about a spiritual transformation of society was increasingly assigned to the depersonalised and anonymous force of industry, an industry elevated into a cultural force as artistic *Gestaltung* permeates each and every one of its products.[47]

The happy accommodation of the artist with industry, however, was not to last. As the Hoechst Building was being constructed Behrens wrote, 'Art is symbolic … The rules of building are mysterious. Its diversity leads towards the infinite, unfathomable … The space is not enclosed, it does not limit, it reaches into incalculable depths and rises into high distances. The lawfulness of this Art is … geometric speculation, mystic number … The spatial boundaries are not the enclosing walls of a prison, but they are precursors, holding the promise of redemption.'[48] The tone of this short text is indicative of a profound change in Behrens' convictions. As for many artists and intellectuals of his generation, the experience of industrialised warfare was profoundly disillusioning, and led to a far less optimistic view of industry's power and the promises of technological progress.[49] Rather than endorsing industry as the most potent embodiment of the modern *Zeitgeist*, a 'return of art' (*Wiederkehr der Kunst*, to quote the title of Adolf Behne's important book of

10.4
**Hoechst Administration Building, Frankfurt; the crystal appears in the floor-pattern and in the shape of all the light-sources**

1919, became the battle cry in the architect's spiritual quest against the predominance of the modern forces of rationality and materialism. This idealist reorientation also gave rise both to the utopian vision of a glass architecture and a renewed interest in *Handwerkskultur* (craft culture).[50]

An exhaustive analysis of the Hoechst Building is provided in Buderath's monograph;[51] we shall look at just a few key elements. One of the symbols that Behrens consistently used throughout his career to illustrate his faith in the possibility of a spiritual transformation is the crystal,[52] which in Hoechst appears in the glass-domed ceiling and is echoed in the floor pattern and in the shape of all the light-sources in the central part of the building. The crystal lights also serve as guides from floor to floor; the rhythm of this movement, a kind of rite of passage from darkness to light, was carefully staged by Behrens.[53] The colour scheme of the glazed brick interior also emphasises this upward movement towards the light, seeming to contradict the material weight of the building's construction, which becomes heavier towards the top of the hall and perspectively emphasises the vertical thrust of the interior. Colour and light are used as symbols of spiritual forces overcoming matter.[54] Other notable features of the building are the emphasis on the contribution of the crafts, represented in the sculptures at the entrance resembling the figures of medieval saints, the gothic lettering of the clock, and the general appearance of the exterior, which seems to owe more to nineteenth century visions of medieval castles and belfries than to modern buildings for industry, an impression enhanced by the parabolic windows that stretch along the main facade.

'Architecture is bound to the earth, but through its spirit it seeks a unity with the cosmic whole, an order that dominates the whole world. This is the drive towards unity, towards totality,'[55] wrote Behrens. The echo of romantic–idealist conceptions of an aesthetic reconciliation is readily apparent in Behrens' environments, but what about the immediate social context and industrial function of his work?

The Hoechst Building could be called a cathedral of industry, as his AEG turbine factory was seen as a temple of work.[56] The religious air of the central space in the Hoechst Building, which feels like a modern interpretation of the gothic cathedral, is the most powerful impression that visitors to this building experience. The experience of spirituality is achieved through an abstraction[57] from the cathedral prototype's traditional symbolic context; the nineteenth century museum as aesthetic church[58] is replaced by the modern industrial building. Explicit references to the gothic and to the cathedral became almost commonplace in the early twentieth century. Not just artists but art historians as well turned the gothic into an ahistorical phenomenon, interpreting the cathedral variously as transcendentalism in stone, the exemplification of a revolutionary spirit, precursor of modern spirituality and a new social order.[59] Walter Gropius propagated the idea of a 'Cathedral of the Future' in the Bauhaus Manifesto of 1919, and Behrens put forward the concept of the *Dombauhütte* (Hut of the Masons' Guild),[60] designing as such an exhibition pavilion for the display of industrial arts at the Münchner Gewerbeschau of 1922. In what was perceived as an age of alienation and predominant materialism, the gothic cathedral came to stand for genuine spirituality and integrity. In the years immediately before and after World War I, expressionism was hailed as the new embodiment of gothic spirituality. It was emphatically declared to be symptomatic of the beginning of a new age of intuition, metaphysics and synthesis, an age that was turning away from the analytical, dissecting spirit of the enlightenment. Expressionism as the artistic heir of the gothic held the promise of cosmic reconciliation and a return to the inner world of the soul. The most prominent and influential reinterpretation of the gothic in this vein was Wilhelm Worringer's *Formprobleme der Gotik* (1911).[61] Worringer's expressionist interpretation of the gothic is based on the same approach of 'style psychology' as his *Abstraction and Empathy* (1908). He interprets 'gothic man' as the relative, the ancestor and soulmate of the modern angst-ridden individual, who suffers from the modern experience of alienation and social fragmentation.

If the nineteenth century secular neogothic movement has been labeled 'gothic without God',[62] and Bahr at the turn of the century speaks of a 'gothic of the soul', I would suggest that the gothic as put forward in German

10.5
*Dombauhütte* (Hut of the Masons' Guild), Peter Behrens, *Münchner Gewerbeschau*,
1922

expressionism by Worringer and which we find exemplified in the Hoechst
Building, can be read as a 'gothic of the murdered God',[63] in the sense in which
Nietzsche had proclaimed the death of God as the birth hour of the new man. In
what Buddensieg has called the 'non-confessional ... cult-building' at Hoechst,
'catharsis and hope through an architecture as Art',[64] we are reminded of one of
Nietzsche's rare direct references to architecture as an *Architektur der
Erkennenden* (architecture of the enlightened ones):

> ... what is missing most of all in our big cities: Quiet and wide spaces
> for thought ... where no outside noise reaches us, where a higher

sensibility would even forbid the priests their loud prayers: Buildings and complexes which as a whole express the higher state of meditation (*Sich-Besinnen*) and withdrawal (*Bei-Seitetreten*). The time is past where the church had the monopoly of thought. As houses of God and seats of the cult of metaphysical traffic, those buildings speak a language of too much constraint for us Godless ones to think our thoughts here. We want to have ourselves translated into stone, we want to amble inside ourselves as we walk through these halls ... [65]

In the Hoechst Building Behrens pulls out all the stops, creating a synthesis of a multitude of time-honoured artistic forms, styles and prototypes to convey a sense of spirituality aspired to, restore a dimension of wholeness deemed lost, overcome the alienation between modern man and his world. The theme of the building can be read as 'per aspera ad astra'. Yet although the aesthetic illusion created in Hoechst was to be so complete that Behrens actually intended the workers to hear the salvation theme from Wagner's *Parsifal* from the belfry on each hour, the reality of industrial work in the early 1920s had little to do with redemption and the quest for the holy grail, of course. It is for this reason that Pehnt criticises the Hoechst Building: 'The business of the everyday is wrapped into the aura of the numinous, the banality of daily chores subjected to the great dictum of fate.'[66] Controversy about the Hoechst Building's artistic merit surrounded it from the start. It was hard for many of Behrens' contemporaries to understand that this seemingly sober artist, whom they had learned to regard as one of the fighters for a modern architecture, had succumbed to such overtly expressionist exuberance and 'aberration'. The younger generation of architects was generally disappointed with the 'Romantic' Behrens of the post-war period, who in their eyes had moved backwards since his AEG turbine factory days.[67]

The theme of spiritual renewal through the creation of a modern *Gesamtkunstwerk* remained a lifelong concern for Behrens. The Hoechst Building, with its splendour and symbolic explicitness, strikes an almost desperate note, for here the theme of redemption is shouted very loudly. Behrens has moved from Zarathustra to Parsifal, from the will to power as exemplified in the fusion of art and technology to the willingness to be delivered from the world of material forces and cold rationality through the hands of the aesthetic saviour. As Walter Benjamin has written, 'The sacral air with which the *Gesamtkunstwerk* celebrates itself is the counterpart of the distraction value which transfigures the consumer good. Both are an abstraction from the social being of man.'[68] Aesthetic universality and industrial anonymity appear as two sides of the same coin.

One of the great achievements of the Hoechst Building from the point of view of the company commissioning it is the representation and interpretation of its product line. *Farben* (the German term combines the two meanings of 'colour' and 'paint', the ideal and its physical manifestation) is a constant theme, and the exploration of the theme and its significance makes for a very successful piece of *Reklamearchitektur* (advertising architecture*)*. However, as with Behrens' work for the AEG, we have to remember that the artist credited with being one of the first to create a corporate identity by means of a unified design programme also holds the curious position of being 'the first artist to devote special care to the beauty of form of peculiarly modern industrial products in terms of some larger cultural conception external to the immediate processes of production and use.'[69]

'Peter Behrens is in part a problematic character; the poet often conquers the practical man',[70] as Breuer, one of Behrens' contemporaries and early critics, observed. It is also the problematic aspects of Behrens' work that make him an ideal exponent of the modern *Gesamtkünstler*. The monumental, celebrational character of Behrens' buildings was apparent throughout his career, and it is the exceptional character of his creations, the idealist quest to lift the work beyond the everyday, which has led to the most pronounced criticism of his architecture. Anderson has argued that 'Behrens showed incessant interest in the *zeitgeist* … with Behrens, architecture answered to "the Time", not to people'.[71] The idealist aspirations to address a higher historical truth are realised at the expense of the architect's engagement with the building's immediate social and functional context. The desire to transcend leads to an involvement with the historical context *ex negativo*, which is one of the issues at the core of the dilemma of the architectural *Gesamtkunstwerk* as idealist aesthetic utopia.

Behrens' Hoechst Building as *Gesamtkunstwerk*, despite its physical integration into the industrial production plant and use for industrial administration, stands like an aesthetic island in the context of this site as well as within early twentieth century industrial culture. This splendid isolation, this self-imposed separation from its context has to be read within the German idealist tradition as an attempt to be the nucleus for the creation of something fundamentally new, rather than as *l'art pour l'art*. Rather than reading the Hoechst Building as either a eulogy to industry or as an aesthetic self-indulgence, it can be seen as a *pièce de resistance* against its own time and culture, an 'untimely meditation' in the Nietzschean sense, which 'aims to work *against* the time and therefore *on* time and thus hopefully *for the sake of* its time'.[72] The artist withdraws from the dominant preoccupations and values of his own culture and time into the aesthetic utopia, his temporary retreat, not in order to evade the outside world and his duties in it, but to represent a more lasting ideal, to restore a lost dimension of unity within a modern culture characterised by spiritual alienation and sociopolitical fragmentation. The Hoechst Building, I suggest, is an example of 'timely untimeliness' in architecture. Like the more overtly utopian expressionist visions of crystal domes, it is most characteristically of its time by being in dialectical opposition to it.

Achieving the balance between the 'timely' and the 'timeless' in architecture was one of Peter Behrens's lifelong preoccupations, and he was careful to equate it neither with the latest formal or stylistic movements nor with modern technological functionalism. In a 1932 speech entitled *Zeitloses und Zeitbewegtes* (The Timeless and the Timely), Behrens speaks of the need for an 'ethical penetration of technology', calling for 'a manifestation of a *gesamtkünstlerisch* awakening and a correspondence of all works [which] would, beyond all aesthetics, weld together the different forms of creation with technology and architecture towards a harmonic unity.' Technology, introduced by Behrens as 'in the idealist sense … the embodiment of a new, still unknown world, a world which is to be thought of as a whole, an all-encompassing totality',[73] thus enters the realm of the 'sister arts' in the early twentieth century. Without it no truly modern *Gesamtkunstwerk* can be conceivable from now on. Though the differences between expressionist visions and the seemingly more pragmatic orientation of the *Neues Bauen* tend to be overemphasised, we need to remind ourselves that although the formal language of expressionism eventually gave way to geometrical simplicity, the socially redemptive agenda of the new architecture had come to stay.[74] The legacy of the romantic–idealist quest for a *Gesamtkunstwerk* permeates the history of modernism long after the disappearance of aesthetic cult buildings.

'Wagner's work hallows our dissatisfation with the modern world, bathing it in the aura of the sacred,'[75] reads a characterisation of romantic theatricality that might with equal justification be applied to the *Gesamtkunstwerk* as realised in Hoechst. The seductive grandiosity of the *Gesamtkunstwerk*, the mesmerising promises of redemption that have been offered from Wagner's Bayreuth to Behrens's Hoechst, continue to attract the art-lover. Pulling the fascinated pilgrim into the realm of an aesthetic counter-world, the apparent conviction with which the *Gesamtkünstler* formulate their absolutist claims as theatres of a new *Weltanschauung* must not obscure the fact that the dream of an aesthetic revolution by means of the *Gesamtkunstwerk* grew out of a state of crisis. While we may admire the singularity and internal consistency of these aesthetic creations, the question that has haunted the *Gesamtkunstwerk* from its beginning has remained. Ever since Nietzsche's ferocious attack on the Wagnerian quest, critics have wondered whether the *Gesamtkunstwerk* can best be interpreted as a kind of aesthetic distraction and escapism, as a manifestation of aesthetic imperialism in a beautiful guise – or, with Behrens, an image of utopia as the dialectical model for the conception of a better world.

# Notes

This chapter is a revised and expanded version of my 'Architecture as "Precursor of Redemption"? Industrial Culture and the idea of the *Gesamtkunstwerk* in German modernism' (Bryant 1999).

1   See for example Buderath, 'Ein Gesamtkunstwerk der Moderne', in Buderath (1990:15–57).
2   Hofmann, 'Gesamtkunstwerk Wien', in Szeemann (1983:84).
3   Behrens (1932:364).
4   Wagner (1983:17).
5   Adorno, 'Die Kunst und die Künste', in Adorno (1977:433).
6   Cassou (1960:25).
7   Klein (1952:298).
8   Michelson (1998:95).
9   How does one approach a phenomenon that seems to be the bane of the art world while at the same time being described as unrealisable, an idea that in the end must remain a utopia? The authors of the 1983 survey exhibition *Der Hang zum Gesamtkunstwerk* (Szeemann 1983) seem to agree that the title had to be *Der Hang* (the desire, tendency or inclination towards the *Gesamtkunstwerk*), since a real *Gesamtkunstwerk*, they argue, does not, cannot – and possibly should not ever – exist.
10  Schelling speaks of art as an emanation of the absolute and 'the true *Organon* of philosophy' (Schelling 1966:380).
11  For a discussion of this idea see Chytry (1989).
12  Schlegel, 'Rede über die Mythologie (Gespräch über die Poesie)', quoted in Gockel, 'Zur neuen Mythologie der Romantik', in Jaeschke and Holzhey (1990:132).
13  Schlegel (1958:269). The German move towards aesthetics in the late eighteenth century has to be understood at least partially as an answer to the failure or rejection of political revolution. Behler refers to the idea of revolution as the 'Angelpunkt im Selbstverständnis der Hauptvertreter des deutschen Idealismus.' See Behler, 'Die Auffassung der Revolution in der deutschen Frühromantik', in Behler (1988:70).
14  A detailed analysis of the background of the idea of the *Gesamtkunstwerk* in early romanticism and German idealism, particularly the relevance of the idea of an aesthetic revolution and its development from early romanticism to Nietzsche, forms the first part of my PhD dissertation *Redesigning Life and Art: The Quest for a Gesamtkunstwerk in Modern German Art and Thought*, (University of Cambridge: Department of Architecture and History of Art).
15  Wagner (1983:67).
16  Vergo, 'The Origins of Expressionism and the notion of the *Gesamtkunstwerk*', in Behr, Fanning and Jarman (1993:12).
17  The formal aesthetic debate about whether the arts should be separated or united, is discussed for example in Böckmann, 'Das Laokoonproblem und seine Auflösung in der Romantik', in Rasch (1970).
18  If the French Revolution of 1789 fired the aesthetic conceptions of German idealist aestheticians, it was the failure of revolution in Germany in 1849 that gave rise to Richard Wagner's explicit formulation of the programme for a *Gesamtkunstwerk* in the nineteenth century, as well as its resurgence as an artistic ambition in the context of the November revolution and the social and political upheaval in Berlin in 1918–19.
19  Hofmann, 'Luxus und Widerspruch', in Hessisches Landesmuseum (1977, Vol. 1, 21).
20  Bisanz (1975:43).
21  Breuer (1995).
22  'Die Verschmelzung aller Künste, um sich im Gesamtkunstwerk der ganzen Wirklichkeit zu bemächtigen und die Kunst zur Lebensform par excellence zu erhaben, ist der unüberbietbare

moderne Mythos des Anfangs.' Jauss, 'Mythen des Anfangs: Eine geheime Sehnsucht der Aufklärung', in Jauss (1989:65).

23  These issues are elaborated further in my essay 'Projecting modern culture: 'aesthetic fundamentalism' and modern architecture' in chapter 5, part I of this book.

24  The similarities of concern for artists like Henry van de Velde, Josef Hofmann, and some aspects of the work of Charles Rennie Mackintosh and Frank Lloyd Wright, surely deserve further exploration, but when discussing the *political* agenda of the *Gesamtkunstwerk* in the early twentieth century, particular attention must be drawn to the Russian avant-garde and the role ascribed to art in the context of the revolution. While a discussion of the similarities and differences would go beyond the scope of this study, the reader is referred to Gofman, 'Gesamtkunstwerk: Ein Leitgedanke der Kunst des 20 Jahrhunderts', in Antonowa and Merkert (1995), and also Groys' controversial *Gesamtkunstwerk Stalin: Die gespaltene Kultur in der Sowjetunion* (Groys 1988). Despite the differences in sociopolitical programme and ideological orientation that exist between different exponents of the *Gesamtkunstwerk*, it is the idealist–ideological charge of the artwork as such that plays a key role. Though Behrens, described by contemporaries as politically conservative by temperament, does not share the revolutionary aspirations of the *Arbeitsrat für Kunst*, the idea of the *Gesamtkunstwerk* as a means towards effecting change guides them all. Another important aspect of the idea of the *Gesamtkunstwerk* and its impact in the twentieth century is its relationship with what Walter Benjamin has called the 'aestheticisation of politics', and the role of the *Gesamtkunstwerk* within fascism.

25  For an introductory survey of Behrens' life and work see Windsor (1985).

26  Posener (1995:15).

27  A notable exception is Anderson's 1968 doctoral dissertation on Behrens' early career, *Peter Behrens and the New Architecture of Germany: 1900–1917* (expanded as Anderson 2000). He provides a detailed discussion of Behrens' view of art as it influenced all his activities as painter, reformer of the theatre, and architect/designer.

28  Behrens (1987:181–184).

29  Behrens (1901:22).

30  Behrens (1901:10).

31  Loos (1962:203): 'The architect meant well … he controlled life … so that no mistakes would be made.'

32  Quoted in Buddensieg, 'Das Wohnhaus als Kultbau', in Schuster (1980:40).

33  As shown by Buddensieg.

34  A more detailed discussion of Behrens' house in Darmstadt, as well as other Nietzsche-inspired works of the same period, is provided in my MPhil dissertation: Häusler (1990).

35  Bahr (1987:57–60).

36  *Das Zeichen* (the sign or symbol) was the title of the opening ceremony in Darmstadt and a reference to Nietzsche's *Zarathustra*. In Darmstadt a huge crystal is presented as 'solidly formed, reflecting change'. The crystal has a long iconographical tradition as a magic symbol of metamorphosis and purity. Bletter argues that when this symbol of transformation, whether spiritual or secular, implies a general social change, it takes on architectural form, but when it stands for individual gnosis alone the image is reduced to the shape of the stone. It is in this later use that it was transmitted to the nineteenth century, from where Behrens takes it. Though I would agree that Behrens' crystal stands for the transformation of self, the alchemical idea of the transmutation of base matter into a noble material, taking place in the individual, implies a 'subjective universality'. When Bletter argues that in Nietzsche's *Zarathustra*, who is himself addressed as the 'Stone of Wisdom', 'alchemical metaphors of transmutation now only stand for narcissistic self-apotheosis,' she disregards the central idea of Zarathustra's mission to be the *Sauerteig*, the seed or nucleus of a total transformation of life (Haag Bletter 1981:20–43).

37  'Dies ist der Jugend edelster Beruf!/Die Welt, sie war nicht, eh' ich sie erschuf.' Quoted in Meyer, 'Faustisches Streben, Zarathustra-Attitüde, Seelentiefe und deutsche Innerlichkeit', in Buchholz (2001:Vol. 1, 113).

38  For a discussion of the ideological aporias of the *Jugendstil,* see Sternberger, 'Panorama des Jugendstils', in Hessisches Landesmuseum (1977:Vol. 1, 3–11); Selle, 'Hoffnung auf eine kultivierte Moderne – Das Darmstädter Kolonie-Projekt zwischen postindustrieller Utopie und Rückzug in die Kunst' (Institut Mathildenhöhe Darmstadt 1990:59–72); and the collection of essays in Hermand (1971).

39  Conrad, quoted in Windsor (1985:41).

40  Meier-Graefe (1987:677).

41  Meier-Graefe (1905:390).

42  Behne (1919:40).

43  Hofmann, 'Luxus und Widerspruch', in Hessisches Landesmuseum (1977:Vol. 1, 27)

44  Buddensieg and Rogge (1980:12).

45  The concept of an artistic synthesis, the integration of technology and the machine as a positive force, is in marked contrast to the ideas of the Arts and Crafts Movement, dismissed by Behrens as 'romantic'.

46  Behrens, 'Kunst und Technik', reprinted in Buddensieg and Rogge (1980:D285).

48  The Werkbund programme of an alchemy of the industrial everyday by means of a fusion of art and industry was criticised by some contemporaries as a *Verkunstung des Alltags,* or *Banalisierung der Kunst.* See Posener (1981:48).

49  Behrens, 'Das Ethos und die Umlagerung der künstlerischen Probleme' (1920), reprinted in Buddensieg and Rogge (1980:D287).

49  E.A. Plischke, who worked for Behrens at the time and was involved in the Hoechst project, later confirmed this fundamental change of attitude: 'The extent to which, with the collapse of 1918, a total spiritual reorientation took place is almost unimaginable today. Not just Bruno Taut, but also Mies van der Rohe, the early Bauhaus and Peter Behrens took a total turn towards German Expressionism.' Quoted in Krawietz (1995:47).

50  Under the title 'The New Romanticism of Handicraft', Behrens wrote in 1920: 'If the ecstasy of work has transmitted soul to the piece, then it is art ... I know that our work has been criticised as not participating in the spirit of the time of the automobile and the airplane; that, on the contrary, our work through its strong emphasis on craftsmanship is a 'romantic movement'. *All right: that's what it ought to be!* We need nothing so much as a little romanticism in order to make our existence more beautiful – indeed, in order to make our life bearable.' Quoted in Anderson (2000:222)

51  See the wide-ranging discussions in Buderath (1990).

52  The symbolism of the crystal and its particular relevance in expressionism have been written about extensively; see for example Prange, 'Das kristalline Sinnbild', in Lampugnani and Schneider (1994:69–97).

53  The careful staging of movement inside his buildings bears testimony to  Behrens' strong dramatic sensibility. His involvement in the reform of the theatre provides another arena for exploring the *Gesamtkunstwerk* idea. For a discussion of this aspect of Behrens's work, see Ch. 3 in Anderson (2000) and also Boehe, 'Das Theater der Darmstädter Künstlerkolonie', in Bott (1977:161–81).

54  For a discussion of the importance of various colour theories on Behrens and his contemporaries see Buderath (1990)

55  Behrens, 'Ethos', in Buddensieg and Rogge (1980:D288)

56  Posener (1981:50).

57  The 'aesthetic differentiation' discussed by Gadamer is the precondition for such an abstraction. The traditional work of art is aestheticised and then new meanings are projected onto it,

as happens with the notion of gothic between the late eighteenth and early twentieth centuries (Gadamer 1986b:88–91).

58  Schrade discusses Schinkel's *Altes Museum* as 'Die ästhetische Kirche' in Schrade (1936).

59  For a history of the interpretation of the gothic in the eighteenth and nineteenth centuries see Robson-Scott (1965). The idea of reviving the gothic cathedral as a Christian temple is one of the most important themes for Karl Friedrich Schinkel after 1810. The gothic cathedral as *Freiheitsdom* is to be a religious, historical and national monument. The claim of the gothic as a truly national style of the Germanic people was the driving force behind many nineteenth century projects such as the completion of the cathedral in Cologne.

60  Behrens elaborates on his concept in Behrens (1923:220–30), affirming the significance of the *Dombauhütte* in terms of an ideal collaboration of all workers and a synthesis of the arts and crafts as well as a higher technology in 'Zeitloses und Zeitbewegtes': 'If we assume that the higher technology as well as ethos have their origin in the intuitive, in the metaphysical, then we must recognise their equality. We must recognise the Gothic age, at least as far as architecture is concerned, as an important technological age. In those times originated the *Bauhütten*, the association of master builders, stone masons and workers of all levels. They were spiritually united in their common work ... they partly aimed at mutual support, and partly they wanted to preserve and enhance those ethical impulses, which were necessary for the creation of such eternal values as the cathedrals ...' (Behrens 1923:363).

61  The reader is referred to the – slightly problematic – English translation *Form in Gothic* (Worringer 1927). For a wider discussion of the expressionist reading of the gothic see Bushart (1990) and Donahue (1995).

62  Kamphausen (1952).

63  This notion is elaborated further in my PhD dissertation on the *Gesamtkunstwerk* (note 15).

64  Buddensieg, 'Architektur als Kunst', in Buderath (1990:65).

65  Nietzsche, *Die fröhliche Wissenschaft*, in Nietzsche (1988: Vol. 3, 524–5).

66  Pehnt, 'Taten und Leiden des Lichts' in Buderath (1990:171).

67  Buddensieg quotes some disappointed voices amongst contemporary artists and critics as well as more recent art historical writings, ranging from puzzlement amongst the audience of a 1923 speech by Behrens, significantly entitled 'Vom romantischen Zusammenhang der Künste' (Concerning the Romantic Synthesis of the Arts) at the Zurich Association of Architects and Engineers, to the dismissive rejection of the building by Albert Speer, who referred to it as 'the kitsch-building of Hoechst' and its labeling as 'degenerate' by other Nazi art ideologues. Buddensieg, 'Architektur', in Buderath (1990:59–73).

68  Benjamin, 'Paris, die Hauptstadt des XIX. Jahrhunderts', in Benjamin (1977:181).

69  (Anderson 2000:127).

70  (Breuer 1910:195–197).

71  (Anderson 2000:260).

72  Quoted in (Krawietz 1995:62). Krawietz' Ch. 4, entitled 'Spurensuche – Peter Behrens und die Philosophie', provides a short discussion of the relevance of Henri Bergson, Eduard Hartmann and Friedrich Nietzsche to Behrens' thought.

74  (Behrens 1932:361–5).

74  For an emphasis on this continuity see also the exhibition catalogue *Expressionismus und Neue Sachlichkeit* (Lampugnani and Schneider 1994).

75  (Harries 1991:29).

**Chapter 11**

# Le Corbusier and the restorative fragment at the Swiss Pavilion

*Dagmar Motycka Weston*

'Quel rôle joue l'esprit poétique dans [les] conceptions architecturales?' ... La raison ne précède jamais, elle intervient ... ma vie toute entière est vouée à l'enregistrement des phénomènes poétiques surgissant à ma portée ... Le monde dans son impassibilité éclate partout en événements poétiques: poésie de la machine, de la raison? Bien sûr! Mais aussi poésie du soleil, des saisons, et des drames de la vie et des batailles que se livrent partout les énergies tendues. Je vois, j'enregistre.

Le Corbusier, 1932[1]

Car de tels mots-notions, on en mettra deux ou dix ensemble. De leur présence, de leurs diverses contiguïtés naîtra un rapport. Ce rapport – écart bref ou immense entre deux notions exactes affrontées (ou confrontées) –, c'est précisément cela que découvre l'artiste ... C'est ... une révélation, un choc.

Le Corbusier, 1938[2]

Poetic analogy has in common with mystical analogy that it transgresses the deductive laws in order to make the mind apprehend the interdependence of two objects of thought situated on different planes, between which the logical functioning of the mind is unlikely to throw a bridge.

André Breton, 1947[3]

One of the most challenging questions for the contemporary architectural discourse is the problem of meaning: can architecture regain its traditional role as an essentially narrative art, or has it become primarily an abstract, autonomous technological discipline? Have the radical changes in modern culture, and in the domain of architectural technology, theory and practice of the last two centuries, irrevocably curtailed architecture's power to communicate about some of the more profound aspects of the human condition, to play an ethical role in our lives? While the twentieth century produced many buildings of great technical and formal virtuosity, powerful examples of an authentically symbolically resonant architecture have been relatively rare. This is due in no small measure to the effective loss in the modern period of the symbolic tradition which had grounded and sustained architecture approximately up to the end of the baroque period. Yet as humans we are embodied, imaginative creatures whose mode of being is fundamentally characterised by its rootedness in the experiential world. We interpret and navigate this world to a significant extent through analogical thought and through metaphor. Our need for meaning in our built environment is undiminished. The postmodernist project has tried to reinject tradition and meaning into architecture with generally disappointing results, since its approach has often been as instrumental as that of the technological modernism it sought to supplant. The need to address the problem of meaning in architecture remains.

One of the fruitful themes in early twentieth-century art and architecture – and one which continues to animate artistic exploration today – is the dissolution of a perspectival understanding of space. Briefly, in nineteenth-century perspectivity the rich phenomena of lived spatiality had become formalised into a neutral, homogeneous Euclidean spatial cage. This was closely linked to an attitude of detachment and instrumentality which has had gravely impoverishing effects in the field of architecture.[4] Successive artists – Cézanne, the cubists and the surrealists – moved away from perspectivity in order to develop more vital forms of representation, better able to evoke various aspects of direct, embodied experience. Their work can be said to manifest a situational understanding which provides an antidote to the somewhat sterile instrumentality of perspectival, beaux arts architecture and academic painting. Through a situational approach – where elements are combined thematically in a topography of meaning – architecture ceases to be a neutral instrumental system and can begin to resume its traditional role as the concrete embodiment of certain typical, stable settings or situations. In this way, tradition can be fruitfully reinterpreted in a contemporary idiom.

The persistent and revitalising role played by the fragment – the fragmentary or disconnected entity found in sometimes incongruous new settings – in modern culture is another point of interest.[5] The fragments used to structure the non-perspectival space of analytical cubist collage are not chosen or combined arbitrarily. Rather they are used to articulate a world, which Merleau-Ponty characterizes as, a 'whole in which each element has meaningful relations with the rest.'[6] The situational setting which is evoked in this case is often that of the Parisian café table. Surrealism is also significant here. The movement developed as a critique of the reductionist rationalism and materialism (what André Breton called 'bourgeois reason') of nineteenth-century culture. Early on, Breton voiced the artists' disdain for the artificiality of perspectival space, and a preference for a metaphorical space structured through collage as a topography of the imagination.[7] In their efforts to reaffirm the validity of such imaginative phenomena as dream and the subconscious as essential parts of human experience, the surrealists focused on analogical thought. In a world where positivistic sciences sought to represent reality as a collection of quantifiable, objective facts, they delighted in revealing the latent, often unexpected connections which exist between things situated in the world of experience, and in making these manifest in their literature and art. Their way of highlighting the mysterious richness of the phenomenal world was through the device of objective chance, the fortuitous conjunctions resulting from the juxtaposition of disparate fragments in a new context.[8] One of their inspirations were the various enigmatic, apparently unrelated fragments making up Giorgio de Chirico's metaphysical world. These pieces often act as what has been called positive fragments; that is to say they are able to import residues of their original worlds into the image, thus greatly extending its meaning. In this way the positive fragment is capable of restoring a rich thematic field to the work of art, and articulating a kind of 'communicative space'.[9] The ability of fragments to act in this way is the result of the essential rootedness of all things we perceive in a world of interconnected meanings. Despite its suppression in our instrumentally dominated technological culture, this world remains accessible to us on a deep level due to our common dwelling among certain stable cosmic conditions.[10] It is this common background which gives us an implicit understanding of what things are like, and what they mean. This phenomenon gives rise to metaphor.

One of the most gifted exponents of a situational understanding in modern architecture was Le Corbusier. Shunning formal *a priori* schemata (in his buildings, if not his cities), he configured his paintings and often his architecture as structures of thematic relationships. His work, right from the early purist years, is informed by a *collagiste* sensibility.[11] The daily, archetypal domestic objects populating the little magical theatre of purist still life become like characters in an alphabet of meaning.[12] These themes or 'topics' (such as the wineglass and its inverse, or the matchbox) soon become emancipated, taking on the character of iconic signs, carrying meanings often quite divergent from those of the original things. Suspended in ambiguous 'implicit' space,[13] the scenes are permeated by mystery and by wonder at the phenomenal world, which reveal a strong affinity with Giorgio de Chirico's strangely animate statues, assemblages of memorabilia, and haunted piazzas.[14] A similar spatial structure, rooted in collage, is also evident, as we shall see, in parts of the Swiss Pavilion.

With the start in the later 1920s of Le Corbusier's cultivation of the *objets à réaction poétique* (objects evoking a poetic reaction),[15] his interest in the primitive intensifies, and the character of his painting (and correspondingly his architecture) becomes more explicitly metaphorical. Where previously the

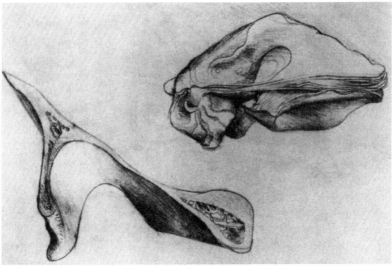

11.2
Le Corbusier,
sketch of
marrow bone
and oyster,
1930

purists had focused on standardised machine-made objects selected for their constancy, the *objets à réaction poétique* were varied and often idiosyncratic, their chance creation and discovery playing an integral role in their power to provoke the imagination.[16] They were mostly organic fragments,[17] admired by their collector for their diversity of plastic qualities as well as for the powerful natural forces (such as rupture or erosion) which over time had helped to bring them into existence.[18] In this way, in contrast to the rather more static character of the purist *objets-type*, the *objets à réaction poétique* manifested a powerful temporality – such as characterises the effect of cosmic conditions on things and on human life. However, both kinds of objects had certain common aspects: on the one hand, both were seen by Le Corbusier as the concrete embodiments of an underlying cosmic order. On the other, they were thematic pieces in a play of situational relationships,[19] the signs in an analogical language.[20] The young Jeanneret had been brought up to revere nature, both through his Jura youth and through the Ruskin-inspired education programme of the School of Art at La Chaux-de-Fonds. Where in his early studies of organic and ornamental motifs, he had seemed primarily preoccupied with form and its inherent geometrical structure, with the later *objets à réaction poétique,* as the name suggests, he became most interested in the objects' allusive and metaphoric potential.[21] Nature now became admired more for the dreamlike strangeness of its meta-morphic power which – like the transforming creativity of the artist, which it nurtures and inspires – can be the source of physical and ethical renewal. Revealing a kinship with contemporaneous surrealist thought, he sees these 'evocative companions' as almost magically animate, metamorphic entities which, through their displacement from their natural habitat and into the realm of man, are rendered strange and acquire a power to evoke and amplify the marvellous dimension of reality.[22] Some of these themes are evident in Le Corbusier's painting *Composition with the Moon,* where the elements of a

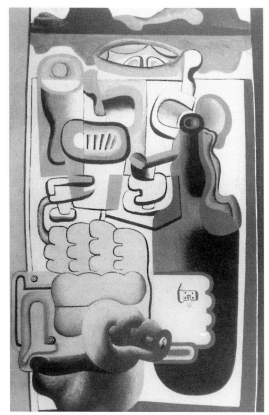

11.3
**Le Corbusier,**
***Composition with***
***the Moon, 1929***

seascape background seem to invade and mingle with a group of domestic objects on a table. The still life, in turn, seems to metamorphose into the figure of an armoured knight.[23] The parallels between Le Corbusier's and the surrealist understanding of the image and collage are evident, the main difference being the architect's focus on the regenerative role of nature. Such evocative natural phenomena are seen as a direct means towards a deeper attunement with its poetic qualities, a vehicle towards man's harmonious dwelling within nature (which it is the role of architecture to facilitate).[24] Around this time he also makes conspicuous use of such surrealistic devices as unexpected ruptures in scale, amplifying the fragments' strangeness and metamorphic potential.[25] His paintings at this time become suffused with an overt eroticism, rooted in the oneiric continuity between the animate and the inanimate.[26] This is also reminiscent of contemporaneous surrealist concerns, and especially of the theme of desire as a central creative and regenerative principle.

Another characteristic of Le Corbusier's work at this time is an increased interest in various manifestations of the primitive, exemplified by the archaic sculptures, sacred and folkloric objects with which he filled the niches and shelves of his own apartment. Some of these pieces can be seen in juxtaposition with his studio rubble wall and modern works by Léger and Laurens in the 'Primitive Arts in Today's Home' exhibition of 1935.[27] The archaic objects which join the *objets à réaction poétique* in Le Corbusier's assemblages are consciously used as temporal fragments, pieces strongly rooted in other times, deployed to create a temporal collage. Again a comparison with de Chirico suggests itself; both artists aim at creating a dreamlike world of temporal indeterminacy. Le Corbusier notes the objects' potential to transcend temporal specificity in what he calls 'anachronisme'.[28] This interest in the primitive, shared by most of the avant-garde movements but probably intensified for Le Corbusier by his travels around this time, evidently relates to the non-illusionism, directness and perceived regenerative power of the primitive artefacts. It is in the context of these themes that we begin to examine the Swiss Pavilion.

# The situational space of the ground floor of the Swiss Pavilion

While the situational play of fragments, relating spatial elements to each other through their meaning, was already apparent in Le Corbusier's early architecture, it becomes a salient theme and a vehicle for the building's metaphorical content at the Swiss Pavilion in Paris of 1930–33.[29] This building has often been seen as a paradigm of Le Corbusier's rational approach to housing and the new city. The reality is, in my view, more complex and interesting. Intended by its clients to be a demonstration of the Swiss position on the cutting edge of

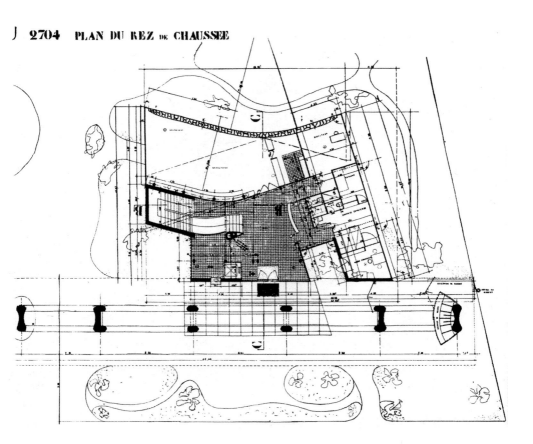

J 2704  PLAN DU REZ DE CHAUSSEE

11.4
**The Swiss Pavilion,
ground floor plan**

intellectual and artistic developments, the building was to embody the modern spirit. It also had a particularly strong ethical agenda.[30] Envisioned as a kind of exemplary home for Swiss youth abroad, the Pavilion naturally lent itself to a range of Corbusian utopian motifs pertaining to the forging of a new ethical order. At the same time this building, which was to combine private study rooms with collective facilities, fitted well with his vision (inspired in part by the Carthusian model and by Russian collective dwelling) of ideal community living. Finally, the Cité Universitaire, with its open greenery, athletic facilities and show housing blocks, was envisioned by its planners as a bold ideological alternative to the old, cramped and unsanitary living conditions of scholars in central Paris.[31] The project thus gave Le Corbusier an ideal opportunity to develop and implement his Radiant City principles, with its programme of salvation which involved modern man's reconciliation with the primitive and regenerative forces of nature. Here, however, this theme is reconsidered and magnified as a representation of the architect's native Switzerland.[32] At the same time his earlier preoccupation with the redemptive austerities of the purist vision was here outweighed by an interest in the power of the fragment to articulate a latent world.[33] The building can be seen as engaging with the fragment in several ways. Firstly, the characteristic Corbusian dialectic of chthonic and celestial motifs takes on a distinctive architectural expression. In a move which was becoming a recurring motif, the building is divided into discrete, functionally and thematically varied parts: the common facilities are housed in the low, shadowy, curved building near the earth, while the individual study cells make up the orthogonal steel-framed structure of the brightly-lit residential block carried atop the massive hull of the concrete slab and muscular *pilotis*.[34] The stair and service tower annex also emancipates itself from the ground pavilion, adopting its own spatial form. A *collagiste* sensibility is further apparent in the building's forms and materials. The white, flat, monolithic surfaces and orthogonality of the purist phase are replaced here by a variety of different treatments – rough, board-marked concrete, smooth ashlar veneer, glass block and curtain wall. Most notably, there is the primitivist curved rubble wall of the refectory, an archaic artefact transposed to a contemporary context. The primitive reference is clearly brought in for its allusive value as a positive fragment (recalling the vernacular stone architecture and even the rocky landscapes of Switzerland),[35] while evoking the chthonic content of the cavelike lower block. An interesting precedent for this collage theme in Le Corbusier's architecture of the time is the little white box of the gardeners lodge perched above its own rubble wall at Poissy. In addition the wall of the pavilion contains a number of other meanings which Le Corbusier weaves into the thematic fabric of the building. One of its most memorable features are the peculiar convex mortar joints which circumscribe the irregular shapes of the stones, so that instead of the great massiveness associated with such masonry, the effect is reminiscent of a lacework veil draped over the wall, giving a sense of strange dematerialisation and lightness.[36] Le Corbusier's motivation seems to be to show that things are not what they seem, and to go beyond that; his transformation of the conventional, objectified meanings opens the way towards wider metaphorical readings. What these might be is suggested by two of the photographs of the building published in the *Oeuvre complète*. The first shows a portion of the north elevation cropped in such a way that the stone wall, framed by two leafy poplar trees, fills its lower half.[37] Above the 'sea horizon' formed by the curving edge of the refectory roof slab rises the orthogonal cladding of the stair tower. The evident intention of the photograph's juxtaposition of the wall and trees is to amplify the latent similarities between the rhythmic, organic pattern of the rubble masonry and that of the trembling mass of the foliage with which it seems to merge.[38] A similar amplification of latent relationships is characteristic of a number of surrealist works, such as René Magritte's *Memory of a Voyage*.[39] In this painting we see an ordinary interior of a comfortable bourgeois house, with French doors opening onto an unearthly, dream landscape beyond. The room is rendered startlingly strange and unfamiliar by being made from the same porous material

which also constitutes the landscape, and which could be either stone (heavy, cold) or cork (light, warm). The painting sets up a thematic tension of affinities and contrasts, challenging the assumed absoluteness of physical qualities, and abolishing facile preconceptions which see each object as a determinate, discrete entity. The stony character of everything in the painting helps to dissolve conventional boundaries between interior and exterior, locating the scene in a kind of metamorphic world of the imagination. The picture sets up a resonant fabric of interconnected meanings and reasserts a phenomenal continuity between things in the world of experience. It thus discloses the surreality immanent within daily reality. In the Corbusier example, a parallel drama is set up between the light, airy leaves and the heavy and immobile masonry, both manifestations of the metamorphic variety of nature. As with the Magritte, clear boundaries between inside and outside, between the man-made and the natural, are abolished, as latent affinities begin to resonate. This serves to restore the sense, suppressed in a rationalist world view, of a primordial kinship between phenomena in the world.

The photograph of the corner of the vestibule in the same *Oeuvre complète* entry[40] reveals another layer of the wall's meaning. It shows a detail of the photo-mural on the airfoil column: a hugely magnified view of organic cellular structure. The oval and polygonal cell boundaries appear white around the dark cavities, an effect hauntingly reminiscent of the wall masonry in the other photographs. Here then, through Le Corbusier's deliberate choice of mural illustration, parallels are drawn between two structures – one pliable and

11.6
**The Swiss
Pavilion, vestibule
with stair and lift**

organic, one rigid and man-made – very different in scales and material qualities, yet both the visually echoing manifestations of certain primordial
processes. This points to an understanding of the rubble wall as one in a series of
*objets à réaction poétique* (the others include the found objects and artefacts
displayed on the shelves and in the built-in cupboards of the building's library
and refectory), consciously combined as different-scaled evocative fragments.
By fragmenting, curving and 'de-materialising' the vernacular Swiss wall, Le
Corbusier avoids the sterility of historicist pastiche. He also elaborates his main
theme of the metamorphic and regenerative power of nature, with temporality
as an agent of transformation. At the same time he alludes to a poetically resonant cosmos of mysterious resemblances accessible not through scientific analysis but on the level of deep, dream-like attunement.

The refectory wall is also an interesting example of Le Corbusier
building in a new way on the traditional iconography of materials. After the
concerns of the purist villas, he now became more preoccupied with tradition,
and with the relationship between his buildings and the cosmic context of the
site and of nature as a source of *poésie*.[41] Thus in the Pavilion the primitivist
masonry is deployed near to the earth as befits its chthonic connotations – this
part of the building sails over the roof of the grotto of a disused quarry.[42] At the
same time, wrapping the northern face of the building, the wall recalls the action
of natural conditions – cold, wind, rain and time – on all things; its organic
texture is vaguely reminiscent of lichen growth around the exposed northern
sides of rocks and trees.[43] The contrast with the relative homogeneity of Le
Corbusier's purist phase is marked. Interior colours, following those of his

paintings, take on richer, earthier tones,[44] with shadows and darkness beginning to feature more deliberately in the thematic content. Most importantly perhaps, where the pristine exterior whiteness and crisp, machine-inspired forms of the purist buildings proclaimed an immunity to the corrupting action of time, the architectural language of the Swiss Pavilion, reflecting the thematics of the *objets à réaction poétique,* begins to engage with temporality as a poetic element. The varied materials begin to have a life and a robustness which allows for a gradual transformation through use and weathering and which, as in the case of the rubble wall or the board-marked concrete of the *pilotis,*[45] are themselves hauntingly evocative of such change. Even glass, conventionally often a rather inert material, is here deployed in a range of thematic ways, evocative of the different processes of its forming, and thematically situated around varying levels of transparency.[46] Clear plate glass is combined with dado bands of textured, semi-transparent glass,[47] reminiscent of certain organic patterns and increasing the sense of being underwater. Le Corbusier's structuring of the shadowy ground level beneath the *pilotis* of his building as an implicitly sub-aqueous realm – the domain of primal waters and chaos which is the source of order (he referred to it as the zone of floods and scorpions)[48] – is rooted in the Corbusian *topos* of the thematic ascent sequence. This metaphorically charged spatial sequence of entry and ascent is common to many of Le Corbusier's works,[49] but reaches a high degree of resolution in the Swiss Pavilion. It can be described as the movement through the building from a dark cave-like, ambiguous space at the lower level, to the full light and orientation of an orthogonally structured space at the building's main or upper level. This corresponds to a symbolic progression from an amorphous domain of latency (comparable to generative chaos), to a luminous domain of fully geometrically formed order, a metaphorical cyclical awakening from the latency of cultural sleep to the creative and ethical action of *droiture.*[50] These entry sequences can be seen as structured to re-enact the drama of awakening to creativity, a paradigm which Le Corbusier practiced at home in his own apartment and which he saw as emblematic of the new life in the Radiant City. At the Swiss Pavilion, the lower, chthonic realm corresponds to the amorphous space of the vestibule, while the geometrically framed, luminous space takes the form of the upstairs study cells in the orthogonal residential block.[51] The iconographic content of this spatial sequence is another reminder of Le Corbusier's sensitivity to the inherent meaning of space (in this case including the distinction between up and down, order and chaos) which, far from being neutral, is for human beings always oriented, heterogeneous and symbolically charged.[52]

This spatial paradigm becomes identified in Corbusian iconography with the sign of the matchbox, in which the unfolded inside sleeve of the box supports its rectangular drawer, divested of its bottom.[53] The configuration thus generates two kinds of space – the lower, concave dark realm surmounted by the

11.7
**Le Corbusier,**
*Sculpture and*
*Nude, 1929*

open, luminous framed space. The artist saw these as two complementary spatial paradigms: that of the earthly, finite domain below, and of the realm of the ideal and the timeless above.[54]

The lift enclosure of the Swiss Pavilion was made of the same semi-transparent glass in metal frames as the vestibule glazing, and resembled a huge aquarium inhabited by a mechanical creature. The sub-aqueous theme was reinforced by gigantic images of crustaceans adorning the inner face of the airfoil column and by the vaguely seashell-shaped upturned light fittings which bathe the ceiling in a diffuse glow.[55] The most striking feature of the ground pavilion, especially as seen from the main south-west approach, is its alternating transparency and reflectivity. In the daytime the glass around the shadowy vestibule reflects the brightly lit surrounding vegetation and the sky, rendering the space of the ground floor pavilion elusive and mysterious. Something of this effect can be appreciated in the widely published photograph of the entrance area with two bespectacled Swiss scholars sitting beneath the *pilotis*. The sky reflected on the vestibule enclosure gives the strong impression that its ceiling, like the surface of a lake, is made of light. At the same time, on bright days the cream-coloured west wall of the caretaker's apartment scoops up sunlight. This provides a dramatically luminous background against which the vestibule, and especially its acute glass corner, become mysteriously transparent, drawing attention to the entrance area and revealing glimpses of the layered spaces, murals and life within. At night the illuminated ground floor pavilion becomes an aquarium of light beneath the dark underside of the housing block, disclosing

11.8
**The Swiss
Pavilion,view
beneath** *pilotis*

its full contents to passers-by. Conscious use was thus made of the architecture of glass to thematise the transformations attending the daily cycle, as well as to evoke the variable character of water. In addition, the Nevada glass block of the stair tower displays the material in its cast, massive and least transparent incarnation, its round lens-like forms making multiple allusions to the organic world which was displayed in the photographs of microscope slides on the nearby column mural. It may also be argued that these photographs, linked as they were with seeing through surfaces, represented another level of implicit transparency in the thematic world of the room. With varying lighting conditions, the two photo-murals engaged in a visual dialogue across the central glass screen.[56]

Le Corbusier's use of architectural fragments and ready-mades, such as the Swiss Pavilion rubble wall, has another significant aspect. By contrast to the mildly oppressive formal homogeneity of the purist years (with their standardised detailing and prescriptive lifestyle), his deliberately *collagiste* approach in the primitivistic buildings can be seen as undermining authoritarian totality. It may be supposed that this apparent loosening of formal consistency and a move towards the bricolage approach reflect an ideological shift in Le Corbusier. As Rowe and Koetter have argued,[57] the diversification of bricolage can be seen as a primitivistic critique of the scientific method, and this seems to correspond to Le Corbusier's general shift in the 1930s toward the primitive, resulting in a more heterogeneous, overtly poetic expression.

Other aspects of the restorative power of the fragment can be discerned in the situational spatial structure of the Pavilion's ground floor

11.9
M. Ernst, *frottage*
'In the Stable of
the Sphinx',
*Histoire naturelle*,
1925

vestibule and refectory. One of the most conspicuous and innovative features of the design is that the entry pavilion, instead of fitting into an extension of the structural grid of the residential block, takes on a form and a life of its own. The places within it are configured out of a primordially simple space. Not conforming to any *a priori* geometrical framework, it is structured as an ambiguous field in which the diverse fragments are made to interact, their relationships giving rise to situational settings. In this way it is closely related to the implicit space of Le Corbusier's paintings and photographs, in which spatial ambiguity was the vehicle for a play of thematic pieces. Permanently in the shadow below the residential block, the ground floor spaces are irregularly shaped and deliberately indeterminate. The orientation of the entry, particularly with respect to the curved glass wall of the refectory (and to the masonry wall beyond) is initially ambiguous. Their structure fits into the Corbusian pattern of thematic ascent sequences, where the lowest stage corresponds to the latent realm of watery chaos. Within this fluid space are positioned several pieces: the double column and radiator beside the entrance, the glass-enclosed lift, the stair bordered by the apparently massive airfoil shaft of the ventilation duct, a reception counter, and two groups of seating. Visible beyond the vestibule, through the glazed partition, was the refectory mural made up of large square black and white photographs. This was foreshadowed in the vestibule by the similar photomural of the airfoil column. The ceiling of both of these rooms is covered by a continuous grid of plasterboard panels which would have echoed the orthogonal structure of the mural (without aligning with it) and which accentuates the irregularity of the plan. However, the apparently loose structure of the fragments is oriented with respect to a series of particular events in which the occupants may choose to participate.[58]

The fragments which constitute the situations are often chosen for their metaphorical content, which they bring to interact within the spatial matrix

11.10
**The Swiss Pavilion, refectory photo-mural and marble table**

of the rooms. In this context it is again relevant that the ground floor rooms were intended for the display in their niches and on their shelves of various *objets à réaction poétique*. This was realised in the library and the refectory, where cupboards and bookcases were designed as part of the central glazed partition. The architect shows them in some of the original photographs as holding a few natural objects very similar to those he collected at this time.[59] The objects were set in dialogue with the collaged images of the photographic mural. The use of magnification and photographic negatives in these images – showing organic, mineral and geological structures juxtaposed with such synthetic things as stacked building materials and architectural fragments – often made them unrecognisable as any particular thing. Instead attention was drawn to the formal or textural relationships between different kinds of structures – large and small, natural and man-made, and so on. On closer inspection, a number of other thematic correspondences emerge.[60] The photo-mural is conceived as an oriented communicative space of meaningful relationships. This work was replaced after the war by Le Corbusier's large mural painting, which is there today.[61]

Le Corbusier's preoccupation with the evocative power of the fragment can be compared with the work of Max Ernst who, perhaps more than any other surrealist artist, developed the art of situational collage. Noting the restorative power inherent in the 'culture of systematic displacement'[62] which is at the root of his collage and *frottage*, Ernst juxtaposed and confronted different realities in the context of his works, creating strange poetic worlds, allowing for the re-situation of meanings usually thought unconnected. Ernst's *Histoire naturelle*[63] can, on one level, be seen as ironically echoing the various nineteenth-century attempts at exhaustive cataloguing and objectification of natural occurrences. It can thus be understood as a poetic critique of the positivistic scientific method. Instead of a factual catalogue, Ernst's is a history of an enigmatic cosmos seen from a distinctly human, situated viewpoint. In it is revealed a latent magical kinship between the animal, vegetable and mineral, between the mythical, cosmic and cosmogonic dimensions. The small is linked with the immense, the trivial with the profound. It is a view of the world suffused with enigma and the sacred – with a dimension of experiential truth. While some of the images make reference to scientific knowledge, the general thrust is to represent the reciprocity of such knowledge with the marvellous phenomena of fantasy, dream and the subconscious. Ernst's whole mysterious cosmos, a hallucinatory quasi-Freudian *Mappa mundi*, is revealed through metaphor to the imagination.

The thematic content of Ernst's *frottages*, with their sense of wonder at the marvellous architecture of nature tempered by time and circumstance, has a close affinity with Le Corbusier's understanding of the *objet à réaction poétique*. In the photo-murals of the Swiss Pavilion, the architect is involved in a similar cataloguing of natural and man-made phenomena which, through photographic manipulation, are rendered unfamiliar, reduced to shadowy textures, estranged from their usual scale, and set in hallucinatory conversations amongst themselves and with the architecture. This aim is reiterated by the display of found objects in this room and by such thematic references as the rubble wall. The combination of evocative pieces suggests a similar aim of disclosing to the imagination the variety and mystery of the cosmos. The most conspicuous examples of his deployment of the situational fragment in this space are the airfoil column in the vestibule and the marble altar-like table in the refectory. The massive shaft, which contains a reinforced concrete column, a heating and ventilating duct and water pipes, was given an airfoil shape in reference to aircraft, setting up the play of meanings situated around air and flight. At the same time, being covered by images of vegetal and marine life, the airfoil seems to glide through the viscous space like a large fish, echoing and amplifying the curves of the depicted marine organisms, engaging them in a poetic dialogue. Finally, this shaft acts as a massive anchor to the staircase, forming a shadowy enclosure before the occupants ascend towards the light flooding in through the glass block of the stair tower. As the airfoil pushes in against the main stair, it squeezes the path (constricted further by one of the outer columns of the partition to the right) leading to the basement stair. The space is compressed, made fluid and indeterminate by the airfoil's buttressing presence. The column thus contributes to the chthonic and sub-aqueous imagery of the vestibule. Its paradoxical attributes of air and solidity, lightness and mass, further set up an interesting tension surrounding the beginning of the journey upstairs. At the same time the column, with its array of imported associations, brings a potent sense of strangeness to its setting in the vestibule. This quality is further accentuated by its somewhat over-scaled proportions (its massiveness contrasting with the transparency of the glass and with other, more delicate fragments in the space), recalling some of the oneiric incongruities of scale characterising surrealist art and Le Corbusier's own contemporary painting.

As we have seen, the Swiss Pavilion is a good example of the power of the fragment to structure a communicative space through its metaphorical potential. This confirms that even in the contemporary context it is possible for architecture to transcend merely formal, technical or aesthetic concerns, and communicate deep thematic meaning. The themes of the fundamental tempo-rality of human life and artefacts, of discipline, creativity and ethical life, and of the regenerative power of nature, are present in some form in most of Le Corbusier's buildings, but are here given a distinctive and imaginative expres-sion. This expression is quite subtle, and easy to miss in casual observation or without acquaintance with Le Corbusier's personal iconographic system. Never-theless, such archetypal themes are ones which we still recognise as essential, even if many aspects of the utopian modernist vision (particularly in the wake of the failure of modern urbanism) now seem misconceived. This pertains to the strongest criticism one can make of the Swiss Pavilion: it continues to suffer from its lack of meaningful context by being part of the fundamentally anti-urban vision of the Radiant City. This is perhaps acceptable in the somewhat artificial and parklike setting of a university campus, but might present more of a problem when placed in a real urban situation. The authoritarian homogeneity and dullness of the Corbusian city is ironic, given his great sensitivity to the restorative potential of collage and the fragment at the scale of the single building. Visiting the Pavilion today, one can level other criticisms against it. The materials and some of the detailing, while less inert than those of the purist phase, still lacked sufficient robustness to enable them (as was, I think, the architect's intention) to evolve and ripen with time – instead some have faded without grace. The original photo-mural ornament, so important in the commu-nication of the building's content, has proved ephemeral, just as the original idealistic vision of communal meditative life has turned out to be at odds with the mundane realities of transient, institutional housing. But the severe budgetary constraints on this building were perhaps more to blame for some of these shortcomings than the architect's judgement. On the whole the building is of great value as an illustration of Le Corbusier's use of the thematic fragment. It reveals its power to restore a dense field of connections lost to objectifying positivistic thought, and to build a space resonant with thematic meaning. Such a space of meanings – a communicative space – can continue to invest the architecture we make today with the necessary sense of recognition and authenticity, and enable it once again to become a source of existential orientation.

# Notes

1  Author's translation: What role does the poetic spirit play in architectural creation? Reason never determines, it informs ... My whole life is dedicated to recording poetic phenomena which spring up within my reach ... The world in its impassiveness erupts everywhere in lyrical events: the poetry of the machine, of reason? Yes, of course! But also the poetry of the sun, of the seasons, and the dramas of life ... I see, I note. (Le Corbusier 1933a)

2  Author's translation: 'For such word-notions, one will combine two or ten together. Their presence, their various proximities, will give rise to a relationship. This relationship – a brief or immense discrepancy between two distinct notions confronting (or complementing) each other – it is precisely this the artist discovers ... It is ... a revelation, a shock.' (Corbusier 1938), Introduction.

3  Breton, 'Rising sign', in Rosemont (1978:281).

4  By perspectivity is meant the mode of thought prevalent in mainstream nineteenth-century culture, an attitude entailing man's supposed separation from the world of objects, and the formalisation of lived spatiality and temporality into formal, conceptual frameworks. The abstract Euclidean spatial model was best suited to the instrumental concerns (such as structural determinism or cost effectiveness) which came to dominate the science of building and architectural production in the nineteenth century. In this approach it became possible to think of the symbolic content of architecture as an issue separate from, and secondary to, that of solving a set of functional and technical problems. The problem of perspectivity and its dissolution in early twentieth-century art and architecture is considered in detail in Motycka Weston (1994).

5  I am indebted to Dalibor Vesely's illuminating work on the fragment in modern culture. See for example his 'Architecture and the ambiguity of the fragment,' in Middleton (1996).

6  Merleau-Ponty (1962:292).

7  'The belief in absolute time and space seems to be vanishing. Dada ... regards submission to the laws of any given perspective as useless. Its nature preserves it from attaching itself, even in the slightest degree, to matter, or from letting itself be intoxicated by words. It is the marvellous faculty of attaining two widely separate realities without departing from the realm of our experience, of bringing them together and drawing a spark from their contact ... and of disorienting us in our own memory by depriving us of a frame of reference – it is this faculty which for the present sustains Dada.' Breton, 'Max Ernst' (1920), in Ernst (1948:177).

8  See Breton, 'Rising sign,' in Rosemont (1978:282).

9  Vesely, 'Architecture and the ambiguity of the fragment', in Middleton (1996).

10  Such conditions include the inherent orientedness of lived space or the cyclicality of day and night, of the seasons and of human life.

11  This manifests itself in his painting, writing and activities as editor of *L'Esprit nouveau*, and culminates later in his *Poème de l'angle droit*.

12  Le Corbusier occasionally refers to them in these terms.

13  Influenced by cubist space, the purists developed an ambiguous 'implicit' space which Le Corbusier was to continue to deploy in both his painting and his architecture. This non-perspectival space of relationships and layered metaphorical meanings is an antidote to the sterile formal space of academic architecture, of which Le Corbusier was frequently critical. It is subtly structured through implication and is perceived at the level of experience through the

embodied participation in its situational structure. As with cubist space, purist space is characterised by tension between two- and three-dimensional representation. The process of perception which animates such space is conceived as a primordial and revivifying drama which has
the power to restore man's full connectedness with the world (in the phenomenon which he
terms *l'espace indicible*) and which thus invests the work with primary meaning.

14  See Motycka Weston (2002).

15  The name was derived from surrealist 'objects of symbolic function'.

16  In *Creation is a Patient Search* (Corbusier 1960:209) Le Corbusier praises the chance occur-
rence which brings such objects as a broken seashell or cut marrow bone into existence. He
collected the *objets à réaction poétique* in various places (on trips or strolls along the beach, for
example) over a period of time, in a manner strongly reminiscent of the surrealists' acquisitions
of enigmatic and evocative objects from Paris flea markets. Both the *objets-type* and *objets à
réaction poétique* embodied for Le Corbusier certain fundamental laws, giving them an archetypal dimension. This made them superior to the excessively personal, trivial knick-knacks
which he had reviled as manifestations of moral decadence in contemporary culture in *The
Decorative Art of Today* (orig. 1925) (Corbusier 1987:83–101 and 132–92).

17  They included water-rounded pebbles and bits of brick, bones, fossils, tree roots, seaweed,
seashells, flints, crystals and pieces of wood.

18  'Ces fragments d'éléments naturels, ... de ces choses martyrisées par les éléments, ramassées au
bord de l'eau ... exprimant des lois physiques, l'usure, l'erosion, l'éclatement, etc., non
seulement ont des qualités plastiques, mais aussi un extraordinaire potentiel poétique.' ('These
natural fragments, these things battered by the elements, collected on the shore ... manifesting
physical laws, wear, erosion, rupture etc. possess not only plastic qualities, but also an extraordinary poetic potential.') Le Corbusier in Charbonnier (1960:107).

19  See note 2.

20  Eduard Sekler was among the first interpreters to note Le Corbusier's deployment of such
object-themes as signs in his iconographic system, and his placing these in networks of meaningful relationships. See for example Sekler and Curtis (1978:229–42).

21  Le Corbusier, in Charbonnier (1960:107).

22  Le Corbusier *'Objets à réaction poétique'*, in Petit (1970).

23  *Composition with the Moon* (1929, Fig. 11.3) is a good example of how Corbusian implicit
space acts as a vehicle for a range of metaphorical meanings. The boat floating on the ambiguously horizontal surface of the sea beyond becomes a container on the table. Its rim provides
the top of the man's helmet, while its handles – in a familiar Corbusian motif – become human
ears. (In a similar way, the moon migrates from the sky onto the surface of the liquid in the
glass, repeating the circle of its rim.) The spout of the siphon/wine bottle, often affiliated with
pipes, here suggests a nose. The most dramatic cluster of transformations occurs around the
fleshy conch shell, which suggests both a corkscrew (connected with the wine) and a hand. A
hand is also echoed by the label on the wine bottle. The image of the castle on the label reads
not only as Chateau LC 1929 (wine being likened to the Dionysian product of the artist, with
the identification of the table and painted canvas), but alludes also to the armoured figure.
Palpable affinities also exist between Le Corbusier's image of the coexistence of the world of
man with time and nature, and certain surrealist images such as the photograph of a locomotive overgrown by forest vegetation published in *Minotaure*.

24  Le Corbusier, *'Objets à réaction poétique'*, in (Petit 1970:178).

25  An incongruity of scale is evident in Le Corbusier's 1931 painting *Léa,* reproduced in Raeburn
and Wilson (1987:Pl. 53), where the three huge objects become almost human participants in
an arcane drama. It is even more startling in *Sculpture and Nude* (Fig. 11.7), where the glass,
bone, and matchbox are aligned with a nude woman like a row of hieroglyphs. Le Corbusier
explained his intentions with respect to the ruptures in scale in his paintings: 'When the

structure of a bone occupies my mind, I try to fill a whole painting with this element and to enlarge the object in proportion to the interest it arouses. I then confront it with other figurative elements which occupy an identical surface but which seem small compared to the object depicted.' Le Corbusier in a letter to Giedion, *Le Corbusier* exhibition catalogue (Zurich, 1938), p. 12, quoted by Moos (1979:307). Also relevant is Fernand Léger's contemporaneous interest in the isolated and radically enlarged object, which manifests itself in his films and his painting.

26   In the painting *Léa,* as in a number of Corbusian works, the oyster shell (drawing on an ancient symbolic motif) represents the archetypal female, while the marrow bone can be seen as a male voyeur at the open door. The giant sardine tin-cum-violin (the latter a cubist analogue for woman) on the table appears to be doing a striptease.

27   See Corbusier (1935:83–5).

28   Le Corbusier, '*Objets à réaction poétique'*, in Petit (1970).

29   The project was carried out as usual in collaboration with Le Corbusier's cousin and partner Pierre Jeanneret.

30   The client committee, headed by Professor Fueter, expressed the hope that the building would contribute to fostering intellectual and moral dialogue between students of all nations. In its ethical aims the project paralleled some of those of the contemporaneous Cité de Refuge in Paris, to which Le Corbusier sometimes referred as an *usine du bien* (goodness factory).

31   The evils and despair of the purgatorial old city of poisonous fumes and gloom (contrasting with the light and air of the radiant city of the saved) are vividly evoked by Le Corbusier in *The Radiant City*. (Corbusier 1967:11–15 and 91–5). The book was completed at the same time as the Swiss Pavilion.

32   In *Creation is a Patient Search* Le Corbusier notes that the original decoration of the curved library/refectory wall were to have been 'pictures of mountain scenery' (p. 98). This prominent display was surely to have set the thematic tone of the building. Mary Patricia Sekler provides an astute examination of the ethical meaning of the tree in Corbusian iconography in her essay 'Le Corbusier, Ruskin, the tree and the open hand,' in Walden (1982). The Swiss Pavilion closely followed the Palace of the League of Nations controversy, giving Le Corbusier, now famous in his native country, a new opportunity to demonstrate his ability to develop an appropriate symbolic expression for a significant public commission.

33   Le Corbusier's interest in the poetics of the fragment is evident in the deliberate combining of archaic and modern, organic and industrial elements in such primitivist houses of the time as the Villa de Mandrot and the Maison de Weekend. It is also palpable in his most explicitly Surrealistic building, the Beistegui penthouse (1930–1). Le Corbusier characterised the primitivistic buildings of this period as important laboratory experiments, (*Oeuvre complète* II (Corbusier and Jeanneret 1964:15–16)), confirming the compatibility of rustic materials – if justified by the local conditions – with the modern spirit. *Oeuvre complète* II (Corbusier and Jeanneret 1964:48).

34   In this respect the pavilion resembles some of the Russian experimental collective dwelling complexes, such as Ginzburg and Milinis' Narkomfin housing block in Moscow, which Le Corbusier knew. See Cohen (1987:122–4).

35   Le Corbusier's use of the masonry wall as a fragment in this way makes it a suitable reference to Swiss vernacular, while preventing the building from descending into a Swiss chalet-type pastiche of the sort apparently favoured by a number of the building's critics. Le Corbusier characterised the primitivistic buildings of this period as important laboratory experiments (Corbusier and Jeanneret 1964:15–16), confirming the compatibility of rustic materials – if justified by the local conditions – with the modern spirit (Corbusier and Jeanneret 1964:48).

36  The treatment of the mortar joints is here very different from that on the contemporaneous de Mandrot and Errazuris houses, or from that of the 'found' rubble and brick party wall of his own studio. This suggests a specific thematic aim.

37  A broader version of this image featured prominently in both Corbusier (1933b:26) and in the section on the building in the *Oeuvre complète* II (Corbusier and Jeanneret 1964:74 and 83), suggesting its significance to the architect.

38  Le Corbusier later suggests something of his aim of revealing the oneiric reciprocities between his building and its natural habitat in *New World of Space* (Corbusier 1948:50), where next to a similar view he notes that the expansive curved wall 'seems to pick up by its concave surface the whole surrounding landscape and to establish a relationship which carries its effect far beyond the actual bounds of the structure itself.'

39  Magritte's *Memory of a Voyage* (1951) is reproduced in Whitfield (1992: Pl. 106).

40  See Corbusier and Jeanneret (1964:85), top right photo.

41  On these shifts in preoccupations and new emphasis on *poésie* see Corbusier (1933b), which contains a transcript of the architect's lecture in August 1933 to the Congrès d'Architecture Moderne and polemically combines a statement of architectural principles with a photo-essay on the Swiss Pavilion.

42  The residential block stands over the quarry on hidden massive concrete piers of its own height, which pierce through dark caves (the ground floor pavilion, by contrast, sits on ordinary foundations on top of the quarry's ceiling). See the two atmospheric section/elevations in Brooks (1982–4:190).

43  A comparison between the open, glazed south elevation and the more rugged, protective north one is invited by the publication of the two long elevations on facing pages of *Oeuvre complète* II (Corbusier and Jeanneret 1964:82–3).

44  'La couleur, expression même de la vie! … L'homme qui vit vraiment, emploie les couleurs.' ('colour, the expression of life itself. The man who is fully alive uses colour'). (Corbusier 1933b:82).

45  The present sturdy *pilotis* are a replacement for the flimsy steel stanchions of the preliminary designs. The outer four are vaguely bone-shaped in plan and were referred to as 'dog-bones' in the architect's atelier (Curtis 1986:105). This analogy is suggestive of the beginnings of an animate or even anthropomorphic reading of parts of the structure, a reading which becomes more explicit with the muscular 'legs' of the Marseilles Unité.

46  Le Corbusier's interest in the variable qualities of glass may in part be due to Pierre Chareau's contemporary Maison de Verre which he admired. It is manifest in several of his own buildings of this time, such as the Cité de Refuge and the Geneva 'Clarté' apartment building, both of which explore the themes of luminosity and transparency, and make significant use of glass block.

47  Such a semi-transparent band also forms the top part of the windows in the vestibule's west elevation. The south curtain wall of the residential block was originally composed in the same way.

48  Corbusier (1967:56).

49  This theme was first identified by Peter Carl. See especially his 'Le Corbusier's penthouse in Paris. 24 rue Nungesser-et-Coli'. (Carl 1988:69). Its beginnings can be traced to the Villa La Roche, but it becomes fully developed at Garches, Poissy, and especially at Le Corbusier's own apartment.

50  Throughout his life Le Corbusier invested the right angle and orthogonal geometry with the connotations of moral rectitude, summarised by the term *droiture*. It appears in section E3 of the *Poème de l'angle droit* (Corbusier 1989).

51  See Carl (1988).

52  Le Corbusier's insistence, against specialist advice, on the south-facing curtain wall on the residential block seems consistent with this interpretation.

53  I am grateful to Peter Carl for this observation. The sign of the matchbox, alone or combined with other fragments, features in a number of sketches and paintings of the 1930s. In the *Poème de l'angle droit*, a circle and an oval are drawn against the upper and lower portion of the matchbox respectively, seeming to suggest the ideal (celestial) and human (earthly) domain.

54  The lower, shadowy cave-like space, receding ambiguously, suggests for him the earthly limitations of perspectival vision. Arising from a limited subjective human viewpoint and particular moment in time, it corresponds to the variable and the incomplete. This view had already been outlined by Ozenfant and Jeanneret in 'Purism', in Herbert (1964:65–6). The perspectively receding, lower space is this earthly, finite domain. By contrast, the geometrically framed, light, orthogonal space of the upper portion of the matchbox is unambiguous and invariable, suggesting a kind of eidetic vision – a timelessness and universality sought by the orthogonal views of purist canvases. To Le Corbusier this kind of space, a complement of the former, is expressive of the ideal order perceptible only to *esprit*. See 'Tracés régulateurs', (Corbusier 1929:18–20).

55  The sense of being submerged is amplified by the narrow strip windows surmounting the east wall of the vestibule and refectory. This motif of the hovering roof is later affiliated with sacred cave imagery at Ronchamp.

56  See photo of entry hall in *Oeuvre complète* II (Corbusier and Jeanneret 1964:85).

57  Rowe and Koetter (1978), esp. Chapter 4.

58  For example the place of entry from below the generous canopy of the residential block is an intimately scaled niche (delineated by the glass elevator enclosure and the acute-angled glass corner) screened by glass. It is marked by a column paired with a pipe stack (the primary structural columns in the entry block are differentiated from those linked to other functions by their colour) and partly enclosed by the free-standing radiator. The empty acute corner – one of the first public glimpses of the life of the building – naturally lends itself to the placement of seating furniture at which to pause and share in conversation. Similarly the area by the west window, defined by the airfoil column, stair and the lift, becomes a natural, more sheltered setting for a table and some chairs, a place for reading and talking with people on their way up to their rooms. These are only two examples of how Le Corbusier has anticipated concrete situational scenarios around which the places making up the vestibule are structured. The architect took some care to illustrate these places in *Oeuvre complète* II (Corbusier and Jeanneret 1964:85).

59  See Corbusier and Jeanneret (1964:77, 85).

60  The inherent orientation of the wall space, for example, is acknowledged: some of the images in the top row show the tops of things, such as the crown of the plant (photographed from below to resemble a tree), the summit of the filigree steel frame of the Swiss Pavilion, or a human head. See the photograph in Brooks (1982–4:152).

61  The 1948 Swiss Pavilion mural represents the culmination of Le Corbusier's collage technique in his art. It is an essential step toward the late-phase iconography of the *Poème de l'angle droit*. One of its constituents, the composite sign of the bull, epitomises the artist's creative process – see Corbusier (1989: Section C1). The most comprehensive study of this seminal painting remains Moore (1977). See also Krustrup (1991) and Becket (1990), both of whom perceptively examine it as part of wider iconographic studies.

62  Ernst (1948:13).

63  Ernst (1982).

# Chapter 12

# The concrete memory of modernity

## Excerpts from a Moscow diary

*Jonathan Charley*

They say that I am old. Ancient. That I was played with in Imperial China and christened in the Coliseum and Pantheon of ancient Rome. They say that wage labour is old too. That on the same building sites of antiquity, workers paid by the piece or the day laboured alongside slaves. But like myself, the idea of a wage as a general condition underpinning everything we do is very modern indeed. I am concrete. I underpin modernity. I think of myself in two ways. As a noun referring to my use as a building material, and in a more philosophical sense. To make something concrete. To materialise an idea. To concretise a revolution.

Forgotten in Europe during the dark ages, it took a philosophical and social revolution for me to be remembered. I was rediscovered in the union of the natural sciences and capitalist economic competition, the crucible of modern technological change. They say it was an ingenuous Englishman, one Joseph Aspdin in 1824, who invented the modern catalyst to building construction by burning chalk and clay together in a limekiln to produce Portland cement. He had no idea of the consequences of his actions. Even more incalculable and unimaginable in its implications for urban development were the visionary activities of a French gardener, Joseph Monier. In an innocuous wire-reinforced flowerpot he inadvertently discovered in 1867 the panacea and pariah of twentieth-century modern architecture. A new opportunity for me to express myself. No longer just concrete, I became reinforced, a single act of genius that opened up a whole new world of possibilities. Stronger than ever before, and capable of being mass-produced under factory conditions.

As reinforced concrete I have been both civilisation and barbarism. I was hope when they used me to make worker's clubs and health centres, shame and tragedy when set in the hardened surfaces of Alactraz and Auschwitz. I was presented as the material of revolution and a container for death rites. Everyone put me to work. I was to be mass-produced on an assembly line with precision. Retrained workers would cast and trowel me. There was to be no mess. I would be bolted together on site in a single flash of a spanner. Housing would be produced in such quantities that politicians would look forward to a lifetime of re-election. And so it was. The nineteen-sixties and seventies were my heyday. The white heat of the technological revolution. How I liked that. The suggestion of speed, urgency, and grand plans. The very stuff of modernity, and I was at the centre of it. At my best I was called upon to concretise modern political ideals in Brasilia and Chandigarh. Solid but expressive. Firmly rooted to the ground, yet exhibiting a wide variety of dance steps. The sinuous curve, lift and somersault. I promised so much more than flat panels and kerbstones.

Scattered across the post-war British landscape, I was the symbol of municipal building in Britain, the chosen material for the physical construction of the welfare state. I was idealised by city planners and architects. There were schools, swimming pools, concert halls, shopping centres, and of course housing. And then there was Glasgow. If the planners had fulfilled their promise and executed the Bruce plan of 1946, Glasgow would have been virtually demolished and reborn as a concrete monument to the technocratic dream of the good city. But then came the fall from grace. I was abused. The word concrete became a synonym for alienation. I stood accused of a lack of poetry and finesse. Branded as anonymous and soul-less, I was held personally responsible for the homogeneous housing that stretched across the horizon of every large town and city. I was facing exile to civil engineering, cast into the unseen world of tunnels, drainage and railway sleepers. But I bear no grudges. I have made a full recovery and have learnt to deal with the constant criticism. Besides, my legacy is too long and deep. Modernity is unimaginable without me. And nowhere more so than in Russia, where I played a key role in each phase of its modern history.

I am like the soldier whose remit is to obey and serve. First I was turned into a commodity and called upon to cement together the capitalist metropolis. I was then summoned as an avant-garde to provide its revolutionary critique. I was made a spectacle of when accused of counter-revolution, and finally rehabilitated as a fetish, the country's saviour. But let me begin at the beginning.

After the great fire of 1812 Moscow embarked on a reconstruction programme. The medieval timber world of Boyars and bears was flattened. First by flame, and then by stone, brick and render. An opportunity beckoned. A neoclassical radial plan was imposed over existing street patterns, and by 1835 ten thousand new houses had been built as a modern city struggled to emerge from the suffocating power of court and church. I waited in the wings for my signal to come on stage. It was between 1817 and 1821, in the midst of the building boom, that E.G. Chelievu conducted a series of experiments. It wasn't exactly alchemy, but nevertheless culminated in the publication in 1825 of the groundbreaking 'Full instructions on how to prepare cheap and better marl and cement'. Cheap and better. He had no idea how important that phrase would become in the twentieth century.

I had yet to mature into full-blooded modern concrete, but as cement I took my place alongside the other indices of the modern metropolis, pavements, stock exchanges and prisons. Russians have a strong claim not just to the modern rediscovery of cement, but also to the whole concept of modernity. Before Moscow was reformatted along with four hundred smaller Russian towns, a pilot exercise was conducted on the Baltic coast. St Petersburg, they say, was the original blueprint for all the dilemmas and contradictions thrown up by the modern metropolis. By encapsulating the enlightenment ideal of geometrically ordered social space, St Petersburg would dispel doubt. It would function as a window on the west and into the future, resplendent with palace, admiralty, fortress, harbour, warehouse, market square, and the Nevsky Prospect. This was a boulevard on which to display, gaze and promenade that would come to rival anything Hausmann would give Paris. A modern stage set whose end was invisible, and which the tourist Bergholz in 1721 noted 'was beautiful by reason of its enormous extent and cleanliness'. Think of it. The modern city as an immense expanse of hygienic space disappearing into the horizon. The dream of a Russian king for a spatially ordered universe which two hundred years later would return to haunt the orthogonal concrete imaginations of architects like Hilbersheimer and Corbusier.

Things did not turn out quite as planned, either in the 1920s or in eighteenth century St Petersburg, for in place of a modernity of certain line and square emerged a modernity of rupture and the unexpected. It started in 1773 with Emile Pugachev, who led the peasants in rebellion. It was followed by Dostoevsky's anti-hero and modern criminal Raskolnikov, fleeing through the courtyards and back alleys, and would end with the murder of the royal family in the cellar. The 'architect Tsar' Peter the Great wanted to be modern, but truly modern cities do not have monarchs. With the final death of absolutism and the proclamation of Red Petrograd in 1917, the workers had wreaked vengeance, not only for the deaths of relatives shot in 1905 but for the ancestors who perished carving the foundations of St Petersburg from the frozen swamps. It was to be expected. Conflict and contradiction are the essence of modern life. Modern audiences love to watch my detonation. In this sense my history mimics that of history in general. A history in which the fixed, concrete things and relationships, which give continuity and structure to everyday lives, can be suddenly broken and swept away. In a single breath I am a future ruin. But forgive me, I am digressing.

In my guise as reinforced concrete I had made my debut on the construction sites of Moscow at the end of the nineteenth century. In what was becoming a thriving capitalist metropolis boasting more than a million inhabitants, I was forced to compete for attention with the new excavators, concrete mixers, and elevators. I was also confronted by the emergence of another peculiarly modern phenomena, the worker's strike. By the time the Winter Palace was stormed in 1917, the smattering of confrontations that had closed the building materials factories and disrupted the construction of railways had developed into something else entirely. As workers poured me into timber moulds I caught mutterings of revolution, strange talk that threatened the concrete foundations of the world that I knew and recognised.

The construction workers said it was nothing personal. They were naturally interested in and becoming increasingly dependent upon cement and concrete. Their quarrel was with those that had appropriated me as private property. This was fascinating. It had always struck me as odd how my constituent parts – lime, clay, sand, metal and stones – could become commodities. It bothered me. It seemed like a heartless and brutal reduction of my intrinsic qualities to the vulgar form of a thing whose existence was contingent on being bought and sold. It bothered the workers as well. They wanted more control over their affairs on site and in the material's factories. To begin with they downed their tools for better wages and improved working hours, but faced with constant arrest and repression it was inevitable that the Trades Unions would become radicalised. In 1917 building workers shouted 'down with oppression'.

By 1918, at the first conference of the All-Russian Union of Construction Workers, slogans had given way to concrete demands. The implementation of workers' control of production in all large firms and for the whole of the building industry to be brought under the 'organs of socialist state power'. Change was afoot, and I looked forward to an unexpected historical role. Demands would soon be made for concrete. How I liked to hear my name associated with the implementation of the grandest of ideals: to make concrete a world where injustice and inequality were to be banished.

It began so well. There were the unfamiliar abstract paintings, youthful flirtations with experimental theatre and literature, amazing new moving pictures starring insurgent workers, bizarre street performances and festivals, all manner of excursions into ideas of non-cities, linear cities and flying cities. Activists talked about setting up communes and exercising their new constitutional rights as workers. There was talk of free labour, free love, and new ways of life. At times the optimism was irresistible, and I confess to having enjoyed my flirtation with the avant-garde in the nineteen-twenties. Theirs was a different take on modernity, or rather a different concept of progress. I had become used to the association of the modern world with capitalist technological and economic revolution, but here I was being called upon to perform a rather different task. Already a hero in my contribution to civil engineering, the illustrious Dnepr Dam, it was in this period that the first modern experiments were conducted to prefabricate and mass-produce me. Like the German modernists Ernst May and Hans Scharoun, the constructivists were convinced of my potential in the mass-production of social housing and other buildings. I was to be the means for fulfilling democratic social objectives by materialising them in built form. It was my simplicity and cleanliness that was so attractive. As a smoothly engineered surface I could be impermeable and hard, yet my structural flexibility allowed me to span large distances. I could be both present and absent, structurally reliable yet freely punctured so as to open up space to air and light, an antidote to cholera and typhoid.

I bristled with anticipation. Manifestos and declarations circulated like Siberian mosquitoes. There was The Association of New Architects who proclaimed 'ASNOVA is working for the masses, which demands standard architecture on a par with a car or a shoe'. That was in 1926. It set my imagination working. How could I be made as a kit? Then there was the OSA, the Union of Contemporary Architects and their first conference in 1928. They were even more ambitious, calling for 'the active and scientific acquisition of all the achievements of world-wide technology in the field of the latest materials, structures, mechanisation and standardisation of building production'. This was stirring stuff. This was my chance.

I went forth to build this new world. There was the dome of the Planetarium in Moscow, the frame of the Dom Narkomfin Housing Commune by Mosei Ginzburg, and one of my favourites, the Club Russakova designed by Konstantin Melnikov. It was the nineteen-twenties, those wonderful few years when everything seemed possible. The race was on to exploit me formally and structurally in the search for what comrade Lissitsky called 'an architecture of world revolution'. As late as 1931 OSA still felt bold enough to extol the virtues of the 'standardisation and typification of buildings which would make it possible to go over from seasonal construction work to mechanised assembly all year round'. But I had already become aware of the looming storm; that particular dream would have to wait for nearly three decades.

Two years earlier, in 1929, The All-Union Proletarian Architects Association (VOPRA) had issued an omen of what was to come. 'We reject Constructivism, which has arisen on the basis of finance capital. The fundamental features of monopolistic capitalism – a bias in favour of capitalist planned arrangements, rationalisation and powerful industrialisation – have determined the nature of this architecture.' A strangulated mouthful of rhetoric though it was, it signalled the end of one history and the beginning of another. I had become used to being a material expression of a more socially just world, but the cracks had deepened into permanent fractures. My adventurous authors, of whom I had become fond of, were placed on trial. For some reason the language of steel frame, glass and concrete put together as geometric assemblage was deemed bourgeois and reactionary. This was more than a little peculiar, for in Germany following the closure of the Bauhaus in 1933 I was accused of being Jewish, Bolshevik and un-German. In the event, VOPRA's programme for the development of a monumental neoclassical architecture came to represent the party line. It had become anti-Soviet to imagine linear concrete housing communes. To the concrete and steel architects of the avant-garde there was delivered a simple choice: assimilate and submit, or face exile.

I was made a spectacle of. With the apparent death of the avant-garde my second brief chapter in modernity had come to an end. I say 'apparent' because, although as a revolutionary critique the avant-garde had been decapitated, many of their ideas and pre-occupations survived and were assimilated. The ideological power of cultural production was one trick the avant-garde bequeathed to the bureaucrats in the Central Committee. The concept of the state regulation and mechanisation of building production was another. The former became one of the ways in which the whole system was propped up; the latter became the organising principle for the construction industry after the Second World War. In a touch of prophetic irony, VOPRA's description in 1929 of the vulgar standardisation of construction captured precisely what the social idealism and technological optimism of the avant-garde would become in the nineteen-sixties, though under very different social conditions.

Though I was to have a central role in this future, my optimism was tinged with sadness as I witnessed the intoxicating idealism of the early years being strangled on the grounds of practicality and authenticity. First, the economic priorities were electrification and industrial reconstruction, not what Stalinists called the 'utopian' flights of fancy of the avant-garde. Second, as a replicated concrete detail I was inappropriate on historical and cultural grounds. Either as a cast-in-situ cantilever on a palace of culture, or in an elegantly mass-produced duplex, my association with the avant-garde branded me an enemy of the people, counter-revolutionary. The party secretaries said that the masses did not understand concrete cubes.

During the thirties I was kept to the margins of the civilian world. My star dimmed. 'The critical assimilation of previous architectural form' became the new political directive for Soviet aesthetics. My experiments with the avant-garde, the concrete communes and worker's clubs had no place in this world. Cast out and mocked, their destiny was to sit as lonely fragments, dismissed by the scornful City of Stalin then growing out of the ground in an ideological spectacular that laid claim to Imperial Rome. Standing on the ruins of Tatlin's tower grew the monumental Palace of the Soviets. The social and technological ambition of the Dom Narkomfin was replaced by the proletarian classicism of prestigious housing blocks.

Despite this blow to my pride I was never very far from the scene. As a consolation I was used extensively for infrastructural and industrial construction. Then of course there was the metro. It might have been an ideological smokescreen to camouflage the gulag archipelago, a projection of empire to impress the outside world, a utopian promise with which to bludgeon the people, but its construction was nevertheless epic. I still remember the day, 14 May 1935: the grand opening addressed by the executioner Comrade Kaganovich. How those words still resound in my head. Announced to the assembled workers, they indicated my rehabilitation. 'In every piece of marble, in every piece of metal and concrete, in every step of the escalator is manifest the new human soul, our socialist labour, our blood, our love our struggle for the new person for a socialist society.' Did you hear that? Me, concrete, an embodiment of the human soul, a socialist soul. This was an unexpected turn of events.

Leading Bolsheviks had always advocated the adaptation of the achievements of capitalist society. The party bosses liked opera, the scientific management of workers, neoclassicism and vertical chains of command. They lifted and adapted the language of bourgeois representation and renamed it socialist realism, something I later discovered was better described as the realism of social deception. They were secretly in awe of Fordist mass production and jealous of Manhattan. Officially they didn't as yet worship concrete, but they would once it was safe. After all, my value and worth had been clearly established.

I visited Moscow in the late forties and was quietly employed in a couple of housing projects, but my desire to become the universal material for all types of construction had to wait for the tyrant's death. March 1953 and he was gone. I had returned to Moscow shortly before the funeral. This time I was going to stay. I got ready, rehearsed my moves. Then in 1955 came the announcement. I loved the wording, subtle so as not to upset those still in awe of the dead Georgian dragon. At last The Central Committee of the Communist Party of the Soviet Union spoke out on the cult of personality in architecture in a document entitled 'On the elimination of over-indulgence in design and construction'. And again in 1957, 'On the development of housing construction in the Soviet Union'. In the small print of the 1955 plan were explicit instructions from the ministries to industrialise the building industry on the basis of the mass production of concrete. I was soon rolling out of the factory gates as large blocks. My mission: to conquer time and space through the speed-up of construction and the territorial and geographical integration of my methods. It was odd. The mechanisation of building production had once been labelled as a grotesquely rationalised expression of finance capital. Now it was being hailed as a brilliant example of Soviet socialist industrialisation, and I was its messenger.

A new ideological programme replaced the monumental propaganda of the metro, the seven wedding cakes, and theatrical boulevards. The party insisted that the class conflicts which had characterised all previous history had been conquered in the Soviet Motherland. Since this was the case, socialism became redefined as technological development, building as scientific planning. In the west, capital and science had long ago merged to form an almost unassailable economic fortress. In the east, science simply became party doctrine. The philosophical foundation for my future victory was laid: an all-consuming technological determinism that was as popular in the west as in the east.

My brief was clear: go forth and solve the housing shortage. For three generations architects drew my facades in simple never-ending patterns, piecing together buildings and neighbourhoods out of books and catalogues of pre-prepared details. Operating as obedient worker technicians, they were subject to the economic and political directives of economists and party secretaries. This was not unique to them, but a fate that befell all other workers labouring for the organisational backbone of modernity, the bureaucracy. Only powerful elites that have successfully concentrated in their hands the means of administration are capable of imagining and simultaneously erecting the same concrete panel on the dotted line of the Arctic Circle and the wandering earth of the Central Asian Desert. I went where I was told. Concrete, like all other property in the construction industry, belonged to the state, but the state belonged to the bureaucracy who, hidden in the Ministries of Construction, controlled every aspect of my daily life.

But there was an obstacle. If the plan targets were to be fulfilled a new type of construction organisation was needed. I seized the moment and formed a DomoStroitelniye Kombinat. The house-building combine. This was to be the machine for delivering housing at previously unimaginable speeds. And what a success it was. At one point Moscow's record-smashing DSK–1 employed twenty thousand workers and built an eighty apartment housing block in fifty-two days. It was simple. All design, prefabrication and onsite erection would be brought within the orbit of a single organisation. Subordinate the trades unions, install a system of one-man management, and introduce bonus schemes and productivity contests between brigades and individual workers. The bosses argued that this wasn't to be thought of as economic competition, but rather socialist emulation. With everything and everyone in their allotted positions, production could be centrally co-ordinated as a continuos flow of information and fabricated objects. The construction process would come to resemble a well-oiled and streamlined machine.

In strictly capitalist terms the planned technological change, yearly increases in productivity, and overall speed-up of construction would have been viewed as an acceleration in the movement of capital aimed at increasing the rate of profit. That is modernity defined as the production of relative surplus value. The words might not have been quite right to describe the process of accu-mulation in a state capitalist economy, but the motivation and consequences were similar to that occurring in the west. For workers it didn't really matter whether construction planning was conducted by the state or the capitalist enterprise. Life in the concrete factory and assembling the kits on site was much the same, as were the results for the users.

Not that you could ever have said such a thing. At the time that sort of criticism was unimaginable. Science was considered objective knowledge, rational planning the practical management of such knowledge. Within such a logic scientific planning was considered beyond any ideological critique by virtue of its political neutrality and innocence. 'Scientific' became the epithet for everything that was deemed to be good and socialist. The further development of socialism was presented as the progressive resolution of technical glitches and problems. Socialism was simply about finding the right solution and perfecting it.

The philosophical justification for such an authoritarian system and the universal remedy for problems in Soviet life lay in the pages of the *DiaMat*, the Soviet replacement for biblical guidance. This concerned the literal application of Engel's essay on dialectical materialism to both the interpretation and organisation of the social world. According to this schema all matter and phenomena, both of the natural world and the social world, including concrete, obeyed three laws: the unity of opposites; the transformation of quantity into quality; and the negation of the negation. Here I have to make a confession. Though they became shibboleths, I found the three laws of the *DiaMat* rather useful for looking at the structure of my life. After all, ideas only become problematic when you believe in them blindly. Take the chemical reactions between my ingredients: clay, chalk, lime, sand, aggregate and steel. Think of the process of my coming into being, from liquid to solid. It is all about the qualitative transformation from one state into another, the extraordinary heat that I generate as I mix and then cool. Then there was the suggestion that these laws should also govern my social and physical application. Well, in a way they did.

The unity of opposites seems to capture rather well my contradictory historical use, as does the second law concerning the transformation of quantity into quality. With philosophical legitimation and the post-war pressures of urban reconstruction, the party's priority was the quantitative increase in the production of things. I was measured in cubic metres, and in a literal sense the yearly increases in my production led directly to a qualitative transformation that could be measured in square metres of housing; quantitative increases in the housing stock represented a considerable improvement in the quality of life for those that had never experienced hot water and heating. This was an important historical battle for the worker's movement to win: the right to live in a home protected from the brittle chill of the Russian winter. But I have always felt that as the goal of socialism fulfilling the targets set in plans was a rather impoverished and utilitarian attitude. These are things that should be taken for granted. Radiators and record quantities of concrete are not the goals of socialism. As a French philosopher commented, 'people do not fight revolutions for tons of steel', or for tons of concrete.

In Russia they did however risk dying for it, and thus was born the cult of the hero worker, an idealised figure of a model Soviet citizen that doubled as an invitation to disable yourself by working beyond the limits of normal physical endurance. There were prizes and a fortnight in a rest home on the Black Sea for the first concrete worker to produce single-handedly a million cubic metres of concrete. The werewolf hunger of the party bosses for greater surpluses knew no bounds. What they called 'socialist accumulation' was a starving beast. The treasure chests were full of bonuses and holidays in which local secretaries were allowed to rummage once the norms had been met. The arms race depended upon it, and so did the acquisition of luxury goods for the family dinner table. But I am rambling again.

In the event the scientific management of labour and industrialised building production became a rather sad parody of itself. Urban construction transformed into the pursuit of statistical targets. This is why I have always liked the third law, the idea that in the clash of opposites and the subsequent resolution of one contradiction a new one emerges. I like to think that this is somehow analogous to my fate. I was both the solution and part of the problem. Hindsight is luxurious, and whatever my subsequent failures I largely resolved the massive shortage of housing. It is just that in the process I was often badly fabricated and inappropriately located in inhospitable climates. But you have to understand why. The economists' and developers' desire for ever greater economies of scale rapidly transformed itself into a pathological obsession. They knew I could perform pirouettes, but I was always ordered to run in a straight line.

Within such a militarised economy my influence and geographical limits knew no bounds. I appeared as a large panel, as a small panel, and as a block. I was prefabricated into cills and lintels, and cast in-situ in all manner of circumstances. My finest early moment was the residential neighbourhood Noviye Cheryomushki, built between 1956 and 1958. A compact residential area that placed low-rise blocks of housing around green squares connected by short walks to schools, shops and cultural facilities, Cheryomushki was in fact a reworking of an idea from the 1920s. It was a simple but successful model, and rapidly became an international hit. Soviet commentators even referred to it as a 'metaphor of modernity'. Every new town had a Cheryomushki. Town planners and concrete firms from western Europe paid visits to witness the forest of cranes swinging the new model homes up in record time. And it wasn't just the layout that was replicated; it was also the large five-storey concrete block houses. Popularly known as the *piati-etashniye*, they were constructed across the Soviet Union. There was even time for the odd experiment, like the attempt to incorporate the Constructivist's concept of the social condenser in the design of the concrete and glass experimental housing commune on Schvernika Street. But none of it was fast enough. The relatively low density of housing and localised facilities was deemed an inefficient and uneconomic use of resources.

Bigger panels would have to be produced, higher blocks constructed. Eight, ten, twelve, sixteen, all the way up to twenty-four storeys. Services like shops would become centralised in larger supermarkets. So was born the *mikrorayon*, wholly prefabricated suburbs and satellites of the capital. This was consecrated in the 1971 plan for the development of Moscow, a vision of a polycentric city with the historic core linked to a new concrete world in which I was to be king, landscape and memory. So followed two decades of relentless industrialised building production. With the invention of automated machines for vibro-rolling, I would dictate the speed of work to the worker. There were even optimists who saw in the 'perfection of industrialised production' the promise of escaping the physiological and psychological unpredictability of human beings. Socialism without heavy labour, or even more ambitious a world where, fully socialised, all the machinery of industrial production would become automated, liberating the Soviet citizen from work altogether. Ah, the dream of free creative labour.

Everything looked good. During the seventies my reign was virtually absolute. New towns like Zelenograd emerged alongside giant planned suburbs like Yasenovo, home to two hundred and twenty thousand inhabitants. Then there was Khimki-Khovrino, even bigger at a quarter of a million. I was all over the city. On into the eighties with Krilatskoye, the further development of Jugo-Zapadnaya, Severnoye Butova and others. There was no limit to the labyrinthine arrangements of tiny blocks of polystyrene shuffled around site models by planners. How many can we fit? How high can we go?

The secret of capitalist modernity has been described as the fetishism of the commodity. But modernity in Russia, conceived of as the centralised state regulation of economic and social life, possessed its own secret, the fetishism of the plan. In 1955 they produced 5.7 million cubic metres of me. By 1960 yearly output was up to 30.2 million; in 1975 114 million; in 1980 135 million; and by 1985 the fabulous figure of 140 million cubic metres of concrete was reached to cap a five year plan in which a record of 552 million square metres of housing was produced. Do you have any idea just how colossal this achievement was? And that was just in the USSR. My techniques and methods were exported to Africa, Cuba, Bulgaria, Germany, Poland – all over the world. Despite the unreliability of Soviet statistical data, I reckon that between 1955 and 1985 the House Building Combines and construction organisations produced around 2,400 million cubic metres of reinforced concrete. This is a monumental modern achievement that has not been fully acknowledged.

By the end of the nineteen-eighties thirteen million workers were labouring in the construction industry. There were two thousand design organisations, and nearly forty thousand construction–assembly organisations. Ninety per cent of all new housing was being constructed out of reinforced concrete. I had come a long way. Reluctantly admitting that there may be a few motivational problems in the industry, the party allowed some room for discussion. Cost accounting, self-management, decentralisation – these were the new buzzwords. Before western management consultants introduced them as mantras of new organisational theory, the Soviet bureaucracy had already institutionalised mechanisms for demonstrating accountability and measuring quality. As an ideological camouflage of deeply rooted contradictions no one really believed in them then, and neither does anyone now. In contrast the Council of Labour Collectives rehung the slogans of the revolution – 'All power to the Soviets', 'All power to the labour collectives'. But it was too late to rescue the administrative command system of economic management.

A new set of problems loomed. There were the boring and predictable accusations of chilling monotony. In response the Institute for Experimental Housing tried to invigorate concrete production, researching how greater flexibility could be built into the system of mass production. Smaller batches, a greater variety of forms. But there was little space for this postmodern shift. Others raised arguments in favour of developing rival technological systems in the building industry. An increased role for brick, steel, aluminium and glass, construction techniques that would threaten my hegemony. But there were more serious issues. The *piati-etashni* began to leak and crumble. Then there was the tragedy of the earthquake in Soviet Armenia, when I collapsed like a deck of cards. But that was not my fault – I had been inadvisably located in a dusty zone of tremors. I was not designed for that. On top of all this there was the feeling that I had fallen out of favour. Now Perestroika and Glasnost gave way to a fierce anti-Soviet anger. Symbols of the old regime, the hammer and sickle, the statues, the placenames, were assaulted. By association so was I. As a concrete panel I was both international and local, a universal language of construction but decisively Soviet, and therefore placed on trial.

The new elite had no time for the *mikrorayoni*. The Noviye Bogati longed for me to disappear along with all of my problems. Like the attempts to rethink democracy, rethinking concrete and dealing with the periphery was of no real interest to the new class of bankers and company directors that had emerged from the shadows of the Communist Party. Entrepreneurs and consultant managers have no use for crumbling leaky twenty-five storey residential towers on the margins of the city. They want the hot stuff, the valuable real estate in the historic core. Eventually someone will have to do something. They can't just leave me. I am the thousand-year city. They can rewrite history, rename streets, remove monuments, rebuild churches, but they cannot erase my presence. I am the legacy of Soviet urban construction. Nobody ever really liked proletarians. They liked their housing even less. Now they don't like it all. Another French philosopher commented that the spectre of Karl Marx would continue to haunt any discourse on the nature of the social world. Maybe, though I feel that's wishful thinking. If the modern world has to live with a ghost, it is much more likely to be me.

Jonathan Charley

# 1920s

# concretising the avant-garde

*"(we call)) for the active and scientific acquistion of all the achievements of world wide technology in the field of the latest materials, structures, mechanisation and standardisation of building production" First OSA Conference 1928*

**Fragments of the avant-garde in Moscow, Gable end of the Dom Narkomfin, Housing Commune of a transitional kind, Ginzburg and Milnius, 1928, Worker's club, Golosov, 1927, Izvestia Newspaper Office, Barkhin, 1925, "We are building socialism", Poster, 1920s, Burevestnik Worker's Club, Melnikov, 1928, Russakova Workers Club, Melnikov, 1927,**

# 1930s

## concrete foundations of an imperial city

*"In every piece of marble, in every piece of metal and concrete, in every step of the escalator is manifest the new human soul, our socialist labour, our blood, our love our struggle for the new person for a socialist society." Kaganovich opening the metro in 1935*

Concrete becomes buried in the foundations and hidden within structure. With the defeat of the avant-garde Moscow emerges as a reborn Imperial Metropolis dominated by the construction after the 1935 plan of new boulevards and an architecture of monumental neo-classicism. Clockwise, Reconstruction of Gorky Street, 1930s, "Neo-Renaissance" Housing, Vaynstein, 1936, Revolutionary Symbolism, Housing, Golosov, 1934, Metro Mayakovskaya, Dushkin, 1936

Jonathan Charley

# 1940s

# triumphant reconstruction

*Victory and post second world war reconstruction was encapsulated by the construction of Moscow's seven "wedding cakes", an attempt to build Soviet skycrapers to dominate the city and compete with the west*

**Concrete structure and details. Ministry of Foreign Affairs, Dushkin et al, 1949, Moscow State University, Rudnev et al, 1949, Apartment Block, Chechulin, 1948, Apartment block, Posokhin, 1950, war-time roof top observation, citizenry constructing barricades across the boulevards**

# 1950s
## *"piati etashni"*
# the re-birth of the concrete metropolis

*"Cheryomuski also became the symbol of modernity within the culture of the late 1950s. Cheryomushki ceased to be a proper name and became a metaphor. Almost every Russian city began to build it's own Cheryomushki, thus not only responding to an urgent housing need, but also asserting it's participation in modernity and its adherence to the new rational patterns of open space arrangement."  Ikkonikov, Russian Architecture of the Soviet Period, Moscow, 1988*

**Industrial reconstruction, Moscow as a forest of cranes, gable ends, detail and aerial view, Noviye Cheryomushki, Five Storey Concrete Panel Housing, 1957, phase one in the nationwide industrialisation of house building production**

# 1960s

## concrete becomes universal

*In the thaw of the Krushchev years in the nineteen sixties, amidst the universal mass production of five storey concrete panel housing, the Schvernika project was an attempt to rework ideas from the 1920s avant-garde. Making reference to Ginzburg and Milnius' Dom Narkomfin, it used an experimental concrete frame construction. Intended for families it was organised as a housing commune that provided communal recreational, educational and sporting facilities*

**Concrete panel factory, concrete tile factory, Soviet Fashion, 1965, propaganda, Gable end of Housing block, House of the new lifestyle, Ulitsa Schvernika, Moscow, 1969, architects, Osterman, Petrushkova et al**

# 1970<sub>s</sub>

# housing combines and mikro-rayoni

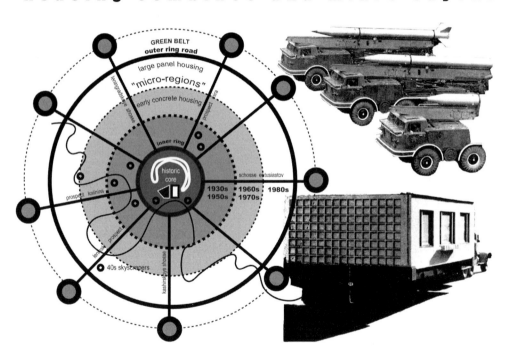

*A diagrammatic plan of Moscow that indicates the form of the radial city inherited from the nineteenth century that was accentuated in the Stalinist plan of 1935. From the 1950s, Moscow expanded outwards in a series of planned housing and administrative zones. These did not appear as neat bands, rather fingers and fragments, but which nevertheless are best represented as rings. This structure was consolidated in the 1970 plan for the city's future development which included new mikrayons and beyond the outer ringroad, the preservation of the green belt and the construction of satelitte towns*

**the mass production and transportation of housing and weapons, pre-fabricated concrete room, missiles on parade**

Jonathan Charley

# 1980s

## concrete determinism

*In the Moscow area by 1989, 90% of all new housing was being constructed out of reinforced concrete. In the last two five year plans of the Soviet era, between 1980-1990, it is estimated that the construction industry which employed thirteen million workers produced across the country as a whole an estimated 1000 million cubic metres of concrete*

**craning a large concrete panel, housing construction magazine, on site on the moscow peripehery, housing block, Mikrorayon Krilatskoye, Moscow, Krilatskoye site lay-out, female construction workers on site. By the nineteen eighties over a third of the construction workforce was made up of women, working in all spheres of the industry as architects, planners, decorators, plasterers, and as the managers of concrete factories**

# 1990s

## modernity's concrete epitaph

*"for a moment the spring sun seemed to pick out the gilt on horse drawn carriages and the whole city seemed to be ablaze as he thought of God, tsars, monasteries, and peasants. He shook his head, blinked, but this time saw nothing but the letter B, for Banks, Barricades, BMWs, Bananas, Bullets and Bandits."*

**1996, concrete foundations are laid for the new luxury shopping mall situated under the former Revolution Square, memorial to a student crushed to death under a concrete flyover by a tank during the uprising of 1991, the "last communists" demonstrate in front of the old Lenin Museum, 1996**

Part Three

# The city

## Chapter 13

# Ildefonso Cerdá and modernity

*Christian Hermansen*

Considering that the aim of this book is to trace the ideas that shaped twentieth century attitudes to architecture and urbanism, and that of this essay is to discuss the work of Ildefonso Cerdá (1815–76) within this context, it is tempting to adopt a model of history that sets parameters against which we can measure Cerdá's contribution. This task would be served, for example, by those historical constructs that adopt the idea of a break or paradigm shift during the period immediately before or during Cerdá's life.[1] Works which come to mind are A.R. Hall's *The Scientific Revolution 1500–1800: The Formation of the Modern Scientific Attitude*,[2] which suggests that the task of the beginning of the nineteenth century was the application of natural sciences principles and methods to industrial production, or Foucault's *The Order of Things*,[3] in which he argues that 1800 marked a fundamental epistemological shift from taxinomia to origins, causality and history. Cerdá's claim to be a prominent figure making significant contributions to such a paradigm shift would allow us to present him as a thoroughly modern figure, a pioneer of the principles and practices that guided twentieth-century urbanism.

While it is true that paradigms change over time and that there is an attraction in the simple elegance of depicting historical change in this way, the reality is that a close reading of Cerdá's work reveals a much more complex and fragmented picture in which paradigms co-exist, overlap, compete and contradict each other[4]. As we shall see, a detailed examination of his work does not negate his role as pioneer of modern urbanism, but it does show that trying to fit his work into a paradigm which was still being formulated during his lifetime offers an incomplete and reductionist picture.

## The historical context[5]

The nineteenth century did not treat Spain well. From its pre-eminent role in the seventeenth century as head of the world's largest empire, in the first decades of the nineteenth century Spain not only lost most of its colonies but was conquered by a foreign power. Charles IV was forced to abdicated in favour of his son Ferdinand VII by the Revolt of Aranjuez in 1808; shortly afterwards French troops occupied Madrid, and Napoleon forced Ferdinand to abdicate in favour of Joseph Bonaparte, Napoleon's brother. Ferdinand was imprisoned in Bayonne where he remained until Napoleon, forced to move his troops out of Spain by the threat to his eastern front, put him back on the throne.

The consequences of the French occupation of 1808–13 were of utmost importance. The disillusionment with the monarchy that had so overtly sold the independence of the country to France fuelled sympathies for the liberal cause, while the distrust of the country's overly centralised government encouraged the regions to organise themselves by raising troops, collecting taxes, promulgating laws, and setting up elected juntas. At national level the elected Cortes (parliament) acquired an importance it had not previously enjoyed, and used its newfound influence to draw up documents like the influential 1812 constitution, liberal in every respect save the guarantee of religious freedom.

Once back on the throne Ferdinand VII, backed by the army, the church and the royalists, annulled the liberal constitution of 1812, arrested liberal leaders, reinstated the inquisition, and ruled like a tyrant for the next six years. By 1820 discontent in the country had turned into a revolt backed by army factions reluctant to engage in the campaign to recover the American colonies. Ferdinand VII capitulated and reinstated the 1812 constitution, at the same time appealing to Austria, Prussia, Russia and France to uphold the principle of absolute monarchy. In 1822, at the Congress of Verona, these nations agreed to help and sent a French army, the 'one hundred thousand sons of St Louis', to invade Spain and impose Ferdinand as absolute monarch. During the next decade Ferdinand's efforts to eradicate all traces of liberalism were so thorough that an address to the King at the University of Cervera contained the phrase 'Far from us the dangerous novelty of thinking'.[6]

In 1830 Ferdinand had his first child, Isabel II, by his fourth wife María Cristina of Naples; to allow her to inherit the throne he proceeded to abrogate the Salic Law. This so infuriated Don Carlos, his younger brother and pretender to the throne, that the first Carlist war broke out in 1833, the same year that Ferdinand died. Maria Cristina took the throne as regent, relying on moderate liberals for support for her daughter's claim to the throne against the conservatives and royalists who supported the claims of Don Carlos. The instability of the regency and the reign of Isabel II can be gauged by the succession of constitutions – 1837, 1845, 1852 and 1855 – which oscillated between the liberal principles of the 1812 constitution and the abrogation of these principles.

The military, being the only well-organised body in the country, emerged as a new political force, and used the military *pronunciamiento* as a means to take power. Political control oscillated between three institutions – the monarchy, the Cortes, and the military. 'Cabinets rose and fell overnight, ministers climbed to power and resigned their portfolios with scarce look at their contents, … and the people found politics a dirty game and stood aside.'[7] By 1868 the unpopularity of Maria Cristina forced her to flee to France to take refuge, much as her father had done sixty years earlier. While the Cortes debated whether or not to bring in a republican system, the royalists pressed for the succession to go to Isabel's twelve-year-old son Alfonso. After two years of interregnum the Cortes decided to make Amadeo I of the House of Savoy the king of Spain. In a climate of extreme political unrest Amadeo reined for two years, resigning in 1873.

The First Republic followed, but the Cortes could not agree on a new federal system and the country soon fell into chaos. The provinces took matters into their own hands, completely disregarding the central authority in Madrid. The more assertive provinces, like Catalonia, declared themselves independent states. The resulting chaos encouraged the supporters of Don Carlos to start the Second Carlist War and take control of much of the north of the country. In November 1874 Alfonso XII came of age while studying at Sandhurst in England, and a month later an army *pronunciamiento* restored him and the monarchy, ending the twenty-two-month-old republic.

Ildefonso Cerdá died in 1876, the year in which Alfonso XII drew up a new reactionary constitution that set up a puppet Cortes to rule alongside the king. His understanding of social structure was greatly influenced by the constant shifts in political power between the liberals, who wanted to realise the potential of progress, and the conservatives, bent on preserving the old ways that stood in the way of progress.

# Ildefonso Cerdá

Ildefonso Cerdá i Sunner was born on 23 December 1815 in the Manso El Serdà,[8] Catalonia, in the secure environment of a long-established minor aristocratic rural family. Being third in line to inherit the estate, he was expected to take up a profession – the church, the military, or one of the 'liberal professions'. His father chose a career in the church, and at the age of fifteen he was sent to Vic to study Latin and philosophy as preparation for the seminary. In 1832, against his father's wishes, Cerdá went to Barcelona to study mathematics and architecture in the school of the Junta de Comercio. In 1835, without finishing his architectural training, he left for Madrid to enrol in the recently reopened Escuela de Ingenieros de Caminos, from which he graduated in 1841.

The staff of the Escuela de Ingenieros de Caminos and the members of the Institute of Engineers were sympathetic to a constitutional, liberal and anti-absolutist political regime, probably because engineers, in contrast with

13.1
Ildefonso Cerdá
(1815–76)

architects, did not enjoy royal patronage, and thus did not feel a strong allegiance to the establishment. In addition, the curriculum of the Escuela de Ingenieros followed closely that of the École Polytechnique in Paris, transformed by Napoleon into a military school for the creation of a technocratic elite with a strong progressive ethos. The École Polytechnique counted the likes of August Comte amongst its illustrious graduates and teachers.

During his studies in Madrid, Cerdá frequented liberal circles and joined the Milicia Nacional, reaching the grade of lieutenant. The *milicias* were volunteer troops under the command of municipalities, and although their role changed according to the current constitution they were generally associated with liberal and radical ideas, playing for example a crucial part in Barcelona's claim to independence from Madrid. The Escuela de Ingenieros and the *milicia* had a profound effect on Cerdá, moulding his ideas about systematic thinking and social concern and providing important contacts for the development of his career. After his graduation Cerdá served as government engineer until 1849, when he unexpectedly inherited his family's estate and resigned his commission to devote his full attention to the creation of a science of the city.

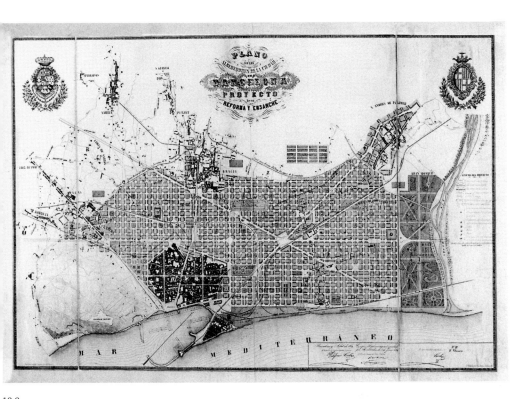

**13.2**
**Competition-winning plan for the extension of Barcelona by municipal architect Rovira y Trias, 1859**

The next six years were devoted to the study of the urban phenomenon and to his role as *Diputado*[9] to the Cortes in Madrid at a time when Barcelona was in the midst of rapid industrialisation. Barcelona was an important and expanding centre for the textile and metal industries, but its growth was restrained by medieval city walls which the military wanted to retain because they made the containment of insurrection easier. The only way industry could expand was by displacing housing, thus aggravating the city's already high density. By the mid-nineteenth century there was total consensus within Barcelona that the remedy to its urban problems lay in the demolition of the city walls and the expansion of the city into its adjoining territory. The demolition of the walls was approved in October 1854, and the *ayuntamiento* commissioned Cerdá to undertake a survey of the territory into which Barcelona would expand.

On 7 June 1859, instigated by Cerdá, the Madrid government issued a royal decree approving an as-yet-uncommissioned Cerdá plan for the expansion of Barcelona. The *ayuntamiento* of Barcelona, who had played no part in the Cerdá expansion plan, tried to have the decree rescinded, but failed. It then proceeded to call for a competition for an alternative plan, and on 20 October 1859 a beaux arts-inspired project entitled *Le tracé d'un ville est oeuvre de temps plutôt que d'un architecte*, designed by the municipal architect Antonio Rovira y Trias, was declared the winner.

**221**

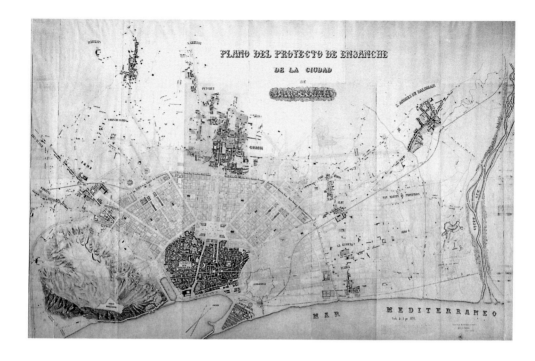

However, another royal decree of 31 May 1860 confirmed Cerdá's plan and put an end to the debate. Attacks against the Cerdá plan continued in the Barcelona press, based on two main issues: the imposition of the will of Madrid against the clear will of the people of Barcelona as represented by their *ayuntamiento*, and the fact that the government backed the plan of a private individual – Cerdá – against the will of democratically elected representatives of the people.[10] In 1867 the Madrid government published Cerdá's *Teoría General de la Urbanización*, a justification of the principles behind the plans for the expansion of both Barcelona and Madrid, and his attempt to create a science of the city.

# Cerdá and modernity

One way of exploring Cerdá's contribution to modern urbanism is to examine four key concepts in the formulation of his theory of urbanism, concepts that have underpinned our attitudes to the city in the twentieth century: movement, communications, form and function, and rurisation.

In 1859 Baudelaire sketched the modern concept of modernity in his *Constantine Guys: le peintre de la vie moderne*.[11] Baudelaire's was not specifically a theory of modernity but rather suggestive fragments of such a theory that would later be developed by Simmel, Kracauer and Benjamin. Baudelaire argued that modernity was that which was new. To be truly new, he argued, you must be in the present; the moment the present become past it has ceased to be new, and thus has ceased to be modern. In consequence, the truly modern is that which is 'ephemeral, fugitive and contingent.' Thus for Baudelaire, fashion has a far greater potential to express the modern than, for example, architecture.

Where does one find this transitory modernity? Baudelaire is unequivocal: 'in the landscapes of the great city' where the *flâneur* roams the metropolis with a 'loftier aim than that of the mere *flâneur*', man-about-town, namely 'seeking out and expounding the beauty of modernity'. Baudelaire continues 'The crowd is his element ... His passion and his profession are to become one flesh with the crowd. For the perfect *flâneur*, for the passionate spectator, it is an immense joy to set up house in the heart of the multitude, amid the ebb and flow of movement, in the midst of the fugitive and infinite.'[12]

Cerdá's understanding of the present closely parallels Baudelaire's notion of modernity.[13] He recalls that his idea about the nature of the present condition of cities came to him in the railway station at Nîmes in 1844, when he watched a great crowd descend from a recently arrived train and work its way into the city.

> What surprised me, in spite of the fact that I had imagined this in my mind many times, was to see those long trains which, loaded with a large quantity of merchandise, a large number of passengers of all sexes, ages and conditions, came and went, appearing to be whole populations hastily changing domicile. After overcoming the surprise of this spectacle, new at that time for me, my thoughts were elevated to considerations regarding the social order, specially when I observed the difficulty with which that mob of unexpected guests penetrated through the narrow doors, scattered in the narrow streets, in search of their shelter in the mean houses of the old city.

This scene suggested two things to Cerdá. The first was a great struggle between a new and an old order. The train and the station, symbolising modernity and progress, represented the new order; the dense city fabric into which the crowd was struggling the old order.

> ... comparing times with times, customs with customs and elements with elements, I understood that the use of steam power to create a motive force signalled to humanity the end of an era and the start of another one. Today we found ourselves in a state of transition, a state which could be longer or shorter, according to the character of the conflict that I realise has already begun, between the past and its

traditions, the present and its vested interests, and the future with its noble aspirations and forward thrusts.

The second was that the locus of the struggle between the old and new orders was going to be within large cities.

> The new era, with its new elements, whose use and predominance is extended every day with new applications, will in the end bring us a new civilisation, vigorous and fertile, that will radically transform the nature and functioning of humanity in terms of the industrial order, the economic order, the political order, and the social order, and will, in the end, take over the whole world. I saw this new civilisation coming at a rapid pace and knocking at our doors, I saw its first manifestations appear in large cities which, because of the nature and circumstances of the war ahead, would become the operational field for the titanic battle between two civilisations for the domination of the world.

Cerdá equates modernity with the present characterised by the confrontation of two orders: the old, and the new that will replace it. This social conflict, which he calls war, is the cause of social change and of progress. Cerdá saw the present as part of a linear development towards perfection: that humanity is inexorably engaged in a historical process of progress, guided by the application of scientific principles to technological advancement. Furthermore he understood that social conflict is the motor for social change, echoing two of his contemporaries: Marx, who suggested that history is governed by a succession of class conflicts over ownership of the means of production,[14] and Darwin, who explained the development of species as a consequence of conflicts between them.[15]

# The science of the city

The industrial revolution had a tremendous impact on nineteenth century society. Productive efficiency shot up, immigration from the country to the city was explosive, and living conditions in industrial cities were worst than at any other time in history. It was the desire to improve the quality of life in the industrial city that motivated most nineteenth-century social reformers, including Ildefonso Cerdá, to address themselves to urban problems.

While other urban reformers directed their efforts to the resolution of local or national urban problems as they understood them, Ildefonso Cerdá, very much in the spirit of the mid-nineteenth century, attempted to create a comprehensive and universal science of the city. He defined the city as the

> ... mare-magnum of people, things, and interests of all sorts, of thousands of diverse elements. In spite of the fact that all elements appeared to function independently of the others, when I observed them closely and philosophically, I noted that they were in constant interaction with one another, often exercising a direct action over each other, and therefore forming a unity.

Cerdá did not understand the term city solely in terms of its fabric, but also in terms of the interaction of all of its constituents.

> My objective was not to express materiality, but rather the way in which, and the system guiding, the formation of groups and how the elements that constitute groups are organised and how they function. In other words, in addition to materiality I wished to show the organism, the life, if one could use this expression, which animates the physical parts.

Cerdá invented the word urbanism as the name for the new discipline he set out to create. He rejected the accepted etymological interpretation of *urbs* (city) as coming from the Latin *urbum*, the curve of the plough blade used in Roman ritual to define the territory of the city. According to Cerdá, in the ancient language precursor to modern European languages *ur* meant hollow or cavity, and by extension a dwelling. From *ur* came *urs*, a group of dwellings, and *urbum*, to open cavities in the earth or to plough. According to Cerdá you would plough both to grow food and to make dwellings, food and shelter being the two basic constituents of human life. *Urbs* and urbanism became the foundations of his new science.

> It became clear that the word city would not do. Maybe some of the words derived from 'city' could have been used, for example *civitas*, which is a word used in modern languages. However, I found that all these words already had meanings that were different to those that I was searching for. In effect, all the words derived from 'city' and 'civil' are applied to express or to refer to the higher states of intellectual and moral development, or to the rights, prerogatives and pre-eminence of man living in the city, or things attributed to him that it was not my intention to address. After having tried and successively abandoned many simple and composite words, I remembered the

word *urbs*, that, because it did not leave Latium, was not passed on to the peoples that adopted its language. The reason for this is that *urbs* was reserved by domineering Rome as a pre-eminent aristocratic title for itself. This made *urbs* useful for me as a virgin derivative, if I am allowed to express it in this manner. A new word for a new subject, so general and comprehensive that it would encompass all the diverse and heterogeneous elements that, harmonised by the superior force of human sociability, constitute what we call a city ... These are the philological reasons why I was lead to, and decided to adopt, the word *urbanisation*.

From Cerdá's formulation of a science of the city I have chosen four concepts whose crucial role in the development of the late nineteenth and twentieth century city has been confirmed by history.

## Movement
The importance of movement in the city is one of Cerdá's most significant insights.

After a quick look at major population centres, I became convinced that, being the product of a passive civilisation, they would present difficulties and obstacles and hinder the new guest who requires and demands more space, more room, more liberty for the expansive character of the unusual movement and feverish activity that distinguishes it. The new civilisation will not suffer these obstacles and difficulties, it will destroy them rather than be condemned to a quietism incompatible with its constituent and essential elements.

My first investigations regarding the requirements of the new civilisation suggested that its distinguishing characteristics were movement and communicativeness.[16] Judging what our old cities, where everything is narrow and mean, could offer to satisfy the new requirements of movement and communicativeness, allowed me to make out new horizons which were wide and vast, a new world for science, towards which I decided to head at any cost.

Changes in patterns of movement have had the most profound effect on the form of the modern city. The expansion of inexpensive mass transport in the latter half of the nineteenth century, in the form of the tram, the railway and the underground, made possible the growth of cities to sizes which had been hitherto unforeseen. For the first time capabilities of pedestrians did not dictate the overall size of urban areas. The impetus for urban growth was further stimulated in the first half of the twentieth century by the rapid expansion of private car ownership.

Cerdá had the foresight to predict the importance of movement before the impact of mechanised and mass transport could be felt, and decades before it led to the explosion of urban form in the latter half of the nineteenth century.

## Communications

The idea of that communication systems are one of the most important components of the city has been decisively confirmed by the development of postal services (1800), telegraph (1837), telephone (1876), fax (1980s), email and the internet (1983).

Since Melvin Weber wrote his article 'The urban place and the nonplace urban realm' in 1964,[17] much has been written about the impact of communications on spatial location. In its most simplistic form it can be argued that communications are so good today that it makes no sense to gather people in a small dense area such as a city centre. People can perform their jobs and social interactions just as well from places remote from cities that are more attractive because of low congestion, low property prices and high amenities.[18]

As with the role of movement within the industrial city, Cerdá had the foresight to understand the importance of communications before most of the technological advances in communications had been developed, and long before the impact they would have on urban form could be felt.

## Form and function

With rare exceptions the word function, meaning what things are used for, was not used by architects until the first decades of the twentieth century. From the eighteenth to the early twentieth century 'function' meant the forces that the building mass exerted on the structure of the building. It is probably Greenough in the1840s and Violet le Duc in the 1850s whose writings can be credited with diffusing the new concept of function as 'use'. Cerdá went further than this. He criticised those who emphasised physical fabric, arguing that the city is made up of a multitude of elements – formal, functional and ideological, and that all these elements form the unity which makes up the city. These elements exert influence and are in turn influenced by other elements.

The idea of the predominance of the relation between form and function was probably taken by Cerdá from recent advances in palaeontology made by Lamarck and Cuvier. Lamarck, proponent of a theory of evolution fifty years before Darwin, published *Philosophie zoologique* in 1809. In it he stated two laws of evolution governing the ascent of life to higher stages: first that organs are improved with repeated use and weakened by disuse, and second that such environmentally determined organ acquisitions or losses 'are preserved by reproduction to the new individuals which arise'. The vital element that Darwin added to Lamarck's theory of evolution was the causes of change in species: 'the survival of the fittest'. Working alongside Lamarck, Cuvier in his *Leçons d'anatomie comparée* of 1800–5 proposed the principle of the correlation of parts, in which the anatomical structure of every organ is functionally related to all other organs in the body, and the functional and structural characteristics of organs result from their interaction with their environment. Species were no longer thought to be static, but to evolve as they adapted to their environment, and were classified not only in terms of their external or skeletal form but also in terms of function, which in the case of extinct species could be deduced through comparative studies with living species. In Paris the displays in the Jardin des Plantes were modified to reflect these new ideas. Cerdá went to Paris while gathering information for the formulation of the science of the city, so it is very likely that he visited the Jardin.

Lamarck argued that the evolution of species resulted from their efforts to adapt and take advantage of the environmental conditions in which they lived, and that the gains of one generation were inherited by the next. In a similar vein Cerdá argued that the evolution of cities occurs because of human efforts to satisfy the basic needs of food, shelter and sociability, and that the lessons learned by one generation are passed on, through culture, to the next. To the idea of evolution through adaptation Cerdá added catastrophes, major environmental disasters sent by God to make humans rethink their relationship with their environment which he saw as useful markers to establish a succession of epochs.

Cuvier had established that there was a relationship between the form of an organism's skeleton and its function, and that the comparative method could be used to deduce the likely function of skeletons of extinct animals. Likewise Cerdá assumed that changes in urban form were a consequence of attempts to improve human relationships with their environment. Cerdá deduced the activities of humans in ancient settlements using a mixture of archaeological research, common sense and rationality, and information emanating from contemporary 'primitive' tribes, using these findings as the basis for understanding the origins and development of urbanism.

## Nature

One of the most important distinctions in western thought is that between the world created by God, nature, and the world created by humans, culture. By the beginning of the nineteenth century there were many theories of nature and its relation to culture. The one that most influenced Cerdá was the sociopolitical concept of nature as freedom, an idea which emerged as a reaction to Europe's absolute monarchies, especially that of Louis XIV. The exponents of nature as freedom saw despotism reflected in the garden designs of the great palaces of the royalty and aristocracy, such as Le Notre's Vaux le Vicompte and Versailles. Such reactions to the cultural domination of nature were based both on the idea of 'paradise as a garden' and on political ideas developed by philosophers such as John Locke. They were first put into practice by William Kent in the late 1730s, who stated that 'all nature is a garden', an idea he implemented in his design of the Elisean Gardens at Stowe, the first garden in which the association of nature and political freedom was explicitly represented.

In his *Teoría General de la Urbanización* Cerdá repeatedly talks about the constraining character of existing dense cities. He explicitly equates nature with individual liberty and the independence of the family, and stresses sociability as one of the advantages to be gained from living in a city. The idea that the most desirable type of urbanisation is one that combines the advantages of the country with those of the city is most often associated with Ebenezer Howard and the publication of *Garden Cities of Tomorrow* in 1898, but this concept was already present in the work of Cerdá, who coined the term 'rurisation' more than thirty years earlier:

> ... ruricemos las ciudades así como vamos urbanizando las campiñas.[19]

Howard was attempting to reverse the large-scale migration of people from rural areas and small towns to large cities which were becoming overpopulated through the control of urban land speculation, which he believed to be the root cause of urban problems. The task Cerdá set himself was different; he wanted to develop a science of urbanism. In order to achieve this he needed to account for the historical development of the city, and it was while researching this history that he came across early urban settlements that combined the advantages of both the country and the city.

According to Cerdá there are two counterpoised sentiments within humanity, whose balance is necessary to achieve a harmonious existence within the city. The first is individual liberty and the independence of the family; the second is sociability, which describes the advantages gained by living within a social group. Cerdá argued that it was in the second phase of his history of development of the city that urban areas came closest to achieving this balance. In this phase each dwelling was located in the middle of a plot of land, thus ensuring privacy and independence, while at the same time being near enough to other such plots of land to make it possible to have the advantages of sociability. In many ways this describes the same sort of suburban development that Howard recommended thirty years later.

# Conclusion

In Cerdá's work we find many ideas that were to become fundamental to a modern theory of the city in the twentieth century and beyond. Though there is no denying that Cerdá was a thoroughly modern figure whose insights into the future nature of the city are remarkable, a close reading of his work reveals a complex and fragmented picture, one in which his pioneering contributions co-exist with ideas that often overlap, compete and contradict each other.

# Notes

All Cerdá quotes are taken from a translation of the *Teoría General de la Urbanización* that I am in the process of completing.

1   In his *Structure of Scientific Revolutions* Kuhn (1962) criticises the cumulative view of scientific development in which science grows step by step by adding new pieces of knowledge to those already in place. Instead Kuhn argues that there are periods characterised by controversies between competing schools of thought, followed by periods in which one of those schools of thought prevails by proposing a paradigm that attracts to it members of other schools, and is sufficiently open to allow further development.

2   Hall (1956).

3   In *The Order of Things* Foucault (1992) analyses the linguistic systems (epistemes) characteristic of certain periods of thought. Although he finds strong similarities between the linguistic systems underlying the classical disciplines of general grammar, natural history and analysis of wealth and their modern replacements, philosophy, biology and economics, he argues that around 1800 there was a fundamental shift in epistemology consisting in abandoning a study of the world based on taxinomia with one based on the study of origins, causality and history.

4   The attitude that I am depicting here is much closer to Carl Schorske's description of European attitudes to the city in the mid-nineteenth century.

5   This short historical sketch is taken mainly from Raymond Carr's (1982) excellent *Spain 1808–1975*.

6   Quoted in Atkinson (1934).

7   Atkinson (1934:164).

8   *Manso* is the Spanish word for an agricultural estate; the Manso Cerdà extended to 150 hectares and had been in the family since 1440.

9   *Diputado* is an elected representative, equivalent to a member of parliament.

10   *Diario de Barcelona*, 5 June 1860. The attack on Cerdá started on this date and continued with increasing venom over the next few days. The only new argument added was the fact that the Junta de Caminos y Canales was the authority that would decide the destiny of the city of Barcelona, thus disregarding the history and cultural traditions of the peoples of Catalonia.

11   Baudelaire (1943).

12   Baudelaire quotes taken from Frisby (1988:11–37).

13   Cerdá and Baudelaire's publications are contemporary. Baudelaire's *The Painter of Modern Life* was published in 1859. Ildefonso Cerdá's main works were *Teoría de la Construcción de Ciudades Aplicada al Proyecto de Reforma y Ensanche de Barcelona* (1859), *Teoría de la Viabilidad Urbana y Reforma de la de Madrid* (1861), and *Teoría General de la Urbanización* (1867).

14  See Marx and Engels (1973), *The Communist Manifesto*, with an introduction by Leon Trotsky.

15  See Darwin (1959).

16  This is when Cerdá introduces what are probably the most far-reaching concepts. He uses the Spanish words *movimiento* (which translates well as movement) and *comunicatividad* (which I translate as communicativity). The word *comunicatividad* does not exist in Spanish; both it and its English translation as communicativity describe the potential of an object to allow the transfer of ideas.

17  Webber (1964).

18  Mitchel (1995).

19  This sentence translates as '… let us make the city rural as we urbanise the countryside'.

# Chapter 14

# 'To knock fire out of men'

## Forging modernity in Glasgow

*Juliet Kinchin*

> It is curious to note how most of the great triumphs of art have been
> won in cities, and in cities, too, whose life was oftentimes of the
> busiest and most complex description ... A civic life would seem to
> knock fire out of men, like the sparks evolved from the contact of
> flint and steel.[1]
>
> Francis Newbery, Headmaster of the Glasgow School of Art, 1897

From the magisterial urban interventions of Alexander 'Greek' Thomson (1817–
75) in the 1850s and 60s to the architecture of Charles Rennie Mackintosh
(1888–1928) and his Glasgow Style associates in the opening years of the twentieth century, Glasgow has been associated with manifestations of the modern in
design, architecture and urbanism. This chapter explores the ways in which
these two generations gave spatial and visual expression to the urban experience
of modernity in Glasgow, capturing the sense of explosive dynamism,
complexity and friction implicit in Francis Newbery's metaphor of civic life as an
industrial forge. The notion of forging modernity also resonates with the transformation of Glasgow in the second half of the nineteenth century into an industrial powerhouse of the British Empire, a transformation as ideologically fraught
as it was economically empowering.[2]

All great western cities in the nineteenth century were experiencing
change at an unprecedented level, but in Glasgow it was felt in a peculiarly
acute and intense way. The city's spectacular growth entailed the constant
renewal and extension of the built environment, along with the creation of
physical infrastructure and transport networks to facilitate the expansion of its
industrial capitalism. During his visit to Glasgow in 1857, the American
novelist Nathaniel Hawthorne was alternately amazed and appalled by the

14.1
**Muirhead Bone,**
*The Forge,*
**etching, 1901**

spectacle of the city. His account vividly captures the sense of simultaneous growth and decay, and the shocking dislocations, both social and visual, that modernisation engendered.

> ... my wife and I walked out, and saw something of the newer portion of the city; and really I am inclined to think it the stateliest city I ever beheld. The Exchange, the other public buildings and the shops, especially in Buchanan-street, are very magnificent; the latter, especially excelling those of London ... Later in the forenoon, we again walked out and went along Argyle-street, and through the Trongate and the Saltmarket. The two latter were formerly the principal business streets, and, with the High-street, the abode of the rich merchants and other great people of the town. The High-street, and still more the Saltmarket, now swarm with the lower orders, to a degree which I never witnessed elsewhere; so that it is difficult to make one's way among the sallow and unclean crowd, and not at all pleasant to breathe in the noisomeness of the atmosphere. The children seem to go unwashed from birth, and perhaps they go on gathering a thicker and thicker coating of dirt until their dying days.[3]

With few planning, environmental or fiscal constraints Glasgow's prodigious growth and congestion continued unabated, creating staggering extremes of wealth and poverty that impacted on the visual and intellectual culture of the city. For the city-dweller the atmosphere of intensified psychic and visual stimulation could be alternately depressing, threatening or exhilarating, and by the very nature of their profession architects and designers were in the thick of it. Psychologically and socially, design skills could be manipulated to reimpose meaning and coherence on the fluid conditions of city life, and to mask the reality of its more unpalatable aspects.

On the one hand, the stylistic assurance, originality and theatrical panache of work by architects like Thomson and Mackintosh seem in tune with the buoyancy of Glasgow's economy, its internationalism, and the climate of thrusting, competitive individualism that pertained in the second half of the nineteenth century. On the other, attempts to impose aesthetic coherence on so many different facets of urban experience can be read in relation to a growing sense of fragmentation and alienation, and as a strategy for reconciling the disruptive potential of the social, economic and technological forces at work. For Alexander Thomson there was 'nothing more repugnant or humiliating than decay', and the task of the modern architect was to counter and transcend the degenerative aspects of the city through the architectural expression of 'divine harmonies' and 'imperishable thought', valuing 'the suggestions of progress, leading upwards into the light of the future.'[4] Likewise Charles Rennie Mackintosh encouraged his contemporaries to clothe 'in grace and beauty the new forms and conditions that modern developments of life – social, commercial and religious' insisted upon.[5]

Glasgow's idiosyncratic manifestations of architectural modernity were bound up with the particular context of the city, its distinctive institutional, commercial and industrial formations, the combination of market forces to which it was subject, and the specific character and form of the local bourgeoisie. The basic contextual framework of the Scottish city was different in detail from its English counterpart. Following the Union with England in 1707, Scotland still maintained a different law, established church, educational system and banking system. Whether considered geographically, racially, socially or spiritually, Glasgow also differed from Scotland's other three cities, despite their shared national history. It was this specific urban context that shaped progressive design in complex and sometimes contradictory ways.

## Expanding mental horizons

To be a middle-class Glaswegian in this period of intensive capitalist expansion was to participate automatically in a cosmopolitan and imperial culture. The

steady stream of colossal ships and locomotives that poured out of the yards along the Clyde provided a tangible metaphor of the rail and sea network that bound Glasgow into the bigger world picture. By 1900 approximately half the world's new shipping tonnage was Clyde-built. Through an advantageous coastal location Glasgow commanded a vast international market, and mental horizons were constantly stimulated by the two-way traffic of artefacts and people through the city. A phenomenal range of industrial skills and products, from textbooks and iron temples for India to carpets for Australia were being exported from Glasgow in the late nineteenth century. Like any other international commodity, fine and decorative arts from overseas were retailed in Glasgow with entrepreneurial flair. The city was famed for its numerous department stores that developed from the 1830s. Alongside these large emporia existed a small but sophisticated network of dealers and decorators which had been fostering an appreciation of continental and oriental artworks among the city's cognoscenti.[6] For designers, manufacturers and retailers, increasing their share in world markets required both aggressive canvassing for sales and staying abreast of political change and technological developments abroad. Daniel Cottier (1837–91), interior designer, stained-glass artist and picture dealer *extraordinaire*, was justly described as someone who 'thought in continents'.[7] By the mid-1870s his business operated between Glasgow, London, New York, Sydney and Melbourne, backed up by a further network of agents in Paris, Vienna and The Hague. Likewise the Blackie family of publishers, who commissioned distinctively modern architecture, interiors and graphics from local designers, had branches in England, North America, India and Australia.

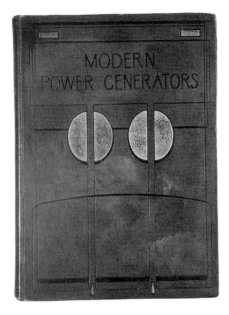

14.4
**Book cover designed by Talwin Morris for Blackie's**

The climate of internationalism arising from Glasgow's trade and industry was amplified by cultural reverberations from ancient political alliances with Scandinavia and Europe. Over many centuries Glasgow's architecture had developed with a strong European flavour. This tradition was bolstered by links with the École des Beaux Arts in Paris, where a large proportion of the British students between 1850 and 1890 came from Glasgow. Marked similarities with Canadian and North American cities were also far from coincidental. Apart from trading connections, the pattern of extensive emigration from the west of Scotland over several generations created extended family networks. Leading architects and engineers frequently looked to American models.[8] With its grid layout, Glasgow qualified as a metropolis in the American or continental sense. This was in marked contrast to London, which the German architect–critic Hermann Muthesius considered 'an immense village', unplanned and haphazard.[9]

From 1851 Glasgow designers and companies, small and large, participated keenly in the many international exhibitions of the period. Cutting-edge design from the city was published and exhibited in centres from Chicago and Melbourne, to Brussels, Budapest, Moscow and Helsinki. On home territory the ethos of voracious imperial expansion was demonstrated in a series of hugely popular and financially successful international exhibitions, which brought a sudden influx of visitors and foreign imports to the city, showcasing Glasgow to the world and the world to Glasgow.[10] The spectacle of James Sellars' exotic 'Baghdad by the Kelvin' in 1888 was followed in 1901 by the largest such event ever held in Britain. Well over eleven million visitors attended, more than double the number that attended London's Great Exhibition in 1851. While historicist in inspiration, James Miller's icing-cake style of architecture on this occasion was a model of prefabricated construction and administrative coordination. More self-consciously modern designers like Mackintosh, George Walton, Jessie King and E.A. Taylor were involved in various smaller displays, but in terms of progressive design the star of the show was undoubtedly the highly coloured Russian Village designed by Feodor Shekhtel, who approached Mackintosh to exhibit in Moscow the following year.[11] Such were the cultural spinoffs of a lucrative trade in timber between Scotland and Russia.

In the well-established tradition of 'the enterprising Scot' many Glaswegians sought work abroad, driven by ambition, evangelism or economic necessity.[12] Among them was George Thomson, Alexander's older brother and partner in the architectural firm, who departed for the Cameroons as a missionary in 1870; Thomson also had relations in North America. An engineering professor teaching in Tokyo arranged an extensive Glasgow–Japan cultural exchange in 1878.[13] In addition to emigration, independent foreign travel for business, education or leisure was becoming commonplace among the

middle classes. For many architects and their associates it was routine to consider a holiday, study or work abroad. On a student scholarship established in the name of Alexander Thomson, Mackintosh could afford an Italian trip in 1891. Later he would oversee exhibitions of his work in Vienna in 1900 and Turin in 1902, and was in regular contact with an international *coterie* by means of letter and telegraph. Before the outbreak of war he and Margaret Macdonald contemplated moving to Vienna, though they finally settled in the south of France. Friends, clients and admirers from far afield also sought the couple out in Glasgow.

14.5
**General view of the Russian Village, designed by Feodor Shekhtel, at the 1901 Glasgow International Exhibition**

By the opening years of the twentieth century this level of networking between international art centres led to talk of a Scotto-Continental School of Design.[14] The mystical symbolism, linear severity and molten qualities of continental art nouveau evidently appealed to many Glasgow Style architects and designers. In particular Mackintosh, Herbert MacNair and the Macdonald sisters (known as The Four) had no compunction about distorting the natural qualities of wood, and developed a highly charged symbolist iconography in their work. In a 1916 survey of British Art Schools *The Studio* magazine noted that Glasgow was alone in employing foreign teaching staff such as the Belgian symbolist Jean Delville, Maurice Greiffenhagen and Eugene Bourdon.[15] Architects in Glasgow evidently knew their English architecture and were abreast of critical debates in the south, not least through the many eminent designers and architects that had migrated south yet maintained links with Scotland (Christopher Dresser, J.J. Stevenson, Bruce Talbert, John Moyr Smith and George Walton to name but a few). Conversely there was a steady flow of eminent English visitors to the city, including William Morris, Charles Ashbee and Walter Crane, many of whom produced designs for Glasgow firms and lectured at the School of Art. From the viewpoint of Glasgow, England was in many senses just another component of Europe. Francis Newbery, himself an Englishman, was struck by the open response of his students to all outsiders: 'To have been born in Glasgow was neither considered of special merit nor a particular recommendation, nor was the welcome anything the colder because a man, other than a Scotch man, was working in Glasgow, as the result of accident or migration.'[16]

Glasgow's design culture exemplified the imaginative freedom, the compressed sense of time and space, that characterised modernity. Thomson never traveled abroad, but this did not prevent him from transcending the visible expanse of Glasgow and dreaming his way into outer space or distant, bygone worlds. 'Philosophers, in explaining the nature of light and endeavouring to give us some idea of the rate at which it travels,' he wrote in 1859, 'tell us that some stars are so distant that, although they may have been created thousands of years ago, their light may not yet have reached us; or that if it were possible for us to fly off into space, we might, as we retire, survey backwards, as it were, all the events that have happened on the planet – that we might, by going a sufficient distance, witness the very first act of its creation.'[17] The monuments of ancient Egypt, India, Assyria, Greece and Rome were all part of his mental geography. A generation later Mackintosh could engage on a sophisticated level with the architecture and design of Japan without ever having been there. Trying to fathom the influences in George Walton's work for Miss Cranston at her Buchanan Street tea-room in 1897, Edwin Lutyens exclaimed 'There is tradition of every country and I believe planet! of the universe – yet 'tis all one.'[18] Such global, even interplanetary, references were all enacted in Glasgow through the local.

## Designing the city beautiful: commerce and culture

The 1901 International Exhibition demonstrated the assurance and practical dynamism of a municipal vision which united industry, art and science. At this point Glasgow's continued economic and cultural progress seemed unstoppable. Physically the form of the city spread ever outwards, and at its periphery surrounding boroughs were enveloped. In the 1891 revision of its municipal boundaries Glasgow had officially mopped up an additional half million citizens; to keep this colossal organism under control required a highly developed municipal machinery. This notion of the city state filled a vacuum in the mental life of the bourgeoisie created by the lack of an aristocracy. Although united by a strong class-consciousness, the middle classes in Glasgow were a large and increasingly stratified group, ranging from a spectacularly wealthy elite to a mass of foremen, clerical workers, shopkeepers and their families at the lower end of the scale. Design played a significant role in expressing their shared values and articulating their positive identification with the city. The industrial and mercantile elite who exerted political control literally built their ideology into the city's infrastructure and institutions through a vigorous programme of improvements which tackled water supply, sanitation, housing, health, arts provision and education. In 1902 *The Times* noted that Glasgow was 'more responsible than any other town or city in the UK for the spread of the various forms of municipal progress which have been developed in the new municipalisation.'[19] The tremendous civic pride in such achievements was paralleled by an aggressive display, in public and private, of individual wealth and cultural prowess, following the precedent set by the city's mercantile elite in the eighteenth century. Competitiveness and materialism characterised public acts of philanthropy or involvement in the arts as well as business affairs. The bourgeois ideal in Glasgow contained no perceived conflict between commerce and culture, nor between promoting self-interests and the greater good of the community. Like the great merchant princes of Italian city-states in the renaissance, Glasgow's elite aimed to excel politically and economically – and culturally.

Citizens were proud of Glasgow's distinct economic and cultural apparatus which functioned independently from national and upper-class power structures. In this respect progressive architecture was certainly in tune with commercial attitudes. 'Fortunately Cockneydom is not Britain,' declared one of Thomson's clients in a lecture to the Glasgow Architectural Society in 1866.[20] As a gas-fittings manufacturer the author went on to express pity for Londoners who had to tolerate 'the most clumsy and ill-fitting gas appliances in their shops and houses – gas so foul and coarse, and street lamps of such imperfect construction, as would not be tolerated in any third-rate town in Scotland'. Without the deadening hand of national institutions such as a Royal Academy or national galleries and museums imposing the stamp of their cultural authority on artistic activity in the city, it was easier for artists and designers to find their own level in the marketplace. No one could accuse Thomson or Mackintosh of replicating London fashions, or of aping aristocratic tastes. While not to everyone's taste, the self-evidently bold and progressive qualities of the new architecture clearly appealed to clients of independent, cosmopolitan and civic-minded outlook, matching their commercial competence and confidence. With occupations and incomes largely dependent on external market forces, a dynamic, assertive and competitive spirit was to the fore. As Francis Newbery observed, 'The business man buys what he likes, or is persuaded to like, or because it pleases him ... Commercialism neither lays down a rule nor demands the following of a tradition. All that is asked is, that the productions of the artist shall be comprehensible to the commercial mind.'[21]

The commissioning of design and craft in both the workplace and the home operated at a largely individual level. Choices in both spheres were often made on the basis of personal recommendations by family, friends and associates. Competitive tendering and committee-based decisions did not become the norm until well into the twentieth century, with the result that similar networks of design patronage spanned public and private life. Consistency of image and status projected across both spheres was important, as demonstrated by the patronage of Miss Cranston and various members of the Blackie family who commissioned houses from both Thomson and Mackintosh. This continuity was strengthened by the importance of business-related entertaining within the home, which was more pronounced in Glasgow than in a city dominated by the professions like Edinburgh.

The modern churches, warehouses, schools and tearooms around the city were public spaces in which progressive style exerted an appeal across the spectrum of the middle classes, both men and women. Glasgow's 'artistic' tearooms, for example, were closely identified with the tenor of life in the city: 'It is not the accent of the people, nor the painted houses, nor yet the absence of Highland policemen that makes the Glasgow man in London feel that he is in a foreign town and far from home', remarked 'J.H. Muir' in 1901; 'It is a simpler matter. The lack of tea shops.'[22] These spaces were renowned for being homely, and it is significant that so many of them were established and run by women. Most public interiors were commissioned by groups of men, but Glasgow's tearooms brought women's artistic patronage and influence out of the confines of domesticity into the visibility of the public arena. Miss Cranston was without doubt the Glasgow Style's most generous and consistent patron. Her chain of tearooms was one of the sights of the city, where thousands of ordinary people had access to a vision of urban chic, fantasy and modernity. Through their startling poster-art and the spectacle of modern pageants, Mackintosh's generation also made more temporary interventions in the cityscape. There has been a tendency to overlook these more ephemeral yet public expressions of modernity, but they formed an important element within a range of creative practice that was rooted in Glasgow's commodity culture and the projection of a distinctive urban identity. In particular the Glasgow Style pageants gave visible expression to an abstract combination of myth, history and allegory, enacting a generalist view of knowledge and civic culture, and showing the interrelationship of art, science, industry and design in the municipal enterprise.[23] These community rituals in allegorical form reaffirmed the concept of the city beautiful, providing a vehicle for aesthetic and political idealism that looked simultaneously to the distant past, the present and the future, to the local and the international.

Design and craft skills were at a premium, applied to both the building and enhancing of the city beautiful, and in the support of a diverse range of industries. Many people's livelihoods depended on an ability to discriminate efficiently in matters of design. As a result of the commercial and industrial profile of the city the bourgeoisie were sensitised to the importance of design. Economic historians point to the interdependent and intensively skill-based nature of the Glasgow economy. Its backbone – ships, locomotives, heavy engineering and textile production – spawned numerous smaller ancillary industries. The kitting out of ships and trains, for example, had ensured the development of a highly skilled and specialised furnishing industry. Although operating on a colossal scale, such industries relied upon a high degree of skilled, labour-intensive specialisation rather than standardised mass production. Each ship was a one-off triumph of art, craft and industry combined. On both a practical and symbolic level there was scope for designers to demonstrate their imaginative transformation of these locally available technologies, materials and skills. Even those architects who had little or no involvement with the city's major industries found their practice locked indirectly into the local skill base. There was an abundance of prototyping expertise, for example, that facilitated the translation of drawings into three-dimensional forms. Architects like Thomson and Mackintosh were able to collaborate with local craftsmen and manufacturers on the production of ironwork, textiles, furnishings, stencilling, plasterwork and ceramics, and to combine these elements in complete schemes detailed down to the last door handle.

The many types of design activity that flourished in the city, whether related to architecture, the fine and decorative arts, engineering or industrial design, shared an emphasis on drawing skills and the conceptual development of three-dimensional forms on paper. Glasgow had a distinguished and diversified tradition in design education dating back to the establishment of the Foulis Academy in 1755, and architects were part of a wider community of designers who were daily to be seen clutching drawing materials on the underground, trams and buses, bound for a draughtsman's office or classes at the School of Art or the Glasgow and West of Scotland Technical College. The latter was more technically oriented than the School of Art, but the two institutions collaborated in the teaching of common areas like architecture and furniture design, encouraging the development of a shared sensibility. Ernest Archibald Taylor (1874–1951), who had connections with both institutions, made the transition from working as a draughtsman in a Clyde shipyard to designing furniture, stained glass and interiors.

The section of society to which modern design appealed was geared to the large-scale production of artefacts. As the international exhibitions so powerfully demonstrated, pride in the city's unrivalled feats of engineering was common to all classes. The skill and precision of the engineer was fundamental to the perception of craft in Glasgow, in contrast with the craft ideal of William Morris and his followers which focused on vernacular and manual traditions. To most Glaswegians the Forth Bridge, completed in 1889, was a marvel and a thing of beauty; to Morris it was 'the supremest specimen of all ugliness'. Large-scale industrial items – ships, cranes and locomotives – were all part of the cityscape and the spectacle of urban living. The river seethed with activity and waterborne traffic, and almost every day locomotives would process through the city to the docks for export. The value attached to technological progress and industrial skills was expressed both in Thomson's abstract and monumental architecture and in the aggressively modern, sleek and stylised forms of Mackintosh furniture. Even where the choice of materials and construction was not technologically innovative, on a visual and more abstract level, Glasgow's modern architecture and interiors alluded to the smooth finish associated with industrial forms, or evoked the formal classical language of engineering.[24] Glasgow's industrial wealth was founded on the manipulation of metal rather than wood, and the fabric of the city was characterised by stone, not timber and brick. The surrounding area contained rich resources of iron and coal, whereas the limited wood stocks in the west of Scotland had been depleted by the rural iron industry in the early nineteenth century.[25] A heightened perception of the qualities associated with metal – a material which could be bent, punctured, welded and moulded with great precision – pervades Glasgow Style furniture and interiors with their streamlined fluid forms, the use of neutral colours, punched motifs and decorative riveting.

Georg Simmel, an academic living in Berlin at the time, was fascinated by the impact of urban living on the mental life of the individual. 'Punctuality, calculability, exactness are forced upon life by the complexity and extension of metropolitan existence', he remarked in 1902. 'The same factors which have coalesced into the exactness and minute precision of the form of life have coalesced into a structure of the highest impersonality; on the other hand, they have promoted a highly personal subjectivity.'[26] The modern architecture and design of Glasgow expressed both of these tendencies in visual, sensual terms. It is difficult, if not impossible, to prove a causal connection between such abstract concepts and style, particularly as such connections would often be made indirectly and subconsciously. Nevertheless the suggestion seems reasonable that designers as sensitive as Thomson and Mackintosh absorbed, synthesised, and on some level translated into physical form Glasgow's collective civic priorities.

# Instability, class conflict and filth

Unlike the middle ranks of society in Edinburgh, the Glasgow bourgeoisie were a precarious volatile group lacking the old indicators of land and family connections. Amid such a great density of diverse people without shared backgrounds or values, visual indicators of status and difference assumed particular importance. To bolster a fragile sense of self the middle classes invested heavily in objectification of their identities. This emphasis on individualism, and on material culture as a physical expression of economic achievement, had been firmly rooted in Glasgow's culture since the enlightenment. The city's economic prosperity was rooted in control of property, capital and labour, which in turn lent itself to dramatic objectification in terms of expenditure on fashionable goods and material possessions. 'The sense of vision would seem to be consulted in the decoration of the homes of people at the expense of all other senses,' noted one guidebook in 1859, 'and in this there is growing rivalship … in a thriving population where the genius of trade and manufacture was continually creating material wealth.'[27] The competitive furnishing of Glasgow homes continued unabated, and nearly fifty years on the English architect Edwin Lutyens picked up on the same qualities of overstatement and theatricality in his description of a visit to the newly opened Buchanan Street Tearooms in 1897. He recounted how James Guthrie had taken him 'to a Miss Somebody's who is really a Mrs Somebody else. She has started a large Restaurant, all very elaborately simple on very new school High Art Lines. The result is gorgeous! and a wee bit vulgar! It is all quite good, all just a little outré, a thing we must avoid and shall too.'[28] While appreciative, Lutyens clearly felt more at ease with the qualities of restraint, gentility and unassuming good taste that were widely perceived as characteristic of English design. Muthesius also commented on this essential cultural difference, with the observation that the Englishman had no urge to impress: 'he even avoids attracting attention to his house by means of extravagant design or architectonic extravagance, just as he would be loth to appear personally eccentric by wearing a fantastic suit.'[29]

Glasgow's unrestrained free market economy was characterised by large-scale capital investment and financial fluctuation. Fortunes were rapidly made and lost. The monumentality of Thomson's architectural vision reflected a desire to instil such urban experience with a sense of moral probity and fixity. Following his death, however, the degree of risk and economic instability in the Glasgow economy was dramatically underlined by the notorious bank crash of 1878, in which thousands were ruined. Because of the integrated nature of the economy, any snags in the wider trade cycle had a widespread knock-on effect. The note of anxiety this introduced to the general optimism was a unifying aspect of the middle-class outlook that came to be echoed in the critical language of instability frequently applied to the Glasgow Style. Phrases like 'extravagance bordering on insanity', or 'lunatical topsy-turvydom', reflected a clearly perceived resonance between design and other areas of bourgeois experience. [30]

There was also increasing disquiet at the appalling social consequences of rapid urbanisation, and the way these impinged on the daily lives of all citizens. By 1900 Glasgow was one of the most intensely urban environments in the world, with more people living nearer the heart of their city than in any other comparable metropolis. At the bottom of the social scale wages were low compared to the national average, and seasonally erratic. In 1887 Glasgow had the highest death rate in the United Kingdom, the highest number of persons per room, and the highest proportion of the population occupying one-apartment homes.[31] This new scale of human abasement created industrial tensions ripe for inflammation, and an unprecedented rise in crime, pollution, prostitution and disease. The middle classes were not immune to the recurrent, life-threatening epidemics that swept the city throughout the nineteenth century. Thomson, for example, lost four children to cholera within the space of three years in the 1850s. The essential fragility of the municipal vision was underlined in 1901 when outbreaks of smallpox and bubonic plague threatened to undermine the massive public relations exercise of the city's great International Exhibition. The spectre of the lower classes – the apparent source of such disease, homicidal violence and moral depravity – loomed large in the middle-class consciousness.

A siege mentality was intensified by the cramped and high-density pattern of building in Glasgow compared with other cities, which meant that different social classes were forced to live cheek-by-jowl. Paradoxically the city's labour force and slums were centrally, inescapably located in the functional heart of the city. In this shared territory the regulation of class interaction was more difficult than in Edinburgh, for example, where the bridges between the Old and New Towns provided a clear physical demarcation. The physical proximity of an underclass in Glasgow served to heighten middle-class anxieties, but also to bolster their sense of group identity. Relations with other social groups were perhaps the most significant element of middle-class value formation.

In search of greater amenity, the middle classes steadily shifted their homes to developments in the west and south of the city. Thomson built tenements terraces and villas in these areas, himself migrating south of the river Clyde, to Moray Place in Strathbungo. Following their marriage in 1900, Mackintosh and Margaret Macdonald had set up home in Main Street near the city centre where the drizzling greyness, the filth and the smog all took their toll. Writing to Anna Muthesius in 1903 Macdonald confided, 'We in Glasgow have been having a terrible winter. It has been most depressing. For two months we have had the gas lighted nearly the whole of the day – the fog has been so thick and black it is just like night. Day after day – it becomes most depressing & it is so very bad for one's eyes, trying to work always by gaslight. We have been trying to find some place we might go and live, so that we should be out of it next winter.'[32] It took them until 1906 to move to the more salubrious and healthy heights of Hillhead in the West End, an area buffered by the University, Kelvingrove Park and the new Art Gallery and Museum, though even here the thudding heartbeat of the city's Clydeside industry could be heard.[33]

On one level the white interiors created by Mackintosh and Macdonald in their new home can be read as a manifestation of an intensified middle-class obsession with health and self-preservation. Viewed in relation to Mary Douglas's classic definition of dirt as 'matter out of place', these interiors made a statement about matter being effectively controlled, and threatened boundaries being redrawn.[34] The advent of modern environmental controls has made it difficult to recreate the starkness of the contrast between the blackened, smoggy reality of the streets and the Mackintoshes' drawing room in Southpark Avenue, described by one visitor in 1905 as 'amazingly white and clean-looking. Walls, ceiling and furniture have all the virginal beauty of white satin.'[35] To preserve such an oasis of light and calm in a grimy, sooty city was an uphill struggle. Nevertheless the constant physical battle against dirt not only helped keep disease at bay, but provided a material analogy for spiritual and sensual values. All references to history, nature and reality were processed and transformed through the modern and artistic temperament. Non-verbally, the uncomfortable forms, impractical colours and self-conscious use of cheap materials in the Mackintoshes' interiors spoke of mind over matter, the suppression of material need through selective inhibition. Social anthropologists like Douglas have pointed to the direct correlation between the growth of urban, industrialised societies on the one hand, and on the other an increased sense of interiority, and a heightened awareness of the human body.[36] The controversial iconography derived from anorexic-looking human figures that was pioneered by the Macdonald sisters in the 1890s powerfully communicated a sense of the pressures on the individual, and on middle-class women in particular.

Design helped to choreograph human interaction and behaviour, and its impact could produce physiological symptoms, as suggested in the following account of entering an 'artistic' household in Glasgow. It was written by the novelist Catherine Carswell (1879–1946), a former student at Glasgow University who moved in social circles centred around the School of Art:

> Even to herself Joanna did not admit how nervous she became in the Lovatts' house. But from the moment the front door closed between her and the street, there was always a tightening of all her nerves. As she passed through the square entrance hall, so unlike any other known to her, with its black-tiled floor, bright blue carpet, and white walls hung with black-framed etchings, her very muscles would

stiffen a little with the involuntary effort which these decorations seemed to demand. In the same way the rooms, though they were neither so large as the rooms at Collesie Street, nor nearly so rich as Aunt Georgina's, imposed a peculiar restraint – these were evidences of a world in which Joanna did not yet move easily, a world where the small talk, like the material furnishings, had its own shibboleths of seeming freedom and simplicity.[37]

The 'seeming simplicity' of many Glasgow Style interiors was indeed elaborately constructed, and at a price. Although there was no flaunting of expenditure in terms of the materials and labour involved in the production of the furnishings, the overall effect could be incredibly labour-intensive to maintain. Like other middle-class households, the Mackintoshes were dependent on domestic servants to sustain their distinctive environment. Muthesius clearly saw the impracticality of their style, and that most people could not, or would not, want to live in this way. Even their most enthusiastic supporters found the Mackintoshes' aesthetic hard to accommodate in its entirety for their own homes, which is hardly surprising given the emphasis on the projection of each respective owner's individuality and taste in the discourse surrounding domesticity. The New Art was certainly more difficult to engage with in the context of intimate, one-on-one encounters in the home than in the relative anonymity of public spaces like tearooms, schools, exhibition spaces and warehouses, where the physical proximity to throngs of strangers in the public areas of the city heightened a sense of mental distance and freedom in the individual. This mental freedom was the essence of a truly metropolitan social life, and it was a freedom which modern architecture and interiors enhanced, providing simultaneously a critique and escape from what Thomas Carlyle saw as a 'murky, simmering Tophet'.

# The spiritual city: rediscovering 'the dear green place'

Glasgow's name derives from the ancient Gallic *glaschu,* meaning 'dear green place', but by the 1840s was Glasgow 'a green flowery world, with azure everlasting sky stretched over it, the work and government of a God', asked Thomas Carlyle, Scotland's dourly Presbyterian prophet, 'or a murky simmering Tophet, of copperas fumes, cotton-fuzz, gin riot, wrath and toil, created by a Demon, governed by a Demon?'[38] Such Old Testament rhetoric set the tone for a popular and persistent vision of Glasgow in which the seething, godless city was pitted against forces of nature. A similarly biblical tone infuses William Morris's reference to Glasgow as 'the Devil's Drawing Room', and Desmond Chapman-Huston's mystical description of the Mackintoshes' home in 1910: 'far away in that mist-encircled, grim city of the north which is filled with echoes of the terrible screech of the utilitarian, and haunted by the hideous eyes of thousands who make their God of gold. Vulgar ideals, and the triumph of the obvious, are characteristic of the lives of the greater proportion of its population; and yet, in the midst of so much that is incongruous and debasing, we find a little white home, full of quaint and beautiful things, with a big white studio.'[39] For modern designers around 1900 the task was to rediscover Glasgow as a 'dear green place', reconnecting on some level with a spiritualised, mythological past. An abstract concept of nature was non-controversial as a source of aesthetic inspiration, and was redolent of a symbolic return to origins.

The themes of evolution, pantheism, symbolism and metamorphosis which characterised progressive design thinking about the natural world were captured in a pageant entitled *The Birth and Growth of Art.* In the final scene, 'The Promise of the Present', figures symbolising the sister arts of architecture, painting and sculpture 'pay their tribute to their dear Mother Nature, and receive their inspiration at her hands. She stands to receive their allegiance under that mystic tree, whose fruit is the gift of the gods, the knowledge of good and evil.'[40] The event was written and directed by Fra Newbery (Head of the Art School) and staged by Mackintosh to mark the opening in 1909 of the newly completed School of Art building. In the introduction the heathen god Pan introduced his companion 'mighty Mother Nature' as the great leveller, an eternal principal uniting all, 'whether rural or city born'. Addressing present and future students of art and architecture he continued, 'Primeval Senses wake from hidden depths,/ And cry for freedom from a shuttered self./ Back to the Land!' In this context 'Back to the Land' entailed the cultivation of nature within the city, 'mid haunts of busy men', whereas for the Arts and Crafts architect Charles Ashbee, 'Back to the Land!' was a rallying cry for the departure of his Guild of Handicraft from London to rural Chipping Camden in 1902, an equally symbolic *rejection* of the modern city.

Newbery's mystical vision was shared by his friend, the social theorist Patrick Geddes. For Geddes a starting point for the creative renewal of urban social life was an historical and geological awareness of place that would enable city dwellers to tune into each city's unique life force. For those prepared to look, the physical evidence of Glasgow's ancient past was omnipresent. Significantly it was to the spiritual and imposing presence of Glasgow's medieval cathedral that Mackintosh turned as a student in 1890, at a time when Glasgow's legacy as a major medieval city of learning, and then as an elegant and vibrant mercantile centre in the eighteenth century, was being overlaid by the city's dramatic industrial expansion.[41] At the same time, in the process of laying the subterranean network of water tunnels and sewers and digging architectural foundations or railway cuttings, the stratification below the city was constantly laid bare, revealing traces of life extending back into geological time. At Moray Place, where a deep sewer was being excavated in about 1860, Thomson's elder brother George described how he and Alexander stopped to observe the trace of fossilised life, the track of an insect embedded in the clay.[42]

Connections with the Scottish land – economic, emotional and recreational – formed a significant element of urban thinking. By 1901 approximately forty per cent of Scotland's entire population was concentrated in the Clyde Valley. Not surprisingly the intellectual and cultural dominance of Glasgow transcended the immediately visible expanse of the city, with the tentacles of its communications network reaching into a huge hinterland spread around the Clyde and north into the Highlands and Islands. Towns in which Thomson and Mackintosh carried out work like Kilmacolm, Bowling and Helensburgh were easily accessible by train or steamer. For cities on the scale and complexity of Glasgow, Patrick Geddes coined the term conurbation to convey the idea of a city which embraced a region. It expressed a new, synoptic view of the city which brought together the previously polarised and distinct concepts of town and country. In a conurbation these elements formed a single organism which, to remain healthy, had to coexist in interdependent balance. Like any living organism the city would show signs of simultaneous growth and decay (as a one-time Professor of Biology at Dundee, Geddes's views were grounded in biology). The provision of public parks, of which Glasgow had more than any other British city, provided a necessary lung and recreational space for the city's inhabitants. On a clear day the hills to the north, snow-capped in winter, were visible, testifying to the closeness of this life-giving relationship. Since 1856 a municipal water supply, so vital to health and amenity, had been piped in from Loch Katrine in the Trossachs. The mighty river Clyde and its Gallic divinity Clutha (celebrated in the murky art-glassware of that name) were central to civic identity, representing both the source of Glasgow's wealth and industry, and the penetration of nature into the heart of the city's industrial culture.

Emotional ties to the land were reinforced by the extended family links of those who had moved to the city. Glasgow's population was in constant flux, and the majority was only one or two generations removed from rustic surroundings. Like other aspects of the middle-class mindset, however, this link with the countryside was fraught with ambivalence. Migration from the Highlands in particular fed the expansion of the industrial sector, and the natural drift of the rural population to the city had been intensified by the agricultural depression in Scotland and the Highland Clearances. For many, the move to the city meant freedom from pre-existing identities, a new start. The upwardly mobile were keen to express visibly their sense of difference from their extended families, and their firm identification with the bourgeois ethos of the city. Even when in the country, in their heads they were living in the city. This was the antithesis of cultural attitudes south of the border. Hermann Muthesius observed that 'In England one does not "live" in the city, one merely stays there.'[43] The last thing most Glaswegians wanted was to mobilise associations of rural living through the houses, possessions and clothes they chose. References to the country and vernacular traditions were important, but needed to be filtered through an urban sensibility. In this sense the Glasgow Style would have struck a chord in that it presented nature processed for the urban palate, with all folksiness and rough edges removed. Through art, nature was rendered modern and civilised, an adornment to the city.

Organicism, synthesis and balance, features central to the concept of the conurbation, were principles that also informed Glasgow Style design and architecture. The nature-based form-language developed through teaching at the Glasgow School of Art in the 1890s could be applied to a wide range of artistic, scientific, commercial artefacts and activities, and executed in varied scales and media. The regenerative, sensual power of nature offered an antidote both to the drab and deadening world of work, and to the tired historicism of mainstream taste. For Mackintosh, entering good modern architecture was 'like an escape into the mountain air from the stagnant vapours of a morass.'[44] Following the South Kensington model, the conventionalised abstraction of natural forms was an important feature of the design curriculum at the Glasgow School of Art. Students were also encouraged to follow Ruskin's injunction 'Go to Nature', by drawing plant-life in the School and sketching outdoors. Some, like Jessie King and Ernest Taylor, spent the summer on the Isle of Arran, a popular artists' haunt. In such designers' work, distinctive colours were distilled from the magnificent scenery around Glasgow – the heathery purples, misty greys, muted pinks and greens – but to express mystical harmonies, rather than the countryside as a place where a rural population lived and worked. This urban view of nature was in marked contrast with the cottagey celebration of rural life and vernacular skills embraced by many members of the English Arts and Crafts movement. To William Morris and his followers, the establishment of craft communities in idyllic English villages and small towns like Chipping Camden offered a retreat from the corruption of the city.

Rural life in Scotland was a rather different matter. Glasgow was within easy reach of sublime expanses of nature, but these had little to do with pretty villages, bosky dells and window-boxes. This was an era of agricultural depression, continued clearances, land reform agitation and mass emigrations. Apart from the frequently inhospitable climate, the whole pattern of agriculture was unlike that in England. Vast tracts in the west of Scotland in the late nineteenth century were characterised by a combination of large estates running sheep and deer, and sparsely scattered crofting communities. The crofting system was not just small-scale farming, but a unique way of life which had been threatened by the brutal dispossessions and forced emigrations of the

Clearances. Many evicted Highlanders not packed onto ships bound for Canada, America or Australia, converged on Glasgow. The 'crofting wars', bitter confrontations between the dispossessed and landowners which came to a head in the 1880s and prompted military intervention, were given prominent news coverage in Scotland. The bitterness and trauma of this episode in Scottish history, and the general severity of the nineteenth-century agricultural depression were not easily forgotten. It was difficult in the context of Glasgow to romanticise rural life. The evocation of vernacular cottage interiors was also inappropriate in terms of most Glasgow housing. Unlike English city dwellers, the majority of Glaswegians of all classes were literally separated from the land by virtue of living in tenemented dwellings (Mackintosh himself was born in a third-floor home). This pattern was the European norm. England's terraced and semi-detached housing was the exception, as Muthesius was quick to point out.[45]

Despite the strong strain of Arts and Crafts in the teaching at the Glasgow School of Art, few attempts were made to flee city life for good. For the most part Glasgow-trained designers like Mackintosh continued to live and work in the city, even though repelled by its squalid aspects. The Scottish Guild of Handicraft was an interesting but isolated and shortlived attempt to emulate Ashbee's enterprise in Scotland. As a co-operative it offered an alternative lifestyle and model of production, also questioning the political status quo through involvement with the new Independent Labour Party in Scotland. Mackintosh shunned such affiliations, and in keeping with the markedly 'individualistic art training' he received at the Glasgow School of Art,[46] exhorted his fellow art workers to look inwards and get on with doing their own thing: 'you must be Independent, Independent, Independent – don't talk so much but do more – go your own way and let your neighbour go his ... Shake off all the props – the props tradition and authority give you – and go alone – crawl – stumble – stagger – but go alone.'[47]

Protestantism, with its complex Scottish ramifications, was such an integral part of Glasgow's culture and history as to be inescapable, and the work of figures like Thomson or Mackintosh is incomprehensible outside this religious tradition. For Thomson the whole cosmos was made meaningful through his firm religious commitment, and an immanent sense of the divine permeated his lectures on architecture. While many of the following generation experienced a crisis of belief, religious values still shaped their architectural thought, their visual ideology. The artistic expression of something as vague as religious temperament is difficult to analyse, but tags such as 'puritan' and 'spiritual' were used both by the designers themselves and contemporary observers. Hermann Muthesius remarked on the peculiarly Scottish 'blend of the puritan and romantic', and Mackintosh himself spoke of the need to respond to the 'religious pressures of modern life', to express 'the ethereal, indefinable side of art'.[48] George

Logan, an active member of the Salvation Army, also described his work in spiritual terms, and incorporated angels, cherubs and fairies in his designs at every available opportunity.[49] His friend Jessie King, a minister's daughter, believed she was in touch with the 'little people'.[50] The figure of John Knox, crowning the Necropolis next to the medieval cathedral, was as important a feature of the Glasgow skyline as the cranes looming over the River Clyde. The spectacular Necropolis, established in 1836, expressed the visual and metaphorical dominance of the Church of Scotland and its adherents – Glasgow's great and good – who inhabited this city of the dead. Although Glasgow also had a large Catholic community, it was largely comprised of recent immigrant Irish or Highland workers who lacked the wealth and political clout necessary to impinge on the world of the Glasgow Style. Potentially the association of Celtic culture with Ireland, and hence Catholicism, was socially problematic, as evidenced in the violence that frequently attended matches between the city's leading football clubs, Celtic and Rangers.[51] Celticism formed an important strand within the Glasgow Style, but in a safely secularised form.

In common with other European centres in the late nineteenth-century, Glasgow had an active Spiritualist Church and Theosophical Society in which many artists and designers were actively interested.[52] One senses a world of association beyond the ornament and forms of the Glasgow Style, ranging from faint subconscious allusion to the full-blown symbolism which earned the early work of 'The Four' the nickname 'Spook and Ghoul School'.[53] The drooping plant forms and emaciated figures are half-dead; the willowy, waistless female figures evoke the inspirational rather than maternal, life-bearing role of women; the soaring attenuated verticals, the muted secondary colours and airy tones all reinforce the ethereal other-worldly aspects of the style; physical comfort and material wellbeing come low on the list of priorities.

In his book entitled *The Evolution of Sex* (1889) Patrick Geddes presented companionable love and cooperation as the highest expressions of the evolutionary process. The principle of 'equal but different', of gendered opposites integrated to form a unity of impression, was also central to theosophical thought. Yeats, a committed theosophist, felt that society was entering one of those rare periods of unity of being, and the mutual interpenetration of the sexes, which would find expression in a bisexual art that had the perfection of each partner without voiding the identity of the other.[54] In architectural terms such ideas found expression in the interiors of 128 Southpark Avenue, in which the Mackintoshes created a circuitous journey of the imagination, aestheticising the daily round of activities within the home in a series of visual and tactile experiences representing different facets of a spiritual transformation. The inhabitants move between the dark nether regions of the masculine preserve (the ground floor hall and dining room) and the heightened spirituality and lightness of the 'feminine' rooms above. Physically and metaphorically they ascend through the house into increasingly intimate spaces, ever more removed from the outer world. There is the sense of a steady casting off of vulgar sensation and the debris of the phenomenological world, starting with the symbolic divesting of outer garments in the transitional space of the hallway, and culminating in the unclothed intimacy of the white bedroom above.[55] Ultimately even this extreme expression of interiority can be seen as a response to the city, a withdrawal fostered by the intensity of social change and the pressures of urban life.

# Conclusion

The recurrent theme of this chapter has been to emphasise that the appreciation, commissioning, design and production of architecture and design are all shaped by the specific culture in which people live. Images, associations, accumulated memories –the very culture of a city – were all fed into the forging of Glasgow's distinctive brand of architectural modernity. The problems of tracking and conclusively defining these elements remain formidable, but little doubt can remain that modern designers, architects and their audience responded to shared patterns of meaning, patterns which resonated across different areas of sensory, emotional and aesthetic experience.

# Notes

1   Introduction to David Martin (1897:xiv).
2   The substance of this essay has been developed from an earlier article entitled 'Mackintosh and the City', which appeared in Kaplan (1996:30–61).
3   Stewart (1962:512).
4   Thomson (1859:41).
5   Mackintosh (1990:222).
6   For an overview of industrialist collectors and dealers see Cumming (1992:9-14); also Fowle (1991:108-11).
7   Wells (1902:145).
8   J.J. Burnet, for example, worked with McKim, Mead and White; Donald Matheson made a study tour of the USA in 1903 to research the remodelling of Glasgow's Central Station.
9   Muthesius (1979:7).
10   See Kinchin and Kinchin (1988) for a fuller account of these events.
11   See Cooke (1984:4-8).
12   See Calder (1986).
13   See Lovelace (1991); Checkland (1989).
14   Jennings (1902:61).
15   Taylor (1916:124–5).
16   Martin (1897:xxi).
17   Thomson (1874) in Stamp (1999:144).
18   Edwin to Emily, 1 June 1898, in Percy and Ridley (1985:57).

19    Chisholm (1902:13).

20    Johnson (1865–7:96–7).

21    Martin (1897:xix).

22    Muir (1901:176).

23    For example the *Masque of the City Arms* (1903), *The Masque of Science and Art* (1905), and *The Birth and Growth of Art* (1909). All were written by Francis Newbery and staged by staff and students at the School. See Kinchin (1998).

24    For a fuller discussion of the workmanship of risk and of certainty see Pye (1968:4–9).

25    Timber was mostly imported from Scandinavia, Russia and the West Indies by the late nineteenth century.

26    'Die Grosstadte und das Geistesleben' in Simmel, *Die Grosstadte: Vortrage und Aufsatze zur Stadteaustellung* (Dresden, 1902); extract reprinted in Harrison and Wood (1992:133).

27    Measom (1859:328).

28    Percy and Ridley (1985:50).

29    Muthesius (1979:10).

30    Jennings (1902:61).

31    Russell (1887:11).

32    Macdonald (1903)

33    Carswell (1920:87): 'From here she overlooked the whole park ... It was a little world to itself, shut in and stuporose. But beyond it to the south ... Joanna could see where the real world began. Nay she could hear, coming from the Clyde across all that distance yet as if it were the beating of her own heart, the dull steady pounding of the yards'.

34    Douglas (1966).

35    Kalas (1933:3).

36    This theme is taken up by Brett (1992), Ch. 4. In Carswell's (1989:31, 38) novel, Joanna (a student at the Glasgow School of Art) is described as 'a fugitive from the realities immediately surrounding her town existence, and her intenser life was lived in her flights'. 'In town Joanna led almost wholly a dream life. The indoor life.'

37    Carswell (1989:164).

38    Carlyle, quoted in Pagan (1847:105).

39    Mountjoy (1910:5), pseudonym of Desmond Chapman-Huston.

40    Newbery (1909).

41    There was a tremendous upsurge of interest in the cathedral, stimulated in part by Walter Scott's romantic description of it in his novel *Rob Roy* (1818). One of the leading figures in the campaign to conserve it was Honeyman with whom Mackintosh was working from 1889. In 1888 Mackintosh sketched the Cathedral Chapter House, and his design 'A Chapter House' was awarded a gold medal in the South Kensington competition of 1892.

42    George Thomson, unidentified letter to his nephew, Rev. James Parlane, in Thomson (1881:33).

43    Muthesius (1979:7).

44    Mackintosh (1990:222).

45    Muthesius (1979:8–9).

46    Taylor (1916:124–5).

47    Mackintosh (1990:223).

48    Mackintosh (1990:223)

49    Logan (1905:118).

50    For a fuller account of her work see Colin White (1989) and Jude Burkhauser (1990).

51    The Catholic 'Celtic' club was established in 1887. The simmering violence between the two teams came to a head at the 1909 Scottish Cup Final, when supporters from both sides invaded the pitch, set pay boxes on fire, and threw stones at the police and firemen.

52  Maurice Maeterlinck was particularly influential in this respect. Also Jean Delville (the Belgian symbolist artist who taught at the Glasgow School of Art from around 1901 to 1907) was deeply interested in Theosophy, as demonstrated in his book *The New Mission of Art* (translated into English in 1910). According to Clifford Bax, many of his Glasgow pupils followed him when he returned to Brussels. The Theosophical Society in Glasgow had five branches in the 1890s.

53  For readings of this imagery and symbolism see Helland (1996), Burkhauser (1990), Brett (1992) and Neat (1994).

54  Yeats (1937) and Gordon (1979).

55  Kinchin (1996:12–29).

# Chapter 15

# The expressionist utopia

*Iain Boyd Whyte*

Expressionist architecture positioned itself, wittingly and unwittingly, as a response to the city – more exactly, the industrial city. In contrast, say, to Britain, industrialisation in Germany came comparatively late. Among the great industrial enterprises Siemens only took off in the 1870s, Agfa was founded in 1872 and AEG in 1887. Starting late, industrialisation and its impact on the social and urban structures in Germany was terrifyingly rapid. As the most extreme condition, Berlin offers the best example of this process. In 1848 the population of the city was 423,000, rising to 774,500 in 1870, 1,900,000 in 1900, and almost 4 million in 1910: a tenfold rise in sixty years.

This population explosion had a predictable effect on the housing stock, as the two- and three-storey housing and corridor streets of the historic city were pulled down to make way for the mean, dark and insanitary Berlin *Mietskaserne* (rental barracks). Courtyard plans were favoured, with apartments lit and accessed from a series of internal courts that ran axially off the principal entrance on the street front. Three courtyards were common; seven were to be found in extreme instances. In the poorest housing the courtyards barely functioned as light-wells. In Berlin yards as small as 5.1 square metres were allowed before the building regulations were revised in 1887, and disease and contagion flourished within these dark wells, exacerbated by minimal sanitary provision. Within the local limits set for street lines and cornice heights, the individual house blocks were free to adopt for facades whichever decorative scheme appealed to the speculator or builder. While this visual free-for-all denied any possibility of creating a coherent architectural scheme on the larger scale, it gave a brash vitality to the city streets that matched the rapacious and self-confident individualism of the *Gründerzeit*, the twenty years of feverish speculation in the 1870s and 1880s. The human cost was high: in 1871, when the new German Reich was established after the Prussian defeat of the French at Sedan, average life expectancy in Berlin was 36.5 years for men and 38 years for women.

With Prussia as the leader of the new Reich, Berlin became the capital city, and Wilhelm I, King of Prussia, became Kaiser of the German empire. The new status of the city demanded appropriate architectural expression, and the Hohenzollern court favoured an extravagant neo-baroque for such

15.1
**Meyers Hof,
Wedding, Berlin,
c.1900**

public gestures as the National Monument for Kaiser Wilhelm I, built beside the Royal Palace in 1895 to the design of the sculptor Reinhold Begas, or for Julius Raschdorff's Cathedral, built in 1894 on the eastern flank of the Lustgarten. Not only was Berlin the undisputed centre of government after 1871, but also the centre for banking, business, transportation, and – in contrast with London or Paris – high-technology industry, both chemical and electrical.

The generation born around 1880 grew up in the fevered atmosphere of the *Gründerzeit*, which witnessed the growth of vast fortunes, massive industrial development and building booms, but also appalling poverty and exploitation. Although a simplification of the endless complexity of the creative process, it is possible and justifiable to see the artistic and intellectual production of the 1880s generation, as they reached maturity before the First World War, as a reaction against the misery and deprivation of the industrial city on one hand, and against the pompousness, vanity and stucco grandeur of the *Gründerzeit* on the other. Artists and intellectuals of all ideological persuasions, from extreme right to extreme left, saw the city in general and Berlin in particular as a festering sore on civilisation. Even Otto von Bismarck, who engineered the unification of Germany, could only condemn the capital city as 'a desert of bricks and newspapers', while the right-wing conservative Friedrich Lienhard coined the phrase 'Los von Berlin!' (Get out of Berlin!), which became the battle cry of the back-to-the-soil movement. From both political left and right came pseudo-Darwinist theories that prophesised the self-destruction of the metropolis. Wilhelm Riehl had already written in 1854 that 'Industrialisation will grow more and more vigorously and in the process and because of it, the modern world, the world of the metropolis will collapse, and these cities ... will remain as broken torsos'.[1] This theme of entropic destruction and collapse brought about by the excessive concentration of power and energy is central to the expressionist analysis of the city. The avant-garde literary journals of the immediate pre-First World War years such as *Der Sturm* and *Die Aktion* groan with apocalyptic premonitions, with titles like *Weltende* (the end of the world), *Die tote Stadt* (the dead city) and *Verfluchung der Städte* (the cursing of the cities). In painting, too, urban destruction was a favoured theme, most famously in Ludwig Meidner's series of apocalyptic landscapes painted in 1913, in which the *Mietskaserne* tumble into vast fissures in the fractured streets in a demonic earthquake of destruction.

Such pessimism seemed justified as the stalemate of the First World War and the Allied blockade brought starvation to the German cities, exacerbated by the terrible epidemic of Spanish flu in 1918, which carried off thousands of victims. The abdication of the Kaiser in November 1918, the German Revolution, and the murderous street fighting between the socialist and Spartacist left and counter-revolutionary right were all seen by the avant-garde as final evidence of the total moral collapse of a society whose institutions and ethics were irredeemably corrupt. Paintings like Georg Grosz's *Woman Slayer* or Otto Dix's savage images of city streets filled with prostitutes and cripples are symptomatic of this perception of the city as the topos of fear and menace. What was to be done against pessimistic reading of the city?

The simplest alternative was to flee. This strategy was favoured by the prosperous bourgeoisie, whose move out to the green suburbs was made possible by the spread of suburban railways in the 1890s. At the same time the bohemians and artists withdrew from the city to set up communes that espoused all manner of social panaceas ranging from teetotalism and vegetarianism to clothing reform and nudism. In the first decade of the new century Berlin was surrounded by communes like the Friedrichshagener Dichterkreis, the Obstbau- kolonie Eden near Oranienburg, and the Neue Gemeinschaft at Schlachtensee in the western suburbs of Berlin. These initiatives in turn spawned some of the most radical cultural movements of the period, including the Freie Volksbühne, which offered radical theatre to the working man, and the Deutsche Gartenstadt- gesellschaft – the German Garden City Society – modelled on its English precedent.

**15.2**
**Wenzel August Hablik,**
*The Path of Genius,*
**1918**

An alternative form of retreat was enacted entirely in the head. The rejection of the materialism of the *Gründerzeit* and of the positivist or mechanist explanation of the world led the expressionist generation into a private, subjective world of the spirit. Man was understood by the expressionist avant-garde not as an integral function of a large mechanism, but as an isolated individual. The images of single figures painted by Edvard Munch, with titles such as *Melancholia*, are symptomatic, and Munch exercised an enormous influence on his younger German contemporaries. The only true world, it was claimed, was the world of the spirit, and the spirit expressed itself in a multiplicity of ways: Nietzschean vitalism, Buddhism, Theosophy and the like. Tangible results that arose from this conviction were Rudolf Steiner's Anthroposophical movement, founded in 1912, and Wassili Kandinsky's manifesto for artistic abstraction, *Über das Geistige in der Kunst,* first published in the same year. As Kandinsky insisted in this text, the redemptive spirit was no longer the Christian God or Christ figure, but the artist. Friedrich Nietzsche lurks behind this insight. In his aphorism 'Wir Künstler' ('We Artists') in *Die fröhliche Wissenschaft,* he insists

It is enough to love, to hate, to desire, simply to experience … we are immediately gripped by the spirit and the power of the dream, and we ascend the most hazardous paths with open eyes and indifferent to all dangers, high onto the roofs and towers of fantasy, without any dizziness, as if born to climb – we sleep walkers of the day! We artists! We concealers of naturalness! We moonstruck and God-struck ones! We deadly silent, tireless wanderers on heights that we do not perceive as heights, but as our plains, our places of safety.[2]

The perfect visual echo of this position can be found in a painting by Wenzel August Hablik, *The Path of Genius* (1918), which shows the single, divinely talented artist climbing through the mountain peaks to the crystalline domes and towers of the New Jerusalem. Behind and below the artist, lesser mortals slump in resignation and despair, unequal to the task.

The yearning for rebirth and renewal took both temporal and spatial forms. Chiliastic dreams of a lost golden age in the German past led to Franz Marc's paintings of eagles, horses, woodcutters and the like: the residents of the primeval German forest. In parallel, the spatial and utopian dreams of a new and better place led to primitive and exotic societies and to works like Kirchner's wooden sculpture modelled on African and Polynesian antecedents. But while it is easy to paint utopian visions for the future, how were these to be achieved in architecture?

A key figure in answering this question is Bruno Taut. Born in 1880, he was the archetypal representative of the expressionist generation, with a seis-mographic sensitivity to the mood of the age. In 1904, for example, he wrote to his brother Max: 'I've read Zarathustra over the last three months – a book of enormous vitality. I've learned a lot from it'.[3] In the years leading up to the war Taut reacted to the consensual Wilhelmine society in an almost programmatic fashion. On one side he espoused decentralisation and the garden city ideals. On the other he argued for purity: pure form, pure colour, pure spirit. As the advisory architect to the German Garden City Society he played a significant role in the housing reform movement. A striking result of this engagement was the Gartenstadt Falkenberg, a garden suburb built on the eastern periphery of Berlin in 1913–14. Falkenberg offered the retrospective utopia, with small, brightly painted terraced housing set along traditional village street patterns in an attempt to regain contact with a lost, pre-industrial past.

For the abstract and more spiritually orientated solution, we can look to the famous Glashaus at the 1914 Werkbund Exhibition in Cologne. This small pavilion gave tangible form to Taut's vision, which was simultaneously published as text in Herwarth Walden's journal *Der Sturm*, of a *Gesamtkunstwerk* (total work of art) in which architecture would join with abstract painting and

15.3
**Bruno Taut, 1916**

15.4
**Bruno Taut, Glashaus at the Werkbund Exhibition, Cologne, 1914**

sculpture to produce an inspirational alternative to the grey facades of the city. In Taut's words,

> Let us work together on a magnificent building! On a building that will not simply be architecture, but in which everything – painting, sculpture – will combine to create a great architecture, and in which architecture will once again fuse with the other arts. The architecture here will be simultaneously both frame and content.[4]

The Glashaus was dedicated to the utopian novelist and poet Paul Scheerbart, who composed the aphorisms incised into the architrave between the supporting drum and dome. As one of his aphorisms insists, 'Glass gives us the new age: brick culture only hurts us'.[5] The fantastic translucent worlds that Scheerbart described in his astral novels found a modest realisation in Taut's pavilion. Under the glass dome was a multicoloured room of glass, accessible via a staircase of glass bricks. At ground level was a water cascade illuminated by a mechanical kaleidoscope. Taut saw in the pavilion 'a well-founded expectation that the eyes and sensibilities of the people will be won for more subtle stimuli. In today's architecture we desperately need to be freed from the saddening omnipresence of stereotyped monumentality.' Through visual stimuli the world was to be educated away from the imperatives of profit towards the world of the spirit. Such expectations, of course, were profoundly historicist. As the title page of Taut's brochure to the pavilion insists, 'The Gothic cathedral is the prelude of glass architecture'.[6]

Taut's aesthetic position was grounded, perhaps unwittingly, in the Kantian model, according to which the purpose of art is not to produce objects of knowledge or use, but rather to research the limits of human possibility. The purpose of art, accordingly, is simultaneously critical and prophetic. Its goal is an ideal human state that might possibly be achieved at some point in the future. The favoured symbol of lost innocence, simplicity and purity, which was to be recovered in the post-industrial age, was the crystal. Here the expressionists consciously linked themselves to a continuum that stretched from Nietzsche's *Zarathustra*, Richard Wagner's *Parsifal*, the German romanticism of Novalis and Goethe's *Faust*, back to the medieval mysticism of the likes of Schwester Hadewich, and ultimately to the biblical Revelation of St John.

Tangible models for the ideal society were drawn from the European gothic and from the orient. Both appear in Taut's book *Die Stadtkrone* (The City Crown), in which images of a harmonious world – be it medieval Strasbourg or oriental Tschillimbaram – are arranged around the great temple or cathedral that acts as the spiritual focus for the whole society. The idea of a city crown appears again and again under various guises in the literature and the utopian drawings of expressionism. Taut himself developed the theme further in a folio

15.5
**Bruno Taut, Glashaus,
internal stair**

SCHNEE
GLETSCHER
GLAS

10

15.6
**Bruno Taut, *Alpine
Architektur*, 1919**

of annotated drawings entitled *Alpine Architektur*, drawn during the First World War and published in 1919. These visionary images of glass cathedrals set high in the Alps and of cosmic constructions of coloured glass speeding through the eternal night were conceived both as a protest against the insanity of the war and as a pointer to a better society, which would devote its energies to peace and understanding rather than self-destruction. As with the Glashaus at Cologne, Paul Scheerbart was the godfather for Taut's vision of a universe too rich and complex to be comprehended by reason alone. Only naïve wonder – the basis of the aesthetics of the sublime – could promote the development of higher forms of understanding. This position, of course, also had implications for artistic production. As one of the characters explained in Scheerbart's novel *Das Paradies, die Heimat der Kunst*

> The concept of art is broadened here; the main issue is no longer representation but rather the invention of things that might be represented ... Our task is not the representation of comprehended perceptions, but a reorientation of comprehended perceptions. In this way we want to make it possible to comprehend new perceptions. You could call this the preparatory work for the artists who come later.[7]

Taut's crystal temples strung from the Alps to the Himalayas had exactly this function. As he admitted in one of the texts inscribed on the drawings,

> The execution will involve incredible difficulties and sacrifices, but will not be impossible. 'The impossible is so rarely demanded of Man' (Goethe) ... But higher knowledge! The greatest work is nothing without the Sublime. We must always recognize and strive for the unattainable if we are to achieve the attainable.[8]

With the end of the war in November 1918, the abdication of the Kaiser and the socialist revolution, Taut and his associates had high hopes that some of their dreams might be realised through the political power of the Workers' and Soldiers' Councils. To this end they established the Arbeitsrat für Kunst (Working Council for Art) in December 1918, with a radical programme for cultural reform based on the benign dictatorship of a self-proclaimed artistic élite. These heady hopes, however, proved to be a chimera, and by the spring of 1919 the attempt to transform the bourgeois democracy into a socialist democracy on the Soviet model had failed. At this point Taut and his associates withdrew into a more private world of utopian and didactic speculation. Walter Gropius, for example, established the Bauhaus at Weimar in April 1919, with a

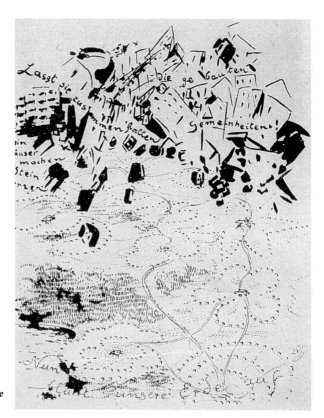

15.7
**Bruno Taut,** *Die Auflösung der Städte* **(The Dissolution of the Cities), 1920**

crystalline cathedral drawn by Feininger on the cover of the founding manifesto, and a programme based very closely on the ideas of the Arbeitsrat für Kunst. Taut himself dreamed of a utopian socialist future based on ownership of the land and mutual aid. His analysis derived not from Marx or Engels, but from Proudhon, Kropotkin and Gustav Landauer. The progressive force in society, according to this thesis, was not the industrial proletariat, but the enlightened artist, working for social reform and a return to the land. This was the theme of Taut's next book, *Die Auflösung der Städte* (*The Dissolution of the Cities*, 1920), in which he proposed – echoing the theories of 1900 – that the cities were doomed and the only hope of salvation lay in a network of self-governing, autonomous communes spread across the countryside. This analysis reflected the trauma of the Treaty of Versailles, which stripped Germany of its merchant fleet and much of its industrial base.

On the astral plane, always present as the complement to the rural and decentralist position, Taut began work in 1919 on his 'astral pantomime' *Der Weltbaumeister* (published 1920). In thirty scenes, Taut depicted the voyage of a gothic cathedral into paradise, where it transformed itself into a crystal palace. At the same time, he founded the *Gläserne Kette* (crystal chain), a group of like-thinking architects and artists, whom he urged to write and draw their

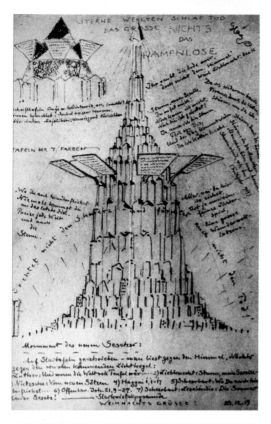

15.8
**Bruno Taut, *Monument to the New Law*, December 1920**

15.9
**Wenzel August Hablik, *Fantasy*, 1920**

visions for the future. Over its year-long active existence the group conducted a fascinating exchange of ideas and images, at the centre of which stood the imperative of a redemptive glass architecture. As the name *Gläserne Kette* suggests, the crystal was the favoured symbol of renewal – the basic, unalterable and irreducible inorganic structure that forged a link with the architecture of the cosmos. The second main argument of the group centred around organic meta-phors, which offered an alternative symbol of primeval, irreducible form, this time in the organic realm. An important source here was Ernst Haeckel, whose popular and inexpensive series of images of single-cell animals and plants had appeared around the turn of the century under the title *Kunstformen der Natur* (*Art Forms of Nature*). These jellyfish-like images of protozoa and protophytes clearly inspired the drawings made around 1920 by Hans Scharoun and Hermann Finsterlin, both mainstays of the *Gläserne Kette* group, and can also be identified in the post-war sketches of Erich Mendelsohn and in his most famous work of this period, the Einstein Tower at Potsdam, built for the physicist Albert Einstein in 1920–24. As a monist Haeckel had argued that all natural phenomena on earth, both animal and organic, are dependent on the same set of natural laws, and the expressionist infatuation with the crystal and with the single cell parallels the contemporary advances in atomic physics, which pointed towards a pure, irreducible source from which all matter proceeds, both organic and inorganic. It is somehow appropriate that the only great work of built archi-tecture to come out of expressionism should be the Einstein Tower, a research laboratory devoted to the atomic particle.

Even as the Einstein Tower was still under construction, however, the expressionist moment had passed. Writing in April 1920 to the *Gläserne Kette*, Bruno Taut wrote that he was 'finished with intuitive works' and wanted to get his feet back on the earth again. A principal reason for this dramatic change of course was the inescapable fact that the formal devices and motifs of the expressionist avant-garde had been taken up by mainstream culture. This is always anathema to the avant-garde spirit, which fights forever on two fronts at once: against the academic establishment on one hand, and against the vulgar taste of the mass, consensual culture on the other. As soon as the avant-garde gesture is adopted for mass consumption the radical impulse is lost, and the avant-garde is forced to stake out new territory and adopt ever more radical positions. This is precisely what happened with the crystals and glass of expres-sionist architecture, which by 1922 had become the very latest design mode, especially in Berlin. A crystalline advertising hoarding appeared on the Avus freeway, for example, designed by none other than Wassili Luckhardt, one of the former members of the *Gläserne Kette*. In a similar spirit Walter Würzbach and the sculptor Rudolf Belling revamped the Skala Tanzpalast in Berlin in an orgy of zig-zag profiles and crystalline forms, which extended even to the ground plan.

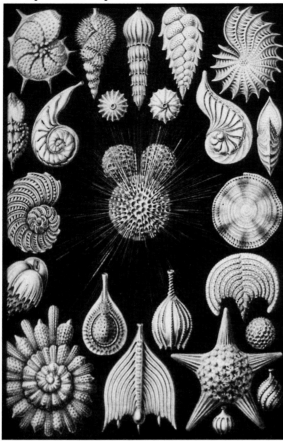

15.10
**Ernst Haeckel, Protozoa,**
**from *Kunstformen in der***
***Natur* (*Art Forms in***
***Nature*), 1899**

Taut concluded *Die Auflösung der Städte* with a poignant farewell
to utopia:

> Can one draw happiness? We can all live it and build it. Utopia? Is
> not utopia the 'certain' and the 'real', swimming in the mire of illusion
> and indolent habit! Is not the content of our aspirations the true
> present, resting on the rock of faith and knowledge![9]

This conscious return to the tangible, concrete and comprehensible
marked one of the most significant turning points in the evolution of architectural modernism. The elements of fear, uncertainty and self-overcoming that
were central to the aesthetics of the sublime were wilfully abandoned in favour
of the mechanical and physical certainties on which functionalist modernism
was grounded. A simplistic view of human existence as something reducible to
the optimal provision of shelter, light and air excluded, by definition, any
attempt to portray absolute greatness, power or awe, either in nature or in architecture. Only in the fledgling German film industry did the aesthetic strategies of

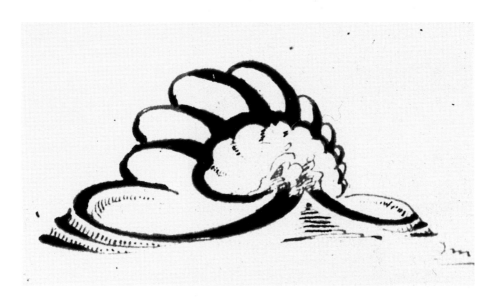

15.11
**Erich Mendelsohn, project
for a garden pavilion, 1920**

15.12
**Erich Mendelsohn,
Einsteinturm, Potsdam, 1920**

the sublime – formerly the domain of the architectural avant-garde – survive, another example of the ever-present risk of assimilation by the mass culture. With expressionism in retreat among the architects there was a brief stylistic vacuum, which was ultimately filled in 1921–2 by the new gospel of functionalism that was arriving in various guises from Holland, France and the Soviet Union.

In the historiography of modernist architecture the victory of functionalism over the emotionally charged architecture of expressionism was seen as a triumph of reason over irrationality, of light over dark. As Sigfried Giedion insisted in *Space, Time and Architecture*, 'Faustean outbursts against an inimical world and the cries of outraged humanity cannot create new levels of achievement'.[10] But devoid of the emotional charge of power, awe, mystery and fear on which expressionist architecture thrived, the white architecture of the modern movement fell prey to the anodyne certainties of the sociologist and the economist. Indeed, the very comprehensibility of the white modernism of the international style – the antithesis of the expressionist sublime – might be seen as a weakness. No uncertainties remained about the contours or dimensions of a building, the repeatability of its elements, its materials, or its economies in space and materials. With the powers and limits of imagination unchallenged, the result was ennui. As Ernst Cassirer noted in the context of Kantian aesthetics, 'Within phenomena themselves the infinite complexity which every organic natural form possesses for us points to the limit of the power of mechanical explanation'.[11]

# Notes

1    Riehl (1851–69).
2    Nietzsche (1906:59).
3    Bruno Taut, letter to Max Taut 8 June 1904; quoted in Whyte (1982:85).
4    Taut (1914 a:174–5).
5    'Das Glas bringt uns die neue Zeit; Backsteinkultur tut uns nur leid.'
6    Taut (1914b:n.p.).
7    Scheerbart (1893:170).
8    Taut (1919: Pls 10 and 21).
9    Taut (1920:30).
10   Giedion (1949:418).
11   Cassirer (1981:346–7).

# Chapter 16

# Walter Benjamin's Arcades Project
## A prehistory of modernity

*David Frisby*

In his prehistory of modernity, Benjamin's intention of reading the nineteenth century as a text that speaks to us in the twentieth should not be taken to imply that a restricted hermeneutic interest lay behind the project. The reality of the nineteenth century was presented to itself as a phantasmagoria, as a dream world, a world of illusions, a mythical world. It was a particular form of 'reason' that would 'clear the entire ground and rid of it of the underbrush of delusion and myth. Such is the goal here for the nineteenth century'.[1] The recognition and subsequent destruction of that dream world was undertaken with the purpose of our awakening through remembrance of the hidden past. Benjamin was impressed by one of the young Marx's aims of 'waking the world ... from its dream about itself'. Like Marx, Benjamin came to realise that this was no easy task for even the most critical method. Benjamin's starting point was the 'profane illumination' of surrealism which confronted 'the world distorted in the state of resemblance, a world in which the true surrealist face of existence breaks through'.[2] Like the work of Aragon, Breton and others it used the city of Paris as its focal point; it was both historical and critical, and not prepared to celebrate the myths of modernity but to undermine them. Benjamin sought to reveal the dreams of the collectivity wherever they were housed – in the arcades and other 'dream houses' – through the process of awakening. As a historical project this meant the unification of awakening and remembrance: 'indeed, awakening is the exemplary instance of remembering: the instance in which it is our fortune for us to recall the most immediate, most banal, most nearby things. What Proust meant by the experimental rearrangement of furniture in the half sleep of early morning, what Bloch recognised as the darkness of the lived-out moment, is nothing other than what is to be secured here and collectively, at the level of the historical.'[3]

Benjamin investigated 'the phenomenal forms of the dreaming collectivity of the nineteenth century'. His critique started not with that century's mechanism and maschinism, but with its narcotic historicism, its craving for masks, a hidden signal of historical existence first recognised by surrealism. 'The present investigation is thus concerned with the deciphering of this signal'.[4] What was masked were material relations under capitalism. For Benjamin 'capitalism was a natural phenomenon with which a new dreaming sleep came over Europe and within it a reactivation of mythical forces'.[5] Capitalism's objects, its technology and, above all, its commodities and social relations, were enveloped in illusions. Whether in the public or private spheres, individuals were surrounded with mythical, illusory phenomena, to the point at which 'collective consciousness sinks into an ever-deeper sleep'. To restore what had passed and create a historical consciousness of what was now occurring required a process of awakening. It required 'the new dialectical method of historiography [which] presents itself as the art of experiencing the present as the waking world, to which each dream, which we term that which has existed, actually relates'.[6]

This was to be achieved by 'the dialectical penetration and the rendering contemporaneous of past constellations'. These past configurations lay in the primal landscape of the arcades, in the phantasmagorias of the panoramas, in the materials of construction themselves (iron and glass), in monuments to transitory ends (railway stations and the like) in the whole world of the commodity (world exhibitions, department stores, fashions), down to the most trivial objects that filled the *intérieur* of the nineteenth century. Above all they lay in the city of Paris around the middle of the century. The whole of 'the explosive material that lies in what has passed'[7] in that city and century was to be brought to the point of being set alight.

This could only take place on the basis of illuminating knowledge of that past and that city, which would enable those interested to penetrate the series of labyrinths that both contained. Benjamin pointed metaphorically to two aids to this task. There exists, he argued, 'an ultraviolet and an ultra-red knowledge of this city, neither of which allow themselves to be confined to the book form: the photo and the city plan, – the most accurate knowledge of the individual element and of the totality'.[8] The deciphering of the 'secret signs' of the dream world of the nineteenth century and its objects was to culminate in the rapid dialectical image. In other words 'it is not the succession from one piece of knowledge to another that is decisive, but rather the leap into each individual element of knowledge itself'.[9] In contrast, the ultra-red knowledge of Paris was perhaps to be gained by the archaeologist who proceeded from a plan. What Benjamin intended was nothing less than a new topography of Paris in the nineteenth century, the excavation of the site of the prehistory of modernity.

The archaeologist of modernity was to investigate the labyrinths of modernity within the Parisian arcades (even the 'catacombs in the arcade'), within the city itself and beneath the city in its underworld of real catacombs. The construction of a topography of the city was essential to his task of producing the dialectical image of antiquity within modernity. The labyrinths of the great modern cities, their most hidden aspect, represented the realisation of the labyrinth of antiquity. This was one of the key features of modernity itself. If Paris in the nineteenth century was the city of arcades that housed the 'mythology of modernity' whose secrets the surrealists had penetrated, then so too, in his projected study of Baudelaire, Benjamin announced that he would deal with 'Paris as *the* city of modernity', as the location for Baudelaire's own 'fresco of modernity'. This study was to develop 'the sublating process by means of which antiquity comes to light in modernity, modernity comes to light in antiquity'.

In his earliest notes to the Arcades Project, Benjamin's intention was to seek out the 'mythological topography of Paris', as Aragon had done earlier on a more limited scale for the arcades with their 'whole fauna of human fantasies, their marine vegetation'. More than this, he recognised the 'affinity between myth and topography, between Pausanias and Aragon (Balzac to be included).'[10] Paris had been rendered not merely mythical but also ancient. This had been achieved a century before Aragon by Balzac, whose *Comedie humaine* represented 'something like an epic record of tradition'[11] and who had secured 'a mythical constitution for his world only through its distinctive topographical contours'. For Balzac,

> Paris is the soil of his mythology, Paris with its two or three great bankers (Nucingen, etc.), with the doctors who appear time and time again, with its enterprising merchant (César Birotteau), with its four or five great courtesans, its usurer (Gobseck), its several military offi-cers and bankers. Above all, however, it is always the same streets and nooks, chambers and corners from which the figures of this circle appear. This means nothing else than that topography is the contour of any mythical sphere of tradition, indeed that it can be its key, as it was for Pausanias in Greece, as the history, situation and distribution of Parisian arcades will be the underworld, sunk in Paris, for this century.[12]

This reference to Pausanias suggests an ancient model for the kind of topog-raphy of Paris in the nineteenth century that Benjamin had in mind.

This ancient model is significant in that 'Pausanias wrote his topog-raphy of Greece in the second century AD as the places of worship and many of the other monuments began to fall into ruin'. Balzac, Aragon, and now Benjamin all gazed upon 'the ruins of the bourgeoisie' in a context in which 'with the upheaval of the market economy, we begin to recognise the monuments of the

bourgeoisie as ruins even before they have crumbled'; in which 'the development of the forces of production had turned the wish symbols of the previous
century into rubble'.[13]

Pausanias's *Guide to Greece*[14] exhibits a distinctive approach to his
subject matter – the ruins of ancient Greece and their mythologies – that is highlighted by Frazer's delineation of Pausanias's work as that of one who

> interested himself neither in the natural beauties of Greece nor in the
> ordinary life of his contemporaries. For all the notice he takes of the
> one or the other, *Greece might almost have been a wilderness and its
> cities uninhabited or populated only at rare intervals by a motley
> throng who suddenly appeared as if by magic* ... and then melted away
> as mysteriously as they had come, leaving the deserted streets and
> temples to echo only to the footstep of *some solitary traveller who*
> explored *with awe and wonder the monuments of a vanished race*' (my
> emphasis).[15]

His topography is that of someone who 'loves to notice the things,
whether worshipped or not, which were treasured as relics of a mythical or
legendary past'.[16]

Benjamin, too, examined the Parisian arcades as primeval landscapes, the city as 'a wilderness', uninhabited except for mythical entities as in
Meryon's remarkable illustrations of the city. In his study of Baudelaire,
Benjamin took such an image of the city with its decrepitude as antiquity. In
contrast, modernity was Pausanias's 'motley throng', the masses. For Baudelaire
'Paris stands as the true indicator of antiquity, in contrast to its masses as the
true indicator of modernity'. Within those masses the man in the crowd experiences the shock of sudden confrontation; Pausanias's 'solitary traveller' is
perhaps the *flâneur* in the metropolis in search of the lost aura of civilisation's
monuments. Pausanias himself could wander around the monuments of ancient
Greece amongst the dead ruins and rubble that still retained a connection with a
mythical past. Indeed, his reconstruction of such mythologies often commenced
with a deserted ruin. Benjamin saw the threshold of mythology in the modern
city in a similar manner:

> In ancient Greece, one was shown places from which the descent into
> the underworld was made. Our waken existence, too, is a land in
> which at hidden places the descent to the underworld commences,
> fully insignificant places where dreams come into their own ... The
> labyrinth of buildings in the city resembles consciousness in broad
> daylight: the arcades (they are the galleries which lead into its past

existence) terminate in the daytime unnoticed in the street. But at
night time, beneath the dark masses of buildings their dense darkness exudes menacingly and the late passer-by hastens past them.[17]

Benjamin saw architecture as embodying the latent mythology of
modernity, embodied in 'the large and small labyrinth' of the metropolis and the
arcade. Beneath them lay 'another system of galleries that extend underground
through Paris: the Metro, where in the evening the lights glow red that show the
way into the Hades of names'.[18] Yet another lay in the catacombs beneath the
city and in the old bed of the river Seine. All represented the labyrinth of antiquity and mythology, and all of these topographical layers of the city required
excavation. Benjamin saw his aim as being 'to build up the city topographically,
ten times and a hundred times over out of its arcades and its gates; its cemeteries
and brothels, its railway stations … '[19] More specifically, part of the planned
Baudelaire book was to be devoted to 'Paris as *the* city of modernity … it brings
the decrepitude of the city into the open insofar as it regards this city as decor.
The poet Baudelaire moves around in it as a play-actor'.[20] This powerful image
of the city as decor was realised in Meryon's etchings of Paris: 'Meryon's Parisian
streets are shafts high above which the clouds pass by'.[21] A not dissimilar image
of the city was evoked by Pausanias centuries earlier: the city of empty, often
ruined, buildings that still held the key to antiquity. This was also Benjamin's
assumption, but for him the city was simultaneously the key to modernity as
well.

'The ultimate and innermost affinity of modernity and antiquity
reveals itself in their transitoriness', which in Baudelaire's work is reflected in
'the frailty and decrepitude of a major city'.[22] Above all, Benjamin detected in
the *correspondences* and allegories of Baudelaire's poetry, drawing their inspiration from mid-nineteenth century Paris, precisely this relationship between
antiquity and modernity. He judged 'the correspondence between antiquity and
modernity' to be 'the sole constructive historical conception in Baudelaire.
Through its frozen armature any dialectical conception was excluded.'[23] The
decrepitude of the city of Paris manifested itself in its drabness: 'the new drabness of Paris … just like the drabness of men's attire, constitutes an essential
element in the image of modernity.'[24] At the same time, however, this modernity
evoked its opposite. It arose out of the fact that 'Baudelaire never felt at home in
Paris. Spleen lays down centuries between the present and the just lived through
moment. Spleen it is which produces the inexhaustible "antiquity". And, in fact,
for Baudelaire modernity is nothing other than the "newest antiquity"'.[25]
Baudelaire, whose fund of images were derived from modern life, drew the
connection between antiquity and modernity in the form of an allegory. This
allegory 'holds fast to the ruins. It offers the image of frozen unrest.'[26] Emancipation from antiquity lay in the allegory; a real emancipation was impossible since

antiquity was so closely bound to modernity. In this sense modernity was both 'anticlassical and classical. Anticlassical: as opposition to classical. Classical: as the heroic achievement of time which stamps its impression'.[27] Yet the allegorical treatment of antiquity was also forced into its opposite: 'the experience of allegory which holds fast to the ruins, is really that of the eternally transitory'.[28] In this respect, 'the image of frozen unrest that allegory represents is ultimately a historical one'.[29] Modernity's constant assertion of the ever-new could not prevent its collapse into the ever-same. It too would decay, its monuments fall into ruin, even when the monuments of modernity least expect it. Modernity, Benjamin asserted, 'possesses antiquity like a nightmare that creeps over it in slumber'.[30]

Such reflections led Benjamin to conclude that 'it is very important that in Baudelaire modernity appears not merely as the hallmark of an epoch but as an energy by means of which modernity is immediately related to antiquity. Amongst all the situations in which modernity makes its appearance, its relationship to antiquity is an outstanding one.'[31] This is as true of Baudelaire's attempt to capture the heroism of modern life in all its forms as it is the fleeting beauty of modernity in the great cities. In relation to the former, 'nothing comes closer to Baudelaire's intention than in his century to give the role of the ancient hero a modern form';[32] the latter is expressed in modernity's 'opposition to antiquity, the new in opposition to the ever-same (modernity: the masses; antiquity: the city of Paris)'.[33]

However significant Baudelaire and his work were in providing a focus for Benjamin's Arcades Project, they by no means exhaust the scope of his project or contain the fullness of his dialectical images of modernity. The archaeologist of modernity might take inspiration from Baudelaire's topographical work, the collector might examine the refuse assembled by Baudelaire's *chiffonier*, and the *flâneur* might recognise himself in Baudelaire's portrait, yet the notes which constitute the Arcades Project are testimony to a much wider conception of the prehistory of modernity. Benjamin's researches extend beyond Baudelaire and the mythological topography of Paris to dialectical images housed within the city and within the nineteenth century itself: the city and its monuments, arcades and streets; the masses as crowds, consumers and revolutionaries; commodities in the arcade, the department store and the exhibition; commodities' ever-new faces as revealed in fashion and advertising; images in panoramas, photography, mirrors and lithography; lighting; human types like the *flâneur,* the idler, the gambler and the prostitute; individual figures such as Fourier, Saint-Simon, Hugo and Marx as well as Baudelaire; changes in historical experience and historical movements; dimensions of modern experience; the changing role of art and the artist – and this does not exhaust the themes which Benjamin's notes suggest he intended treating. At different stages in his Arcades Project Benjamin saw many of them coalescing in his treatment of

Baudelaire. The project of a prehistory of modernity and its dialectical images has a wider scope, perhaps one that was too wide to encompass within a single work. Nonetheless, it is possible to illuminate some of the dialectical images of modernity that this wider project contained.

The labyrinth of city streets, the city's architectural monuments, the masses who populate the city, the world of commodities and its illusions, the illusory retreat from that world in the *intérieur* and the illusions of historical tradition – these are some of the most important of these images. To traverse these labyrinths is to become aware not merely of the dream world of the nineteenth century, but of the changes in perception and experience that were their counterpart. The labyrinths were to be illuminated not merely through the topographical vision derived from the ultra-red knowledge of the city plan, but also through the shocking image derived from the ultra-violet knowledge of the photograph. Benjamin likened the activity of the materialist historian assembling images of the past to someone operating a camera, who is interested not merely in the inverted reality of the photographic image of bourgeois society – as that society wishes to see itself, but in what the camera actually produces, negatives in which what is light is dark and vice versa. Such a person can choose a close-up of the fragment or a 'larger or smaller extract' from the whole, 'a harsh political or a filtered historical lighting' for the images. Such images or fragments affirm the discontinuity of the past that is handed down to us, the fragments of the oppressed, repressed in order that history may appear 'as the continuum of events' whilst hiding the fact that 'the continuity of tradition is illusory'. This implies nothing less than the fact that 'the conception of discontinuity is the foundation of genuine tradition'.[34]

The significance of Benjamin's metaphor of the camera and its images is not an isolated reflection on his method. The Arcades Project was explicitly concerned with the production of images of modernity in the nineteenth century, not merely in art forms such as Baudelaire's poetry but also in the concrete sense of changes in images of the city brought about by their architectural transformations – and the construction of the arcades, Haussmann's rebuilding of the centre of Paris, and the Paris Commune's destruction of some of the city's monuments are of prime significance. Benjamin was concerned with the transformation of perception and experience in the artistic realm. This included the decline of what he termed 'auratic' experience, primarily though not exclusively of art works. It also consisted of a concrete examination of the production of images themselves – in photography, lithography and mirrors, in forms of lighting, and in building materials such as iron and glass. For Benjamin, technology was never reducible to 'the mastery of nature'. New techniques transformed the objects of perception and human beings' relations with them. The study of all these things was an essential prerequisite for the investigation of the mythical dream world of the nineteenth century, and out of these images, these fragments of the past, could step the future. Benjamin suggested for instance that we search the early photographs of individual subjects 'for the tiny spark of contingency, of the Here and Now, with which reality has so to speak seared the subject, to find the inconspicuous spot where in the immediacy of that long-forgotten moment the future subsists so eloquently that we, looking back, may rediscover it'. Photography reveals the secrets of motion by telling us 'what happens during the fraction of a second when a person *steps out*'.[35] Each fleeting moment could now be made to endure. In fact, 'a touch of the finger now sufficed to fix an event for an unlimited period of time. The camera gave the moment a posthumous shock, as it were.'[36] Baudelaire's response to the early daguerreotype was that it was 'startling and cruel'. This shock element was crucial to Benjamin's account of the transformation of modern experience. But photography could also reveal something else, namely 'the physiognomical aspects of visual worlds which dwell in the smallest things, meaningful yet covert enough to find a hiding place in waking dreams, but which, enlarged and capable of formulation, make the difference between technology and magic visible as a thoroughly historical variable'.[37]

The smallest, often unregarded, things are precisely what interest Benjamin. Enlarged, they take on a new significance. Benjamin praised Atget's photographs of Paris, which never took as their subject-matter 'great sights and the so-called landmarks', but seemingly insignificant aspects of the city's streets. Atget's streets, like those of Pausianas, are almost always empty: 'The city in these pictures looks cleared out, like a lodging that has not yet found a new tenant.' Such photography's virtue lies in the fact that 'it gives free play to the politically educated eye, under whose gaze all intimacies are sacrificed to the illumination of detail'.[38] Benjamin's own 'politically educated eye' had earlier cast its gaze over his native Berlin after his visit to Moscow in the winter of 1926–7. His image of Berlin was one shared at times by Kracauer, a city not merely 'cleared out' but cleaned too. As Benjamin expressed it, 'For someone returning home from Russia the city seems freshly washed. There is no dirt, but no snow, either. The streets seem in reality as desolately clean and swept as in the drawings of Grosz ... Berlin is a deserted city ... Princely solitude, princely desolation hang over the streets of Berlin. Not only in the West End ... they are like a freshly swept, empty racecourse on which a field of six-day cyclists hasten comfortlessly on'.[39] Like Kracauer, Benjamin turned to Paris as a city that retained for him an image of the human labyrinth which constitutes the city. The 'most vivid and hidden intertwinings' of people he discovered – so Benjamin informs us in *A Berlin Chronicle* – in Paris

> where the walls and quays, the places to pause, the collections and
> the rubbish, the railings and the square, the arcades and the kiosks,
> teach a language so singular that our relations to people attain, in the

solitude encompassing us in our immersion in that world of things, the depth of a sleep in which the dream image waits to show the people their true faces.[40]

By 1935 the illuminations of Paris in the Arcades Project had moved into a less rhapsodical phase, impelled by an explicit concern for social history, Marxism, and 'new and radical sociological perspectives'.

The materialist physiognomy of Paris was to proceed from the topographical layers of illusion to reveal the true face of the city and of modernity in images. Nineteenth century Paris was, for Benjamin, the location of modernity and its images. At its centre were the arcades, the collective 'dream houses' that included the panoramas and the dual aspects of the streets, the symbol of modernity in Haussmann's destruction of old Paris and 'the home of the collectivity' of the masses whose own symbol was the barricade. The threat of social movement expressed in the barricades strengthened a process that was already underway: the more rigid separation of public and private spheres and a retreat into the *intérieur*. The masses constituted the dark side of modernity (as in Baudelaire's image of the 'sickly population'): they could, as crowds, form a veil through which the *flâneur* and the idler passed; they could constitute a revolutionary movement; and as consumers they constituted an advantageous mass to those concerned with the circulation and exchange of commodities. Especially from the mid-1930s onwards, Benjamin asserted that the key to modernity lay in the fetish character of the commodity, writing in 1938 that 'the fundamental categories' of the Arcades Project would 'be in agreement with the determination of the fetish character of the commodity'.[41] The commodity had already revealed its secret life in the arcade; it was to take on a more public role in the department stores and world exhibitions. Its endless transformations were made possible by its ever-new face, enhanced by fashion and advertising. As a key to the experience of modernity, Benjamin hoped to link this ever-new face with the frozen historical world of the recurrence of the ever-same. The consequences for modern experience were outlined in *Some Motifs in Baudelaire* under the rubric of the shock of the ever-new.

The notion of the secret life of the commodity had first been revealed to Benjamin in Aragon's account of the Passage de L'Opera in which the arcade, which was then falling into disuse and had already taken on the mystique of the archaic, contained a dream world of a past age. Benjamin sought to go beyond the mythology of Aragon's treatment of the arcade through the investigation of its historical and sociological foundations in the early nineteenth century. Benjamin quotes a contemporary Parisian guidebook which describes the arcades as

a new contrivance of industrial luxury … glass-covered, marble-floored passages through entire blocks of houses, whose proprietors have joined forces in the venture. On both sides of these passages, which obtain their light from above, there are arrayed the most elegant shops, so that such an arcade is a city, indeed a world in miniature.[42]

The preconditions for their emergence in Paris during the third and fourth decades of the nineteenth century were, Benjamin argued, 'the boom in the textile trade' and 'the beginnings of construction in iron' which was 'made use of for arcades, exhibition halls, railway stations – buildings which served transitory purposes'. The arcades were also 'the setting for the first gas-lighting' as well as indicative of the increased use of glass as roofing material. Economically they might be viewed as 'the temples of commodity capital', the forerunners of the early department stores (which also often displayed their wares beneath an elaborate glass canopy). Metaphorically, Benjamin described them as 'constructions or passages which have no outside – like the dream'.[43] Their capacity to function as dream worlds was enhanced by the arcade's affinity to 'the church nave with side chapels'.[44] The passer-by and *flâneur* could enter their quiet refuge from the streets dominated by traffic into an environment in which one of the two components of the street – trade and traffic – fell away. This suggests that 'what is really at work in the arcades is not, as in other iron constructions, the illuminations of the inner space but rather the subduing of the external space'.[45] This paradoxical relationship between *intérieur* and *extérieur* constituted the 'complete ambiguity of arcades: street and house'.[46] This 'ambiguity of space' within the arcade was enhanced by 'its wealth of mirrors which extended spaces as if magically and made more difficult orientation, whilst at the same time giving them the ambiguous twinkle of nirvana'.[47] For all these reasons, 'something sacred remains, a remainder of the nave of this series of commodities which is the arcade'.[48]

The arcade symbolised a storehouse of latent mythology, a more secret labyrinth within that of the city. Its entrance was a threshold to the dream world, originally to the 'fairy grottos' of the Second Empire, and with their decline to the 'primal landscape of consumption'. The arcade was an interior landscape, which even in its period of decline clung to its secret dream world. For Benjamin it represented not the mythology of modernity, but its prehistory:

> Just as Miocene or Eocene rocks in places carry the impression of monsters from these earth periods, so the arcades lie today in major cities like caverns with the fossils of a subterranean monster: the consumers from the pre-imperialist era of capitalism, Europe's last dinosaur. On the walls of these caves there grows as immemorial flora the commodity and, like the tissue of an ulcer, it enters into the most irregular connections.[49]

In this primal landscape of consumption 'the commodity hangs and forces itself unrestrained like images out of the wildest dream'.[50] This juxtaposition of disconnected exchange values, the early trace of commodity fetishism, has a different significance for Benjamin. The disorder is embedded in the dream world of the nineteenth century whose microcosm is the arcade. The 'landscape of an arcade' consists of an 'organic and inorganic world'. Amongst the former are 'the female fauna of arcades: whores, *grisettes,* old witch-like saleswomen, female second-hand dealers, *gantieres, demoiselles* – the latter was the name for female-attired arsonists around 1830'.[51] The 'souvenir is the form of the commodity in the arcades'.[52]

The heyday of the arcades was a short one, and Benjamin started out from Aragon's surrealistic image of them just as they were disappearing. Their decline had been under way for decades; the *flâneur* stepped onto the boulevards after Haussmann's reconstruction of the centre of the city. Benjamin gave as reasons for the decline of arcades 'broadened pavements, electric lighting, the ban on prostitutes, the fresh air cult'.[53] Just as significant, however, were changes in the lighting of the arcades.

As long as gas and even oil lamps were burning in them, they were fairytale palaces. But if we wish to think of the high-point of their magic then we imagine the Passage des Panoramas around 1870 when on the one side gas lights hung and on the other there still flickered the oil lamps. The decline began with electric illumination. But basically it was not really a decline, rather more accurately an abrupt transformation'.[54]

There followed 'the epoch of forms and signs', and their names remained a filter of our knowledge of their past. The centre of commodity display moved elsewhere. This transformation signified for the arcades that 'at a single blow they were the hollow form from which the image of "modernity" was cast. Here, the century smugly reflected its absolutely newest past'.[55] The absolutely new had fallen into decay, and the glitter of the commodity shone elsewhere.

The arcade was not the only repository of the dream worlds of the nineteenth century that Benjamin sought to illuminate. They were also housed in the early dioramas and panoramas, sometimes located within the arcades, whose preparation 'reached its peak just at the moment when the arcades began to appear. Tireless efforts had been made to render the dioramas, by means of technical artifice, the *locus* of a perfect imitation of nature.'[56] Yet while they 'strove to produce life-like transformations in the nature portrayed in them, they foreshadowed, via photography, the moving-picture and the talking-picture'.[57] The dioramas and panoramas portrayed towns and cities from far away, landscapes, classical events, decisive battles and the like. They also gave the spectator a view of his or her own city, so that 'in the dioramas, the town was transformed into landscape, just as it was later in a subtler way for the *flâneurs*'.[58] Hence 'the interest in the panorama is in seeing the true city – the city indoors. The true is that which stands in the windowless indoors.'[59] This is also the case for the arcade. The panorama's effect, however, was produced by standing high up on a circular platform in an enclosed building in which 'the painting ran along a cylindrical wall, roughly a hundred metres long and twenty metres high'.[60] The development of lighting techniques enabled the configuration of images to change, from diorama to nocturama and the like.

The phantasmagoria of the city as interior landscape was paralleled by the emergence of a 'panorama literature', and anthologies of 'individual sketches which, as it were, reproduce the plastic foreground of those panoramas with their anecdotal form and the extensive background of the panoramas with their store of information ... They were the salon attire of a literature which fundamentally was designed to be sold in the streets'.[61] The world of bizarre figures displayed in these physiologies 'had one thing in common: they were harmless and of perfect bonhomie'. The menacing dimensions of the crowd in the city's landscape could be transformed in this harmless view of the world.

It was not merely the landscape of the city and its population that could be rendered harmless. The same could happen to history too, trapped in its own dream house: the museum. The thirst for the past could be controlled in the museum so that 'the inside of the museum appears ... as an *intérieur* elevated into a mighty person'.[62] As to its contents, 'there exist relations between the department store and the museum, between which the bazaar creates a mediating link. The massing of art works in the museum approaches that of commodities which, where they offer themselves to the passer-by in masses, awaken the notion in him or her that in them too a portion must fall'.[63]

Yet if the entrance to such structures as museums, arcades, panoramas and railway stations represented the threshold to the dream world of the nineteenth century, to the labyrinth of dreams, they all existed within the context of a more diffuse labyrinth of the street. Like the ambiguous dream-houses, the streets of Paris also exhibited a dual significance. They too could appear, at times, to be monuments to the bourgeoisie. Above all this was true from mid-century onwards of Haussmann's 'urbanistic ideal ... one of views in perspective down long street-vistas ... The institutions of the worldly and spiritual rule of the bourgeoisie, set in the frame of the boulevards, were to find their apotheosis. Before their completion, boulevards were covered over with tarpaulins, unveiled like monuments'.[64] This was, however, only one side of Haussmann's attempt 'to ennoble technical exigencies with artistic aims'. His other, 'real aim ... was the securing of the city against civil war. He wished to make the erection of barricades in Paris impossible for all time ... The breadth of the streets was to make the erection of barricades impossible, and new streets were to provide the shortest route between the barracks and the working-class areas'.[65] The open perspective of the new boulevards suggests that the streets were to be cleared of all except admirers, spectators and above all, with the inclusion of the *grand magasins* in the contours of the boulevards, consumers. All this was accomplished on the basis of a limited technology. Haussmann 'revolutionised the physiognomy of the city with the most modest means imaginable: spaces, pickaxes, crowbars, and the like. What measure of destruction had been caused by even these limited instruments! And along with the growth of the big cities were developed the means of razing them to the ground. What visions of the future are evoked by this!'[66] During the Spanish Civil War a note by Benjamin reads 'As the Spanish war shows, Haussmann's activity is today set to work by totally different means'.[67]

There is another dimension to Haussmann's rebuilding of Paris which is implicit in the potential of the grand boulevards to become a new *intérieur* for the bourgeoisie. One of Haussmann's contemporaries recognised the mechanism by which that same bourgeoisie dealt with the housing question: 'That method is called "Haussmann"'. By this Engels meant

> the practice, which has now become general, of making breaches in the working-class quarters of our big cities, particularly in those which are centrally situated, irrespective of whether this practice is occasioned by considerations of public health and beautification or by the demand for big centrally located business premises or by traffic requirements, such as the laying down or railways, streets, etc. No matter how different the reasons may be, the result is everywhere the same: the most scandalous alleys and lanes disappear to the accompaniment of lavish self-glorification by the bourgeoisie on account of this tremendous success, but – they appear again at once somewhere else, and often in the immediate neighbourhood.[68]

This process was accelerated in Paris after the Commune. Whereas on occasion Haussmann 'expressed this hatred for the rootless population of the great city … this population kept increasing as a result of his works. The increase of rents drove the proletariat into the outskirts. The Paris *quartiers* thereby lost their characteristic physiognomy. The red belt appears'.[69]

For many Parisians, Haussmann 'had alienated their city from them. They no longer felt at home in it. They began to become conscious of the inhuman character of the great city'.[70] Some, like the artist Meryon, were able to capture the earlier Paris before it crumbled under the instruments of the *artiste demolisseur*. Baudelaire praised Meryon's engraving of Paris that 'brought out the ancient face of the city without abandoning one cobblestone. It was this view of the matter that Baudelaire had unceasingly pursued in the idea of modernity' and which he found in Meryon's 'interpretation of classical antiquity and modernity'.[71] Meryon's engravings were to have appeared with texts by Baudelaire; all that exists, however, is Baudelaire's appreciation of Meryon's work.

> Seldom have we seen the natural solemnity of a great city depicted with more poetic power: the majesty of the piles of stone; those spires pointing their fingers to the sky; the obelisks of industry vomiting a legion of smoke against the heavens; the enormous scaffolds of the monuments under repair, pressing the spider-web-like and paradoxical beauty of their structure against the monuments' solid bodies; the steamy sky, pregnant with rage and heavy with rancour; and the wide vistas whose poetry resides in the dramas with which one endows them in one's imagination – none of the complex elements that compose the painful and glorious decor of civilisation have been forgotten.[72]

Yet as Benjamin points out, Baudelaire's image of modernity and its affinities with antiquity, with eternity, could not survive. Even the fate of that which Benjamin took to be the secret subject of Baudelaire's poetry, the city of Paris, has been transformed:

> To be sure, Paris is still standing and the great tendencies of social development are still the same. But the more constant they have remained, the more obsolete has everything that was in the sign of the 'truly new' been rendered by the experience of them. Modernity has changed most of all, and the antiquity that it was supposed to contain really presents the picture of the obsolete.[73]

A measure of how fragile the modernity of Haussmann's boulevards was came during the Paris Commune, when the 'burning of Paris was a fitting conclusion to Haussmann's work of destruction'.[74] The Commune did not survive to erect its own monuments, though 'the barricade was resurrected … It was stronger and safer than ever. It extended across the great boulevards, often reached first-storey level, and shielded the trenches situated behind it'.[75] The anonymous masses again took on a definite form and entered the public sphere not as an anonymous mass but as a revolutionary, proletarian movement. They were the ever-present threat to the Parisian bourgeoisie in the nineteenth century, as the events of 1830, 1848 and 1870–1 testify. In response to this threat, the mass's *intérieur*, the streets themselves, were transformed by Haussmann only to be re-transformed during the Commune into barricades. The masses symbolised one of the essential features of metropolitan modernity: the fact that the phantasmagoria of the bourgeois world could be more transitory than had been dreamed of; the possibility that the nightmare of Marx's image of transformation in which 'all that is solid melts into air' could come to pass.

One way of dispelling this nightmare was to prevent the proletariat from entering the public sphere at all, whether in the guise of a formal political party or, more informally, as organised labour. For the private citizen, however, another strategy for relieving the burden of this nightmare lay in the retreat into the *intérieur*. This presupposed that the living-space can be distinguished from the place of work. When this occurred that living space

> constituted itself as the interior. The office was its complement. The private citizen who in the office took reality into account, required of the interior that it should support him in his illusions. This necessity was all the more pressing since he had no intention of adding social preoccupations to his business ones. In the creation of his private environment he suppressed them both. From this sprang the phantasmagorias of the interior. This represented the universe for the private citizen. In it he assembled the distant in space and in time. His drawing-room was a box in the world theatre.[76]

This *intérieur* was populated with a whole array of objects, from furniture to everyday utensils. Benjamin's aims were 'to decipher the contours of the banal as a picture puzzle'. They too were a part of the dream world of the nineteenth century: 'Picture puzzles as the schemata of the world of dreams has long been discovered by psychoanalysis. We, however, are not so much on the track of such certainties of the soul as on the track of things. We seek the totem-tree of objects in the thicket of prehistory. The highest, the ultimate mask of this totem-tree is kitsch.'[77] But yet again there was no security against the transitory nature of modernity or the waking dreams of the exterior.

As an inward retreat, the *intérieur* too was populated with a labyrinth of dreams and mystery. Rather than being a retreat from the world of dreams outside, 'the *intérieur* of this period is itself a stimulus to intoxication and the dream'.[78] To live within it was to be trapped 'within a spider's web, that dispersed the events of the world, hung up like the dried out bodies of insects'.[79] Hence, the *intérieur* did not recommend itself as a way out of the layers of dream world that enveloped it. Instead, it provided the casing for a reified world of individual lived-out experience (*Erlebnis*) that could blossom out in all its variegated forms. It supported the inwardness which Adorno claimed was 'the historical prison of the prehistorical human essence'.[80]

The inner space of the *intérieur* was filled with furniture that retained the character of fortification, embattlement against the outside world and its transitory nature. Its complementary aspects lay in the masking and encasing of the dwelling's contents. The drive to mask that extended throughout the nineteenth century was a consequence of the fact that the relations of domination became insecure. Bourgeois power-holders no longer possesed power in the place where they lived; the very style of their homes provided little more than false immediacy. 'Economic alibi in space. *Intérieur* alibi in time'.[81] The masks in which the interior was clothed were directed towards the dream world, 'furnished for the dream'. The procession of styles – gothic, Persian, renaissance – meant that 'the bourgeois dining room became a festival hall from Cesar Borgia, out of the boudoir of the housewife arose a gothic chapel, over the study of the master of the house there played iridescently the apartment of a Persian sheikh'.[82] Such costumes hid what lay beneath them: 'they exchange glances of agreement with nothingness, with the trivial and the banal. Such nihilism is the innermost core of bourgeois cosiness'.[83] In his notes Benjamin also cited Simmel on the plurality of styles at the turn of the century.

Benjamin recognised that the strategy of encasing the contents of the *intérieur* was a complex one, extending to the living space itself. Within the notion of the dwelling

> on the one hand, the primal – perhaps eternal – must be recognised, the reflection of human being's stay in the womb; and ... on the other side, disregarding this prehistorical motif, dwelling in its most extreme form as a state of existence of the nineteenth century must be grasped. The primal form of all dwelling is existence not in the house but in the casing. The nineteenth century, like no other, was addicted to the home. It conceived of the home as human beings' casing and embedded them with all their accessories so deeply into it that one could liken it to the inside of a compass box where the instrument, with all its replacement parts lies in deep, most often purple, velvet recesses.[84]

The bourgeoisie's compensation for 'the inconsequential nature of private life in the big city' was sought 'within its four walls. Even if a bourgeois is unable to give his earthly being permanence, it seems to be a matter of honour with him to preserve the traces of his articles and requisites of daily use in perpetuity'.[85] The bourgeoisie found a casing for everything,

> for slippers and pocket watches, thermometers and egg-cups, cutlery and umbrellas ... It prefers velvet and plush covers which preserve the impression of every touch. For the Makart style, the style of the end of the Second Empire, a dwelling becomes a kind of casing. This style views it as a kind of case for a person and embeds him in it

together with all his appurtenances, tending his traces as nature tends dead fauna embedded in granite.[86]

Yet the casing never proved to be as secure as granite. By the turn of the century it received its first major shock: '*Jugendstil* fundamentally shattered the nature of casing. Today it has died out and the dwelling has been reduced: for the living by the hotel room, for the dead by the crematorium'.[87] Benjamin spent the majority of his exile existence in hotel rooms; Kracauer's early architectural commissions were for post-First World War cemeteries.

The casings had served to hide the traces of the transitory, symptomatic of an unconscious recognition of their purpose. 'Dwelling as the transitional – in the concept of the "lived out life", for example – gives an inkling of the hasty actuality which is hidden in this process. It lies in the fact of moulding a casing for ourselves.'[88] In the recesses of the *intérieur* the bourgeoisie could, for a while, create the illusion of their heroism by surrounding themselves with the costumes of greatness. The *intérieur's* physiognomy was best seen in the dwellings of the great collectors of the nineteenth century, where the interior provided the casing for a museum.

Benjamin detected a further dimension to the significance of the traces of living:

> Coverings and antimacassars, boxes and casings, were devised in abundance, in which the traces of everyday objects were moulded. The resident's own traces were also moulded in the interior. The detective story appeared, which investigated these traces ... The criminals of the first detective novels were neither gentlemen nor apaches, but middle-class private citizens.[89]

This literary genre 'concerned itself with the disquieting and threatening aspects of urban life'. One of these was the absence of traces of individuals in the metropolis, especially within the masses. Individuals sought asylum not in the *intérieur* but in the crowd, where 'the masses appear as the asylum that shields an asocial person from his persecutors. Of all the menacing aspects of the masses, this one became apparent first. It is at the origin of the detective story.'[90]

If the city was 'the realisation of the ancient human dream of the labyrinth', then the masses were 'in the labyrinth of the city the newest and least researched labyrinth'.[91] The masses constituted an essential element of one of Baudelaire's recurring images of the city: 'For Baudelaire, Paris stands as a testimony of antiquity in contrast to its masses as testimony of modernity'.[92] Though the streets could be viewed as a deserted labyrinth of buildings, they were not always merely an empty decor. Rather the streets 'are the home of the collectivity. The collectivity is an eternally unquiet, eternally moving entity that lives, experiences, recognises and feels between the walls of houses just as much as individuals in the security of their four walls.'[93] For them, even the arcades acquired a new significance. For the collectivity 'the arcade was the Salon. More than in any other location, the street revealed itself in it as the furnished, lived-out *intérieur* of the masses'.[94]

On the streets this permanently moving and changing collectivity appeared as the crowd which fascinated nineteenth century commentators and writers before becoming the threatening masses. The crowd constituted itself in a peculiar manner; they gathered,

> but socially they remain abstract – namely, in their isolated private interests ... In many cases, such gatherings have only a statistical existence. This existence conceals the really monstrous thing about them: the concentration of private persons as such by the accident of their private interests. But if these concentrations become evident – and the totalitarian states see to it by making the concentration of their clients permanent and obligatory for all their purposes – their hybrid character clearly manifests itself, and particularly those involved.[95]

Engels – often quoted by Benjamin – spoke of 'the brutal indifference' of the crowd in the great cities, concluding that 'one shrinks before the consequences of our social state as they manifest themselves here undisguised, and can only wonder that the whole crazy fabric still hangs together'.[96]

# Notes

This chapter is an extract from David Frisby's *Fragments of Modernity* (MIT Press, 1986) which outlines the main features of Walter Benjamin's ambitious and incomplete Arcades Project, an exercise that he worked on from the late 1920s until his suicide in 1940. Intended as a 'prehistory of modernity' focusing on 'Paris: capital of the nineteenth century', the project was recorded in an elaborate and cross-indexed note system. It was first published in English in 1999 as *The Arcades Project*. For an analysis of the project see Susan Buck-Moss, *The Dialectics of Seeing*, MIT Press, 1989.

1   Benjamin (1982:571).
2   Benjamin (1969b:234).
3   Benjamin (1982:491).
4   Ibid:493.
5   Ibid:494.
6   Ibid:491.
7   Ibid.
8   Benjamin (1980, IV:357).
9   Ibid:425.
10  Benjamin (1982:1031).

11    Ibid:134.

12    Ibid.

13    Benjamin (1973a:176).

14    Pausanias (1971).

15    Frazer (1900:22).

16    Ibid:34.

17    Benjamin (1982:135).

18    Ibid.

19    Benjamin (1980, I:1173).

20    Benjamin (1982:419).

21    Ibid:423.

22    Ibid:422.

23    Ibid:423.

24    Benjamin (1980, I:666).

25    Benjamin (1982:377).

26    Ibid:439.

27    Ibid:463.

28    Ibid:470.

29    Ibid:309.

30    Ibid:405.

31    Benjamin (1980:681).

32    Ibid:164–5.

33    Ibid:1236.

34    Benjamin (1979:243).

35    Benjamin (1973a:243).

36    Benjamin (1979:243–4).

37    Ibid:251.

38    Ibid:177–8

39    Ibid:318.

40    Benjamin (1982:*1166*).

41    Benjamin (1973a:158). For a full examination of their architectural significance see Geist,
        *Passagen*, Munich 1971.

42    Ibid:513.

43    Ibid:86

44    Ibid:668–9.

45    Ibid:1030.

46    Ibid:1050.

47    Ibid:280.

48    Ibid:670.

49    Ibid:993.

50    Ibid:617.

51    Ibid:1034.

52    Ibid:140.

53    Ibid: 1001–2.

54    Ibid:1045.

55    Benjamin (1973a:161). For a study for the panoramas and their variants see Oettermann, *Das
        Panorama. Die Geschichte eines Massenmediums*, Frankfurt 1980.

56    Benjamin (1973a:161).

57    Ibid:162.

58    Benjamin (1982:661).

59   Ibid:656.
60   Benjamin (1973a:35).
61   Benjamin (1982:513).
62   Ibid:522.
63   Ibid:513.
64   Ibid:174–5.
65   Benjamin (1973a:85).
66   Benjamin (1982:208).
67   See Engels, 'The Housing Question' in Marx and Engels, *Selected Works*, Moscow 1962, pp.557–635, esp. pp.607–8. Quoted by Benjamin (1982:206).
68   Benjamin (1973a:174).
69   Ibid.
70   Ibid:87.
71   Ibid:88–9.
72   Ibid:90.
73   Ibid:176.
74   Ibid:175.
75   Ibid:167–8.
76   Benjamin (1982:281).
77   Ibid:286.
78   Ibid.
79   Ibid:289.
80   Ibid:288–9.
81   Ibid:282.
82   Ibid:286.
83   Ibid:291–2.
84   Benjamin (1973a:46).
85   Ibid.
86   Ibid:46.
87   Ibid.
88   Benjamin (1982:292).
89   Ibid.
90   Benjamin (1973a:169).
91   Ibid:40.
92   Benjamin (1982:559).
93   Ibid:437.
94   Ibid:535.
95   Ibid.
96   Benjamin (1973a:62–3).

# Chapter 17

# Impromptus of a great city
## Siegfried Kracauer's
## *Strassen in Berlin und Anderswo*

*Graeme Gilloch*

> The worth of cities is determined by the number of places in them devoted to improvisation.
>
> Kracauer (1987:51)

> We Jews are always seeking a promised land, provided it be elsewhere.
>
> Eco (2002:464)

He has a child's face, we are told, and its changing expression confirms our suspicion that although he is here, seated at the grand piano in a fashionable Berlin bar, he is utterly lost in his thoughts, his dreams, his memories. Playing the merest 'murmur' [1] of music to accompany the hum of conversation, shifting seamlessly from one melody to the next, his hands and fingers move effortlessly over the keyboard as if they have a life of their own, as if wholly independent of the man's middle-aged, rather corpulent body. The pianist plays absent-mindedly. Yes, he is certainly here, but he is also, unmistakably, 'elsewhere'. [2]

Our musician is one of the motley assortment of characters found in the 'Figures' section of Siegfried Kracauer's *Strassen in Berlin und anderswo* (*Streets in Berlin and Elsewhere*), a collection bringing together a few of his *feuilleton* pieces originally published in the *Frankfurter Zeitung* between 1925 and 1932. After training and working as an architect in Frankfurt, Kracauer began writing for the newspaper in early 1921, initially as a local reporter and then as a salaried member of the journalist staff before finally becoming a full editor in 1924. Following a ten-week initial visit to Berlin in the late spring and

early summer of 1929, during which time he collected the materials for his celebrated study of white-collar workers, *Die Angestellten: Aus dem neuesten Deutschland* (1929–30),[3] Kracauer moved permanently to the city to assume the
position of editor for the *feuilleton* section and the various film reviews and
literary features forming the paper's Berlin pages. He was to spend the next
three years in the German capital before wisely fleeing to France immediately
following the Reichstag fire of 1933.

Kracauer was a prolific writer, producing nearly two thousand pieces
for the *Frankfurter Zeitung*[4] dealing with the widest possible array of subject
matter: observations on the everyday street life of Frankfurt and Berlin; tales
recounting particular occurrences and memorable encounters; pen portraits of
eccentric and otherwise remarkable street figures; discussions of urban architecture, planning and design; film and literary reviews; reports from exhibitions,
shows and premieres; and occasional dispatches from abroad providing for
reflections on conditions 'elsewhere'. While he was fully aware that the vast
majority of these writings would experience the usual fate of newspaper articles
– penned one day, read and forgotten the next – Kracauer nevertheless
harboured hopes that some of them might prove of more lasting significance
than mere journalism, a term he always considered to be pejorative. He was – so
he claims – incapable of treating such texts as mere *Brotarbeit*, as matters of
financial expediency, and instead composed them with the same spirit and
meticulous care as his more substantial literary, sociological and philosophical
writings.[5] The reason for this attentiveness is clear: Kracauer recognised the
felicity with which textual miniatures were able to capture exemplary and decisive moments culled from the concrete reality of quotidian metropolitan existence, raw material shot through with social, political and existential
significance. For Kracauer, as for his colleagues Walter Benjamin and Ernst
Bloch, the surface manifestations of the cityscape were 'traces', 'hieroglyphs',[6]
'dream images' to be recovered, scrutinised and deciphered by the critical theorist.[7] For them, this was the stuff from which modernity was made and the basis
of its legibility.[8]

It is not surprising, therefore, that the idea, first muted in the 1930s,
of bringing some of his writings together into a single volume appealed immediately to Kracauer. It was an opportunity to construct a kind of literary mosaic
that would capture the image of the kaleidoscopic city, and penetrate its reality
in a way forbidden to mere reportage.[9] Kracauer could have modelled his
proposed *Strassenbuch*[10] on Benjamin's 1928 *One-Way Street*, a seemingly exemplary instance of fragmentary composition and one similarly preoccupied with
the critical rendering of snapshots of metropolitan culture and the 'profane illumination' of modernity. Benjamin's observations of the cityscape serve as convenient points of departure for philosophical-metaphysical speculations and
political enunciations, the text being strewn with his own, often enigmatic,

aphorisms, dream images, jokes (sometimes witty, sometimes lamentable) and other asides and anecdotes. Kracauer envisaged his *Strassenbuch* rather differently: its building blocks were not fragments but rather self-contained literary miniatures – much more akin to Benjamin's *Denkbilder* – and these texts, although certainly not bereft of wider philosophical and sociological import, displayed a greater loyalty to the material itself, to the concrete particulars and specific moments of the urban environment. The book was intended to give voice to this cityscape, rather than simply to utilise its nomenclature as captions for the presentation of abstractions. Accordingly there was to be little if any reworking of the forty-one *feuilleton* pieces Kracauer selected for inclusion, and certainly no overarching commentary. The thematic and conceptual repertoire would emerge from the material itself, and from the organisation and juxtaposition of texts. These were to be divided into three sections based on spatial, rather than chronological considerations: 'Auf der Strasse' ('On the Street'), 'Neben der Strasse' ('Beside the Street') and 'Figuren' ('Figures').

Kracauer's hopes for the publication of his *Strassenbuch* remained unrealised for many years. In 1963 Suhrkamp Verlag published *Das Ornament der Masse* (*The Mass Ornament*), a volume Kracauer dedicated to Adorno containing a plethora of the *feuilleton* pieces[11] along with a number of more substantial writings. *Strassen in Berlin und Anderswo* appeared the following year. It would be easy to overlook what is remarkable about the Berlin book. Despite the most traumatic thirty-year historical interval imaginable, it is extremely similar in structure and content to Kracauer's original conception: most of the earlier selection is retained, there are few additions, and the texts are divided into four sections under a new rubric: 'Strassen' ('Streets'), 'Lokalen' ('Localities/Bars'), 'Dinge' ('Things') and 'Figuren'.[12] In short, the writings Kracauer deemed most significant in the 1930s have, in his view at least, maintained their relevance through into the post-war period. The question arises: How are these texts, written as contemporary pieces addressed to a contemporary readership, supposed to speak to a subsequent generation? Seemingly Kracauer saw no need to address this issue. He provides no foreword or introduction to these writings elucidating how, for example, they might be read as intimations of the catastrophic events to follow, as analyses of a class so soon to swap its white collars for brown shirts. There is no revision, no attempt to give the pieces a new inflection or orientation. This is especially puzzling given Kracauer's abiding sense that his work had never received proper scholarly recognition. Here was an opportunity to bring his writings to the attention of a new and highly receptive generation of German readers, one already schooled in the critical theory of the Frankfurt Institut für Sozialforschung. It is surprising how little he chose to make of it.[13]

The new title is the only pointer. The inclusion of the name of the city 'Berlin' – the fact that this needed to be specified at all – is perhaps indicative of

Kracauer's own distance from the city, geographical, temporal and intellectual, and that of the book's potential readership. In 1930s Germany, *Strassenbuch* would surely have sufficed as a title. Where else, when else could this be other than the here and now of contemporary Berlin? The titular reference to Berlin serves to specify the focus of the book, which otherwise might not be so obvious given that no fewer than fourteen of the thirty-three pieces[14] date from the period (1925–9) when Kracauer was in Frankfurt. The reader is left in no doubt: Berlin is what matters, Berlin rather than 'elsewhere', an 'elsewhere' that is always defined in relation to Berlin, an 'elsewhere' which Berliners might nevertheless chance upon in the course of their travels – whether real, physical journeys, or spiritual-intellectual 'wanderings'. In the final analysis, everywhere else – even Paris[15] – is simply *anderswo*.

This is instructive. For Kracauer Berlin came to constitute the definitive modern metropolis, the 'newest Germany', an environment in which the tendencies and prospects of contemporary political, social, cultural and economic transformation were most manifest and legible. And at the heart of this newest Berlin, crowding its streets and squares, frequenting its myriad shops, enjoying its diverse attractions, was an ascendant class produced by changes in the character and organisation of labour, a series of new socioeconomic strata developing their own distinctive practices, perceptions, experiences and aspirations: the office workers, the white-collar employees, the salaried masses. Kracauer recognised with great prescience that contemporary Berlin was *their* city. *Die Angestellten*, which itself appeared in instalments in the *Frankfurter Zeitung*, was conceived as an exploration of this increasingly dominant class, one whose habits and dispositions had gone hitherto unobserved precisely because of their apparent visibility, their very proximity and familiarity.[16] As an intrepid urban ethnographer, Kracauer set out to map the *terra incognita*[17] of these clerical and service staff, not only with respect to their experience of the occupational structure, but also in terms of their fundamental role as customers, consumers and clientele, as shoppers and browsers, as spectators, audiences and readers. The countless distractions spawned by the modern metropolis – the glistening display windows of department stores and fashionable boutiques, the showrooms and exhibition halls presenting the latest gadgets and devices for modern living, the glossy magazines, brochures and catalogues, the dazzling neon signs illuminating the nocturnal boulevard, the brilliant interiors of polite cafés and the moody half-light of less reputable bars and dancehalls, the splendour of picture palaces and variety theatres, the rides and attractions of the early amusement parks like the Lunarpark, forerunners of contemporary theme-parks – all these elements of the modern cityscape and mediascape vied for the attention of these salaried masses, all sought to separate the masses from their salaries.[18]

And it was not long before many of these petty bourgeois employees were to part company with their earnings. The *Strassen in Berlin* collection presents images of this class subject to the economic and political crises of the early 1930s. The growing numbers of unemployed come to constitute an atomised mass,[19] one which has to endure petty indignities in the crowded labour exchange, to negotiate its labyrinthine corridors leading not back to the world of work but only to the back of beyond, elsewhere.[20] The sombre gloom of the welfare centres providing temporary refuge for the penniless contrasts with the spectacular illumination of the cityscape of consumption.[21] Anxious investors queue forlornly outside banks in the vain hope of recovering their precious savings and assets.[22] Patrons assemble at dismal morning and matinee cinema performances in search of cheap respite from the boredom of useless days.[23] Even the Tiller Girls, the exemplary Taylorised mass ornament spawned by the distraction industry, have had their day – nothing could be more out of kilter with the paralysed industrial production system than their upbeat, dynamic, rhythmic routines.[24]

The figures of poverty and destitution, unwanted reminders of a collective bad conscience, stalk the cityscape and imbue it with a sinister atmosphere, even where the lights shine brightest and the mood seems most convivial.[25] For Kracauer the city is haunted, not just by the shadows of economic misery and the growing spectre of political terror, but by something far less tangible. The city is haunted by the alienated modern condition itself, by its inhabitants' inner emptiness and loneliness, by the absence of social solidarity, community and consolation, by the evacuation of meaning and hope for which no upturn in the stock market could compensate. Such a bleak vision of a disenchanted, functionalised, mechanised modernity, a social form whose obsession with abstraction and quantification blinds people to the real qualities of things, finds repeated expression in Kracauer's early writings,[26] and unmistakably suffuses the Berlin studies.[27] No wonder: the salaried masses are, for Kracauer, the 'spiritually homeless' *par excellence*. With their futile, individualistic bourgeois aspirations to culture, education and self-improvement, with their inability to make common cause as a class for themselves, let alone ally themselves with the working class they disdained, these strata owe their very existence to a rationalised, bureaucratised, impersonal and inhuman modern world and, as its functionaries, find themselves the main bearers of its stigmata: melancholy, frustration, resentment, repression, boredom, fatalism. This is perhaps why Kracauer saw so little need to revisit and transform the material from the original *Strassenbuch*. Certainly his writings for the *Frankfurter Zeitung* bear witness to, and are marked by, the particular crises of the time – how could they not be? But they also seek to penetrate below the surface of things, to distil what is more enduring (sociological and philosophical insight) from the ephemeral (the material of reportage), and thereby capture metropolitan modernity as, if nothing else, the fully recognisable (pre-)history of our own times, a modernity which is now for us both 'elsewhere' and here and now, a modernity which is nothing other than continuing alienation, reification and catastrophe.

Doubtless the unsalaried masses would not figure among the well-heeled and well-connected clientele entertained by our pianist, but 'Der Klavierspieler' ('The Pianist') serves as a point of conjuncture for a number of central themes and experiences in Kracauer's Berlin studies: distraction; transition and transience; asynchronicity and improvisation. The first of these, *Zerstreuung* (distraction), is most apparent here, playing a double role: the pianist, as a professional musician, is part of the 'entertainment industry' and a provider of distraction for others; moreover, and this is perhaps most significant, he is himself presented as a distracted figure, one inattentive both to his musical labours and his otherwise preoccupied audience. The pianist plays in a state of distraction to an audience that is equally distracted: he is elsewhere and so are they. This is music to be heard, not listened to. In fact, it is scarcely to be heard at all. This music is intended to be almost inaudible, music as acoustic décor, music as melodious wallpaper covering over any unsightly cracks in conversation. The pianist's non-audience have other, more enticing distractions: gossip and chatter.[28] The music must not distract them from these more serious, more salacious delights. Accordingly, there is little pleasure to found in this music – indeed, perhaps this is why no one listens, not even the pianist himself. For him, of course, it is work. The merging of the spheres of 'entertainment' and labour is central to Kracauer's studies. He understands the proliferation of 'entertainment', the massive burgeoning of amusements and 'leisure' activities as a new and pernicious extension of capitalist domination, in which the rhythms and logic of the industrial work process are transposed into the cultural domain. Leisure and entertainment involve the systematic orchestration and arrangement of 'free time' into a mass phenomenon spuriously experienced as individual expression. Rationalised, alienated labour finds its corollary, its after-image, in the mechanised synchronicity of the mass ornament and in the 'organised happiness' of the amusement park.[29] Kracauer's miniatures bear critical witness to the dutiful character of distraction: at weekends the Berliners decamp and head off in search of pleasures elsewhere, a mass exodus which leaves behind an eerily deserted Sunday cityscape;[30] hordes of day-trippers venture into the country to sample the pseudo-rustic delights of the *Mittelgebirge*;[31] fatalistic crowds flock to performances of clairvoyants and mesmerists;[32] in the cinema, they laugh in the right places.[33] Autonomy, spontaneity and distinction have no place in the modern world of heteronomous entertainment and de-differentiation. The Berlin variety theatres, providing for all tastes and none, present banal popular singers and chorus line routines on the same bill as classical virtuosos and talented but impecunious soloists, a mingling of commercial entertainers and serious artists that squeezes out any genuine cultural variety.[34] These 'motley (*kunterbunt*) offerings'[35] fail to disguise the fact that the white-collar world of distraction lacks contrast and colour. Our pianist, a figure who like his music fades imperceptibly into the background, and his audience, prattling pilgrims to the 'cult of boredom', are symptomatic of this grey indifference, this dull insensitivity to the unique qualities of things.

Kracauer's critique of mass culture and distraction clearly prefigures Adorno and Horkheimer's 'culture industry' thesis. The all too 'easy on the ear' music of our pianist is precisely that which Adorno was to lambaste in his writings on tin-pan alley jazz. The corresponding inattention of the audience provides the clearest possible evidence of their diminishing capacity for concentration, of a pernicious and pervasive 'regression of listening'. Kracauer's response is more equivocal than this, however, and more sensitive to positive moments in popular cultural forms and the experience of distraction.[36] His work involves what one might term a 'dialectics of distraction', one which anticipates Benjamin's famous 1935 essay 'The Work of Art in the Age of Mechanical Reproduction'. Benjamin's principal argument, that the advent of photography and film lead to the withering of the aura of the unique, original, authentic work of art, needs little elaboration here. His convoluted attempt to develop a positive understanding of distraction is, however, of relevance to us. Firstly, Benjamin claims that film and photography involve reception in a state of distraction. These new media neither demand intense concentration nor allow leisurely contemplation. Rather the distracted cinema audience is relaxed and receptive, at ease with and aware of its own expertise, conscious of itself and critical of what is presented on screen.[37] This is because, and this is his second main point, the inattentive apprehension of works of art – Benjamin's model here is the everyday architecture of the cityscape – provides for an intimate and privileged familiarity with the object that is the very antithesis of the fetishistic adoration of the auratic artwork.[38] Habitual acquaintance integrates the artwork into everyday life. Familiarity with the cityscape fosters a sense of confidence and composure which are in turn preconditions for a form of mastery – not mastery of the object and the environment, but rather, as Benjamin puts it in *One-Way Street*, of one's relationship with them.[39] Lastly, distraction points in the direction of a new sensitivity to the world, of new forms of recognition and receptivity. Habit may threaten to dull the senses, to produce amnesia and indifference, but distraction also transforms the everyday environment. Distraction is not simply inattention, a failure to give due consideration to what is at hand, but rather it is a paying attention elsewhere, the diversion of one's perceptual faculties to that which is not of immediate importance, to what is commonly overlooked or neglected. Distraction is, then, an attentiveness to that which lies at the edge of our conventional field of vision or which fleetingly crosses it. Distraction involves a particular openness to the marginal, the liminal and the transient, to that which escapes the everyday perceptual realm. It is here that film emerges as the medium of distraction *par excellence*. It is film that focuses our attention on the hitherto inconspicuous, that identifies and penetrates the 'optical unconscious',[40] that discloses the *terra incognita* of the modern cityscape. Film promises distraction for the distracted: it engenders a heightened appreciation of the urban environment for an accomplished, expert audience.

The images of metropolitan modernity presented in Kracauer's *Strassen in Berlin* are indebted to the constellation formed by distraction, film and urban space. In many respects this collection of miniatures finds much of its theoretical armature in Benjamin's 'Work of Art' study and, later, in Kracauer's own *Theory of Film: The Redemption of Physical Reality*,[41] a book containing numerous echoes of Benjamin's essay.[42] For Kracauer, film has an elective affinity for the cityscape, revealing that which normally goes unnoticed, capturing the unstaged flow of movement on the urban streets. Film enhances our perception of physical reality, and in so doing restores to the city precisely those qualities and colours which have faded in alienated, everyday perception. For both Benjamin and Kracauer, film gives the dreary, quotidian metropolitan environment an electrifying charge, an explosive volatility.[43] But while Benjamin advocates both film and architecture as media of distraction, Kracauer sees a distinction and tension between them. Given Kracauer's architectural training and the numerous pieces on particular buildings, architectural exhibitions and design competitions he wrote for the *Frankfurter Zeitung*, the absence of architecture in the Berlin collection is noticeable and instructive.[44] For Kracauer, it is not in conventional architecture that the modern is to be found. The monumental, the planned, the static forms of the city are uninspiring. Kracauer's interests lie elsewhere: modernity as flux and the fortuitous, as instability and indeterminacy, as contingency and happenstance.

> One can distinguish between two types of cityscape: those which are consciously fashioned and those which come about unintentionally. The former spring from the artistic will which is realised in those squares, vistas, building ensembles and perspectives which Baedecker generally sees fit to highlight with a star. In contrast, the latter come into being without prior plan. They are not, like the Pariser Platz or the Place de la Concorde, compositions which owe their existence to some unifying building ethos. Rather, they are creations of chance and as such cannot be called to account. Such a cityscape, itself never the object of any particular interest, occurs wherever masses of stone and streets meet, the elements of which emerge from quite disparate interests. It is as unfashioned as Nature itself, and can be likened to a landscape in that it asserts itself unconsciously. Unconcerned about its visage, it bides its time. [45]

Chance configurations of light, of traffic and of crowds in motion, fleeting constellations and conjunctures, the physical fabric of the cityscape in perpetual transformation, Berlin subject to time – these spatio-temporal, cinematic aspects of modernity fascinate Kracauer. In the city 'Everything moves, everything stirs (*Alles regt sich, alles bewegt sich*)'. [46] Only film is adequate to this sense of motion. His collection of miniatures is like a cinematic odyssey around the metropolis, a series of filmic vignettes, cut and juxtaposed to create a montage of images setting up a series of tensions, bringing the 'surface-level expressions' of the city into sharp and critical focus. [47] Kracauer's book is conceived as cinematic images of the city [48] – not as moving pictures, of course, but as pictures of movement – for those urbanites for whom the cinema had become the pre-eminent form of modern mass media: the white-collar workers themselves, the little shop girls, who went to the movies. These office and retail employees, one should not forget, were the original readership for these journalistic writings. The texts in *Strassen in Berlin* are cinematically-inspired offerings for cinematically-inclined and experienced audiences, distractions for the distracted.

In Berlin *alles regt sich, alles bewegt sich*. The city is a rich hunting ground for those in search of sites of transition, objects in transit and transient forms. Kracauer's eye was caught by those spaces and places that we habitually pass through and fail to notice, the loci of the optical unconscious: anonymous squares serving as junctions for tramlines; unnoticed expanses hurried across by busy commuters and shoppers; the subterranean world of the underpass; the railway bridge spanning Friedrichstrasse affording the driver and passengers a fleeting initial impression or last parting glimpse of the city centre; the corridors of the labour exchange. Kracauer's eye was also caught by settings where we must linger unwillingly, hang-outs where we must hang about: cinema foyers and hotel lobbies, cafés and bars, the welfare refuge. In reality those environments we move through and those where we must patiently bide our time are one and the same: limbo in different guises. Kracauer carefully brought together all these dismal non-places, these thresholds, these in-between spaces (*Zwischenräume*). For him they were so many waiting rooms, temporary shelters for the spiritually homeless of the modern metropolis.

The 'Lindenpassage' (actually the Kaisergalerie) is exemplary here. The arcade is to be walked through, a passageway connecting streets, yet it is also an interior which invites dawdling, which encourages us to tarry awhile. The luxurious and fashionable goods which filled the arcade during its heyday [49] are now long gone, and it has come to serve as a temporary resting place, as a refuge, for the curios and remnants of a now-vanished world of commodities. Like Benjamin and the surrealist Louis Aragon before him, Kracauer observed the metropolitan shopping arcade as a ruin on the point of demolition, as it was about to disappear irrevocably, at last sight. In *Strassen in Berlin* we find not only a dialectics of distraction at work, but also a complex dialectics of disappearance. Spaces, things and figures are captured only as they are about to vanish, or are remembered only after they have already gone. This tension is made explicit in 'Strasse Ohne Erinnerung' ('Street Without Memory').[50] In this key text Kracauer recalls how, before catching a train, he strolled along to one of his favourite cafés on the Kurfürstendamm only to discover with surprise that this had closed. He has to make do with another bar nearby, one far too garish for his taste. A year or so later Kracauer is suddenly struck by the disappearance of this second café – a sign in the window merely informing passers-by that the premises are now available to rent. He reflects upon this seemingly banal, commonplace experience: 'Elsewhere, the past clings to the places where it resided during its lifetime; on the Kurfürstendamm it departs without leaving so much as a trace. Since I have known it, it has changed fundamentally again and again in no time at all. The new businesses are always brand new and those they expel are always wholly obliterated'.[51] On the Kurfürstendamm – and it is not insignificant that this is a locus of consumption, fashion and leisure – the new eradicates the old without residue, and it does so with ever greater rapidity. For Kracauer the fate of the cafés brings into focus distinctive features of our experience of the modern cityscape: how 'perpetual change erases memory';[52] how the endless quest for novelty merges into the flow of undifferentiated, empty time; how the past is consigned to oblivion by the present, and perhaps most importantly how it may fleetingly reappear as a disturbance that gives a shock to today's passer-by. For it is paradoxically the act of obliteration, the present absence of the former cafés, which brings them so vividly to mind. Demolition and erasure bring with them a sudden appreciation of what is no longer there.[53] The Kurfürstendamm is without memories, but Kracauer is not.

The peculiarly memorable character of the transient and of the in-between – Kracauer provides a particularly neat illustration of this drawn from the distraction industry: the 'number girl' at the Scala variety theatre.[54] Between each act it is her task to walk across the stage in front of the lowered curtain carrying aloft a number corresponding to the programme number of the next turn. Her performance, such as it is, is a brief interlude between the performances. However charmingly she executes her 'delicate mission'[55] – and for Kracauer her every appearance is nothing less than 'a genuine solo'[56] – she herself does not figure in the programme, she is allotted no number. But we ignore the number girl at our peril. For who is to say in later years what will prove significant and what will prove trivial? After all, Kracauer observes, warming to his theme, dates may acquire an importance for later historians long after the actual events of those years themselves have faded into the background. The numbers may be all that is left to us. To exemplify this, all that remains of Kracauer's evening in the Scala are his memories of their smiling bearer. Of the acts to which they referred, not a trace.

Our pianist is a complex figure of memory: firstly, his playing involves a particular combination of memory and improvisation as familiar, well-rehearsed melodies are connected and embellished with a few new accents and flourishes; secondly, freed from the need to concentrate on his music, the pianist is preoccupied with his own memories;[57] thirdly and most significantly, the pianist prompts and sets in train Kracauer's memories. Kracauer interrupts his account of the Berlin bar and its pianist to recall another piano player, one whose task it was to accompany the silent films shown in the local cinema many years before, and one so remarkable and memorable that he figures twenty-five years later in Kracauer's *Theory of Film* as 'the drunken pianist'.[58] This musician was, so it was said, a genius who once trained at the conservatoire, and whose talents had now gone to waste. Seated at the piano in the cinema in various states of inebriation, this pianist could see little – his hazy view of the screen was completely obstructed by the instrument itself. Thus unable to match his music to the events of the film unfolding on screen, he contented himself by playing music at random: old melodies and familiar songs played time and again; improvised passages and flights of fancy; military marches and popular dance tunes; whatever gave expression to his stupefied senses. Film image and impromptu music accordingly bore no particular relation to one another. It was precisely this asynchronicity which charmed Kracauer: 'This lack of relation between the musical themes and the action they were supposed to sustain seemed very delightful indeed to me, for it made me see the story in a new and unexpected light or, more important, challenged me to lose myself in an unchartered wilderness opened up by allusive shots'.[59] By coincidence there were odd occasions when the music did correspond in some way with the cinematic events presented, occurrences which gave the impression that some sublime, secret connection might have existed all along, 'a relationship which I considered perfect because of its accidental nature and its indeterminacy. I never heard more fitting accompaniment.'[60]

Kracauer's anecdote is highly suggestive. It clearly privileges the spontaneous, the improvised and the contingent in opposition to the predictable, the predetermined, and the routinised, precisely those features characteristic of modern mass metropolitan culture. There could be no greater contrast between the idiosyncratic music of the drunken pianist and the precision-engineered soundtracks produced by the distraction industry, with their formulaic, standardised themes and calculated emotional manipulation. Out of time, out of step, out of kilter – what could be less like the mass ornament, indeed what could debunk the mass ornament more completely than this piano music? The asynchronous music of the pianist becomes a wry comment upon – and destabilises – the meaning of the events on screen: melancholy music transforms comedies into solemn occasions, just as the earnestness and intensity of serious drama is subverted by frivolous tunes. This is important because Kracauer extols the comic and critical powers inherent in things out of time, asynchronous – or more precisely anachronistic – objects. He writes satirically of certain objects of the recent past which once attested to the orderly world of the upstanding, independent, self-assured bourgeoisie, objects whose subsequent fate serves as an index of the present crisis: a three-page 'historical study' chronicles how the dignified fashion of wearing braces has waned in a period of belt-tightening;[61] an essay ponders the sociological significance of the umbrella's apparent fall from favour;[62] another recounts the comic adventures and humiliations of an upright piano, an instrument increasingly out of favour and one of the first things sold to the second-hand shop when economies become necessary.[63]

The last of these narratives is perhaps the most interesting, and provides a counterpart to the figure of the pianist. Kracauer's anthropomorphic tale of the upright piano is an allegorical story ridiculing the social and cultural pretensions of the lower middle classes, the white-collar workers. The upright piano, symbol of familial musicality and *Bildung*, stands cramped and rarely played in a corner of a typical petit bourgeois home. It aspires to cultural elevation, considering itself the very equal of a grand piano, only less showy. The piano dreams of being elsewhere, of taking its proper place in the order of things, of being the object of attention and admiration. Its special qualities and capacities are unused and unappreciated. As is only proper, it prefers solitude to any alliance or acquaintance with the other objects of the interior. It has no wish to be part of the furniture. Such snobbery is cruelly exposed, however, when the piano finally ventures into a nearby bar – a far less sophisticated establishment than that of our piano player, indeed more the haunt of the drunken pianist – and finds itself playing together with a number of other instruments in an impromptu band. But things go awry – the music is banal and inferior, the democratic spirit is not to the piano's liking, and conflict ensues. Its lid is slammed shut and it is sent packing. In *Das Klavier* Kracauer presents a far from subtle satire of the utterly individualistic sensibility and bourgeois orientation – the product of envy, deference and resentment – which ensures the failure of the salaried masses to develop any sense of class consciousness. They pin everything on their personal upward mobility and are left dumbfounded when real life circumstances catapult them in the opposite direction. They are a lonely crowd searching in vain for solace, for satisfaction, for compensation and consolation in the remaining pleasures of the cityscape. In the streets of Berlin and elsewhere the metropolitan multitudes fail to find companionship in misfortune.

It is clearly no coincidence that *Das Klavier* is followed immediately by the 1927 piece *Das Schreibmaschinchen* (The Little Typewriter), for here is another figure who, like the pianist, sits lost in thought as his fingers move swiftly and lightly over the keys in front of him: the writer. It comes as no surprise that our piano players – sober and sozzled – should serve as opportunities to reflect upon the position and practice of the contemporary writer, the journalist and the critic. The musicians are highly ambiguous, highly ambivalent allegorical figures, yet they offer a utopian vision of textual production. To be able to write as they play – effortlessly summoning from the keyboard a smooth, endless flow of words – would be, as Benjamin observed, nothing less than to be in the Arcadia of one's writing.[64] In the ease and expertise of the pianist, the painfully staccato rhythm of writing has been transcended. This positive inflection is particularly evident in the case of the intoxicated musician – spontaneous, improvised, unpredictable, his creative spirit and unpredictable output run counter to and debunk reality, providing thereby its most perfect critical accompaniment. The drunken pianist brushes the contemporary against the grain, as Benjamin puts it, and in so doing acts as an intemperate, unintentional critic, an accidental theorist. But he is also an absurd and ridiculous figure who, having frittered away any real talent, is reduced to eking out a living in a miserable movie-house. The café pianist has fared little better. He is the centre of inattention, a distracted daydreamer whose principal aspiration is to be elsewhere. He is a non-entity whose mediocre music falls on deaf ears. Our baby-faced piano-player corresponds to a writer without readers, a contemporary Cassandra whose voice goes unheard amidst the chatter and gossip. Perhaps it is the scream of this Cassandra that echoes through the streets of Berlin and elsewhere, and which makes Kracauer's blood run cold.

# Notes

I would like to thank the Leverhulme Trust, whose generous financial support facilitated the research for both this chapter and the next. I am very grateful to the Deutsches Literaturarchiv in Marbach am Neckar for their kind assistance in providing access to previously unpublished materials from the Kracauer archive, and to the Archivzentrum of the Stadt- und Universitätsbibliothek of the Johann Wolfgang Goethe University, Frankfurt am Main, for the use of materials in the Max Horkheimer archive and the Leo Löwenthal archive.

1   Kracauer (1987:105).
2   See Kracauer (1987:105).
3   Translated as *The Salaried Masses: Duty and Distraction in Weimar Germany*, London: Verso 1998.
4   Ward (2001:33).
5   In a letter to Adorno of 22 July 1930, Kracauer claims to devote the same loving care to these short lived minatures as to his major texts. (See: Deutsches Literaturarchiv 72.111/5).
6   Kracauer makes clear his concern with the 'hieroglyphics of space': 'Spatial images are the dreams of society. Wherever the hieroglyph of a spatial image is deciphered, the basis of social reality presents itself' (Kracauer 1987:52).
7   In the opening to Kracauer's 1927 essay 'The Mass Ornament' we find the most explicit formulation of this: 'surface-level expressions … by virtue of their unconscious nature, provide unmediated access to the fundamental substance of the state of things. Conversely, knowledge of this state of things depends on the interpretation of these surface-level expressions. The fundamental substance of an epoch and its unheeded impulses illuminate each other reciprocally' (Kracauer 1995:75).

8   Ward (2002:81) rightly describes Kracauer as 'the archanalyst of Weimar surface culture.' For a discussion of Kracauer's technique of critical decipherment see Stalder in Volk (1996:131–55).

9   Kracauer writes 'Reality is a construction. Certainly life must be observed for it to appear. Yet it is by no means contained in the more or less random observational results of reportage; rather, it is to be found solely in the mosaic that is assembled from single observations on the basis of comprehension of their meaning. Reportage photographs life; such a mosaic would be its image' (Kracauer 1998:32).

10  Deutsches Literaturarchiv 72.3496a.

11  Indeed there is some overlap: the *Mass Ornament* collection includes some of the same pieces as the *Strassen in Berlin* selection.

12  All of these miniatures along with many of Kracauer's other *feuilleton* writings have subsequently been published in the three volumes of his *Schriften* 5 (Kracauer 1990), and in two further collections: Kracauer (1996) and (1997b). Many of the film reviews have also appeared in Kracauer's 1974 *Kino* collection, and in the appendix to the German translation of the Caligari study (Kracauer 1979).

13  Perhaps he hoped that Adorno's retrospective essay, 'Der wunderliche Realist', written to mark Kracauer's 75th birthday in 1964, would perform this task. Of course it did no such thing, and served only to provoke another acrimonious exchange between them.

14  Fifteen if one includes the extract from Kracauer's 1928 novel *Ginster*, published in the *Frankfurter Zeitung* in 1931.

15  See 'Von Berlin aus gesehen', the first part of Kracauer's 1931 'Pariser Beobachtungen' (Kracauer 1990:5.2:25–6).

16  Kracauer compares the situation of this class with that of the purloined letter in Edgar Allan Poe's famous story, a missive that eludes detection because, rather than being hidden, it remains in full view amongst other letters casually left on a writing desk.

17  See Kracauer (1998:28–9).

18  For a full discussion of Kracauer's writings on this employee culture see Band (1999) and Hofmann in Kessler and Levin eds (1990:87–104).

19  Kracauer (1987:56).

20  See 'Über Arbeitsnachweise', published 17 June 1930 in Kracauer (1987:52–9).

21  See 'Wärmehallen', published 18 January 1931 in Kracauer (1987:59–63).

22  See 'Kritischer Tag', published 15 July 1931 in Kracauer (1996:54–6).

23  See 'Kino in der Münzstrasse', published 2 April 1932 in Kracauer (1987:69–71).

24  See 'Girls und Krise', 1931 in Kracauer (1990:5.2:320–1).

25  See Kracauer's description of the Christmas market in 'Weihnachtlicher Budenzauber', published 24 December 1932 in Kracauer (1987:30–2).

26  Such ideas find their fullest elaboration in Kracauer's copious unpublished 1917 study 'Über das Leid des Wissens und der Sehnsucht nach dem Tat' ('On Suffering under Knowledge and the Desire for the Deed') (Deutsches Literaturarchiv 72.3538) and his partially published 1919–20 study of Georg Simmel (Deutsches Literaturarchiv 72.35.23), a sociologist – and of course Berliner – whose writings both recognise this modern condition and exemplify it. Kracauer returned to this theme of the modern indifference to qualitative distinctions in his final chapter for *Theory of Film* (Kracauer 1997a (1960)), 'Film in Our Time'.

27  The enigmatic piece 'Schrei auf der Strasse' is a good example of this all-encompassing urban anguish: the terrifying scream which pierces the tense atmosphere and which sends a shiver down Kracauer's spine emanates neither from political disturbances on the street nor from the unfortunate victims of everyday violence, but seemingly from the very streets themselves. See Kracauer (1987:21–3).

28  See Kracauer (1987:108).

29  In 'Organisiertes Glück: Zur Wiedereröffnung des Lunarparks', 7 May 1930 (Kracauer 1996:73–5), Kracauer writes: 'unseen organisation ensures that pleasure assails the masses in a prescribed order. Perhaps people want it that way; after all, during the day they are guided by signals, party manifestos and associations' (Kracauer 1996:73).

30  See 'Sonntagsausflug' in Kracauer (1996:43–5).

31  The term *Mittelgebirge* means a low mountain range. Kracauer uses the term to describe an urban hinterland, the site of weekend excursions. He plays on the notion of the *Mittelstand* (the middle classes) who turn the Mittelgebirge into a leisure setting. See Kracauer (1987:90–1).

32  See Kracauer (1996:29–30).

33  Kracauer notes 'The places in which there is laughter are indicative of the audience' (Kracauer 1987:70).

34  See Kracauer (1996:31).

35  Kracauer (1996:31).

36  Kracauer's enthusiasm for the performance of the Andreu-Rivel clowns at the Scala Variety Theatre is the clearest example. See 'Akrobat – Schöön' (Kracauer 1987:101–4).

37  See Benjamin (1973b:236 and 242–3).

38  See Benjamin (1973b:241–2).

39  See Benjamin (1985:104). Kracauer's pianist is an exemplary instance of such mastery amid distraction. Our musician plays fluently and flawlessly, demonstrating a professional expertise developed through many years of practice. He exhibits a technical, tactile control of the keyboard and pedals borne of familiarity and habit. His work no longer requires his attention. His complete proficiency enables him to be both here and elsewhere in the same moment.

40  See Benjamin (1973b:239). Compare Kracauer (1997a (1960):46).

41  Kracauer (1997a (1960)).

42  For this insight I am indebted to Barry Langford's 'The Strangest of Station Names: Changing Trains with Benjamin and Kracauer' (unpublished paper presented at the Siegfried Kracauer Conference at the University of Birmingham, 13–14 July 2002).

43  Compare Benjamin (1973b:238) and Kracauer 1997:48).

44  For a discussion of architecture in Kracauer's *feuilleton* pieces see Hess in Volk (ed.) (1996:111–29).

45  Kracauer (1987:40).

46  Kracauer (1987:30).

47  As some commentators have observed, the book may be seen as a counterpoint and corrective to the abstract and contrived use of montage in Walter Ruttmann's 1927 *Berlin – Symphony of a Great City*, a film Kracauer condemned as portentous, contrived and vacuous. See Kracauer's review of 17 November 1927 in Kracauer (1979a:404–5). See also Kracauer (1947:182–8) and Kracauer (1997a(1960):207).

48  Ward (2002:160) astutely notes this cinematic quality of Kracauer's texts in the *Strassen in Berlin* collection.

49  Built in 1873, the initial fashionable period of the arcade was already over by 1888. See Levin's editorial notes in Kracauer (1995:388–9).

50  First published in *Frankfurter Zeitung*, 16 December 1932.

51  Kracauer (1987:17).

52  Kracauer (1987:17).

53  This moment of recognition and redemption, in which an object or edifice appears to us for the first and only time, becoming memorable in the very instant in which it is about to vanish irrevocably, lies at the heart of Benjamin's Arcades Project. It corresponds to the notion of the 'dialectical image', that conjuncture and mutual illumination of past and present which constitutes Benjamin's decisive historiographical category. Kracauer's observations on the railway bridge

over Friedrichstrasse, offering at one and the same moment a first and a final image of Berlin, serve an allegorical function in this regard.

54   See Kracauer (1987:113–4).

55   Kracauer (1987:113).

56   Kracauer (1987:113).

57   As Kracauer suggests, the images that arise before him are 'without doubt memories, which would like to hold on to him. Paralysed he confronts the past and his child-like face shows only too clearly that he has never come to terms with it' (Kracauer 1987:107).

58   See Kracauer (1997a(1960):137–8).

59   Kracauer (1997a (1960):137).

60   Kracauer (1997a (1960):138). As if to echo *Der Klavierspieler*, Kracauer immediately follows this anecdote in *Theory of Film* by quoting Aaron Copland's comment that film soundtracks consist of music 'one isn't supposed to hear, the sort that helps to fill the empty spots between pauses in a conversation', and Igor Stravinsky's observation that such music 'has the same relationship to the drama that restaurant music has to the conversation at the individual restaurant tables' (Kracauer 1997a (1960):138).

61   See 'Die Hosentraeger: Eine historische Studie', 30 October 1926 (Kracauer 1987:86–8).

62   See 'Falscher Untergang der Regenschirme', 7 April 1926 (Kracauer 1987:88–90).

63   See 'Das Klavier', 23 February 1926 (Kracauer 1987:77–81).

64   See Benjamin (1985:71).

**Chapter 18**

# Orpheus in Hollywood
## Siegfried Kracauer's Offenbach film

*Graeme Gilloch*

## Long shot

Paris, 1864. A crowded boulevard on a late spring evening. Ludovic Halévy and Henri Meilhac, Jacques Offenbach's librettists, are strolling towards the composer's house, exchanging pleasantries and witticisms with acquaintances encountered *en route*. The imperial carriage rolls past, pursued by the usual procession of vehicles occupied by a motley array of notables, fashionable figures, courtesans and other hangers-on. Among them is Hortense Schneider, Offenbach's capricious leading lady for the last decade. Her carriage pulls up. She 'calls the two men over and, in a vulgar outburst, rails against Offenbach'. The carriage moves off. Meilhac's response to this petulant outburst is telling. He turns to Halévy and declares that Hortense was born to play the part of Helen of Troy in an operetta.[1]

This comic incident opens Siegfried Kracauer's *Jacques Offenbach: Motion Picture Treatment*, an unpublished 22-page German text now held as part of Kracauer's literary estate in the Deutsches Literaturarchiv. With this rather peculiar script, written around 1938, Kracauer hoped to interest Hollywood film companies, producers and directors[2] in acquiring the motion picture rights to his then recently published *Jacques Offenbach und die Paris seiner Zeit* (1937), a book already translated into English,[3] the merits and critical acclaim of which Kracauer was quick to extol. Had the *Motion Picture Treatment* succeeded, the ensuing sale would not only have generated much-needed funds for the impoverished exile but also, more importantly, have initiated an important contact with – indeed a potential lifeline to – influential figures in America. It failed. Although Kracauer later notes in a letter to Max Horkheimer[4] that Metro-Goldwyn-Mayer did take out an option on the American edition of the book, nothing came of it. The rights remained unsold, the film unrealised. Kracauer remained poor and in Paris.[5]

The prospect of Kracauer, then nearing his fiftieth birthday, becoming a Hollywood screenplay writer may – but should not – surprise us. The *Motion Picture Treatment* was certainly not Kracauer's only attempt to write for the cinematic medium which so fascinated him. Sometime between 1933 and 1936 he had written a five-page 'thematic sketch (*Ideen-Entwurf*) for a short film' to be entitled *Dimanche*;[6] and in the mid-1940s was involved in writing the script of *The Accident*, later retitled *Below the Surface*, a so-called 'test-film' intended to elicit and examine audience reactions as part of the Frankfurt School's studies in prejudice and anti-semitism.[7]

Kracauer had reason to be quietly optimistic: in a letter of 22 August 1939 he points to the production of a Fox film biography of Offenbach starring Lily Pons.[8] Offenbach's music was enjoying fresh acclaim on the East Coast: the 1937 New York Metropolitan Opera House performance of *The Tales of Hoffmann* is still considered one of the best recordings. In 1938 MGM released Julien Duvivier's *The Great Waltz*, a musical film biography of Johann Strauss, Offenbach's great rival and eventual successor. If this production could be a box office hit, then why not an Offenbach movie? Duvivier's film was to prove a mixed blessing for Kracauer. On the one hand it was an exemplary instance of the cinematic realisation of a musical film based on the life and work of a composer, demonstrating the reconciliation of apparently uncinematic subject matter with the inherent realism of the film medium.[9] On the other hand, Kracauer observes, its financial failure was decisive in MGM's decision not to take up the option on *Orpheus in Paris*.[10]

The Offenbach *Motion Picture Treatment* is neither a brief overview (like *Dimanche*) nor a detailed, finalised film script (like *Below the Surface*), but rather a halfway house, a sketch of the main scenes of the film as a series of vignettes culled from the book with an indication of the tenor and direction of the dialogue, and directions – sometimes general, sometimes specific – regarding the use of Offenbach's music, both as background accompaniment and as actual performance. From the outset Kracauer makes it clear that the text itself was not to stand alone, but should be read in conjunction with the book. This elaborated the essential context and delineated the principal characters on which the *Motion Picture Treatment* drew. In short, the book is the key to the *Motion Picture Treatment*: this is Kracauer's view. I want to reverse this formulation and suggest the following: to appreciate the book, one must read the *Motion Picture Treatment*. My argument is this: viewing Kracauer's Offenbach book through the lens of his film sketch brings into sharp focus its intentions and inventiveness, and facilitates a new appreciation of its critical power, complexity and playfulness. The book appears in its true, vibrant colours only when understood as the literary product of a profoundly cinematic imagination.

As its title suggests, *Jacques Offenbach and the Paris of his Time* was intended to be something other than a conventional biography detailing

Offenbach's life and musical achievements. Kracauer's study was conceived from the outset as a *Gesellschaftsbiographie* (biography of a society),[11] a critical attempt to discern, decipher and reconstruct a particular society and historical epoch – the Paris of the Second Empire – through the lens of one particular, exemplary individual and opus: Offenbach and the operetta. Such a biography was for Kracauer both necessary and legitimate. It was necessary because, like Walter Benjamin, Kracauer recognised Paris as the capital of the nineteenth-century, as the site of 'diverse motifs that persist today' and as 'the immediate precursor of modern society'.[12] Kracauer's biography was intended as a history of the origins of the present, of modernity itself. It was a legitimate undertaking because, in his view, both composer and music were fully in tune and in step with the 'spectacle of wealth and brilliance',[13] the very phantasmagoria of modern commodity culture that stood at the heart of the city.[14]

> Many great artists have been comparatively independent of the times in which they lived; not so Offenbach. He had to be in perpetual contact with the world about him in order to be creative at all. All who knew him bear witness to the fact that he was the very personification of sociability. He plunged into social life because it alone supplied him with the necessary tensions. He lived in the instant, reacting delicately to social changes and constantly adapting himself to them. The speed with which he made a name for himself was largely due to the fact that at the moment of his debut a society was crystallising that satisfied the cravings of his being. He had only to be himself for success to be there for the asking.[15]

The fashions of the arcades and pomp of the world exhibitions; the vitality of the boulevards and cafés; the intrigues of the salons and boudoirs – these provided the essential themes, rhythms, tempi and language for Offenbach's music.[16] The operetta could set the cityscape to music because it corresponded so perfectly with it. The Paris of Offenbach's time was theatrical, fantastical, artificial, nothing less than a dream world,[17] but the operetta harboured other dreams too, which disturbed the sleep of the dreamers. For Kracauer Offenbach's music possessed a utopian moment or promise[18] that was wholly irreconcilable with the frivolity, complacency and self-deception of imperial Paris. Offenbach's levity and wit, while seemingly in accord with a society which invented and privileged the *bon mot*,[19] were born of a melancholy recognition of the hypocrisy, triviality and inhumanity of the Second Empire. The operetta was imbued with a satirical spirit ridiculing and mocking the folly of the very milieu from which it sprang. Offenbach's music was an enchantment that disenchanted his world – unfolding this paradox was central to Kracauer's study, for it was the composer's ironic sensibility and critical distance that

lifted the operetta above the banality and mediocrity of its time and transformed it into an art form worthy of critical appreciation and contemporary
redemption. Kracauer's 'biography of a society' was dedicated to this twin task of
re-evaluation and recovery.

Offenbach's music is not the only thing in need of redemption: so too is
Kracauer's study. Kracauer's claims as to the critical success of his book were not
based on its reception by his closest colleagues. Theodor Adorno, Walter
Benjamin and Leo Löwenthal were unanimous and unequivocal in their condemnation. In his letter to Benjamin of 4 May 1937, Adorno writes with utter dismay:
'It has far exceeded my worst expectations'[20] Adorno denounced the wholly superficial and simplistic treatment of Offenbach's music as 'crassly erroneous',
Kracauer's preface and the idea of a 'societal biography' is dismissed as 'shameless
and idiotic', and the book's supposed 'social observations' are described as 'old
wives' tales, the foolishness and superficiality of which find their only equivalent
in that blinking petty bourgeois look with which, half admiringly and half resent-
fully, he squints at 'society' and indeed the demimonde'.[21] Adorno concludes 'if
Kracauer really does identify with this book, then he has definitely erased himself
from the list of writers to be taken at all seriously',[22] a list which, knowing Adorno,
was probably extremely short and did not always feature Benjamin. Benjamin, a
not wholly disinterested figure given his own ongoing and never to be completed
work on the Paris of the Second Empire, was quick to endorse Adorno's damning
verdict – 'I cannot believe that our judgements about the book diverge in any way'
– and is equally forthright in his critique: 'He has simply made an example of the
thing'.[23] Kracauer's naive and apologetic stance apparently lacked any insight:
'Lovely as many of the things in his sources are,' Benjamin laments, ' they only
appear shabby and mean in the text itself. And hardly any of the numerous anecdotes make a proper effect when they are reproduced.'[24] It is, ironically,
Benjamin who makes the most telling comment, one whose spirit Adorno was to
communicate to Kracauer in a letter of 13 May 1937.[25] Interpreting the book as a
desperate act of financial expediency, Benjamin writes 'with this book Kracauer
has essentially resigned himself. He has composed a text that only a few years
ago would have found its most ruthless critic in the author himself'.[26] The
consensus was that with the Offenbach book Kracauer had betrayed not only
Adorno and Benjamin but, most of all, himself. He must be rescued from his own
folly.[27] By 1937 Adorno knew best. He was fast outgrowing his former mentor
and, dare one say it, his own boots too. He would soon take it upon himself to
rewrite Kracauer's work for him.

The Offenbach book failed to find a defender even in Löwenthal,
Kracauer's closest and most enduring friend at the Frankfurt Institut für
Sozialforschung. As editor of the *Zeitschrift für Sozialforschung*, Löwenthal not
only readily agreed to the publication of Adorno's – albeit more muted – critical
review,[28] he also suggested a less ambivalent, more strident critical conclusion.

For Löwenthal, the fact that Kracauer's book was reviewed at all in the *Zeitschrift* did the study an undeserved honour.[29] It is little wonder then that subsequent scholars have paid little attention to Kracauer's Offenbach writings.[30] Even as sympathetic a critic as Martin Jay passes quickly over the Offenbach book, noting Adorno's 'mixed review' and acknowledging the greater 'conceptual daring and breadth of vision' of Benjamin's *Passagenarbeit*,[31] praise indeed for a text that was never actually finished. The *Motion Picture Treatment* seems to constitute a particular embarrassment and the surest sign of Kracauer's destitution, both economic and intellectual. Had it been made, Jay remarks, the Offenbach motion picture would have been precisely the kind of film that Kracauer himself would later condemn in *Theory of Film* as uncinematic: a mere historical costume drama.[32,33] Jay's conclusion echoes Adorno and Benjamin. But even if undertaken as *Brotarbeit* – a charge Kracauer utterly refuted – the Offenbach studies demand more sustained consideration than this. As far as the *Motion Picture Treatment* is concerned there is a clear precedent for the critical reassessment of minor media works undertaken by impecunious intellectuals – a series of scripts for radio broadcasts for children in the early 1930s written and presented by a certain Dr Walter Benjamin.

Jeffrey Mehlman[34] among others has persuasively argued that Benjamin's radio scripts – among them children's radio dramas, tales of extraordinary figures and events, historical and contemporary stories of Berlin – be understood as 'theoretical toys',[35] playful opportunities to test out the author's fundamental philosophical ideas and theological motifs. If Benjamin's radio writings are really intellectual experiments, allegories for adults disguised as enlightenment for children, surely we should not exclude the possibility that Kracauer's *Motion Picture Treatment* might be something more than schmaltzy, waltzy entertainment for the 'little shop girls' who go to the movies. *Brotarbeit* or not, Kracauer's *Motion Picture Treatment*, read as a 'theoretical toy', as the articulation in miniature of the Offenbach book's critical themes and motifs, may provide new insights and fresh appreciation. To detect in the fragment or minor work the kernel of and key to the greater one – such a way of reading is not only reminiscent of Benjamin's monadological practice but also imitates a conventional scientific and cinematic technique for the exploration of physical reality: the long shot–close up–long shot sequence.[36] If the book forms the panoramic long shot, the context, the background, the whole, then the film treatment constitutes the close up, scrutinising and penetrating the detail. What might such a close up reveal?

## Close up

Let us take Jay's accusation that Kracauer's proposed Offenbach film would have been of an uncinematic nature which shuns the fundamental qualities and possibilities of the medium. There is a degree of truth in this charge: in its very artificiality and closure, Kracauer argues, the historical drama does indeed compromise film's predisposition for capturing unstaged reality as an endless continuum.[37] 'Whenever a film maker turns the spotlight on a historical subject or ventures into the realm of fantasy,' Kracauer warns, 'he runs the risk of defying the basic properties of his medium'.[38] Given such dangers, Kracauer is keen to emphasise and explore how filmmakers seek to 'mitigate the inherently uncinematic character of films that resuscitate the past',[39] to compensate for and alleviate their inevitable 'staginess and finiteness'.[40] Far from compromising Kracauer's cinematic imperatives, the *Motion Picture Treatment* might better be understood as an exemplary cinematic compromise. Even a cursory consideration of the proposed film substantiates this thesis.

It is significant that the film's opening scene is the Parisian boulevard since, as Kracauer repeatedly insists in *Theory of Film*, the metropolitan street with its crowds, distractions, contingencies and ephemera constitutes the most cinematic of subject matter.[41] Film has an 'elective affinity' for the city, a special and 'unwavering susceptibility to the street'.[42] Moreover, in the dawdling duo of Halévy and Meilhac we are immediately presented with the image of the *flâneur*, the observer of the flow of modern urban life whose predilection for the visual, for the fleeting and fortuitous, make him an exemplary cinematic figure.[43] The next scene finds Halévy and Meilhac entering Offenbach's house in the rue Laffitte. As they climb the stairs to his apartment a growing din is audible. People are milling on the staircase. The hallway is crowded with journalists, theatre types, artists and dilettantes. Pushing through this throng into the salon we notice Offenbach, surrounded by yet seemingly oblivious of the noisy hustle and bustle around him. Unperturbed, he sits at a table composing. In accordance with Kracauer's insistence on the primacy of the visual in film, these first images of Offenbach tell us much: he is at home here. From the boulevard to the salon this is his social milieu, this is the vital atmosphere and essential raw material for his musical labours.[44] He is at the height of his powers and popularity amid the superficial 'joy and glamour' of the Second Empire.[45] The operetta is inspired by this world – petty, precious and pretentious, frivolous and fashionable, insincere and eccentric – and provides its signature. This is *la vie parisienne*.

These loud and lively opening scenes provide an interesting contrast with a sequence towards the end of the film which envisions the waning popularity of Offenbach's operettas amid the transformed social, cultural and political conditions of the 1870s.[46] Images of the 1878 World Exhibition appear in *Schnellmontage*. The figure of a sandwich man is seen, a figure who according to Benjamin is the final ruinous incarnation of the *flâneur*. The camera follows him as he trudges along the boulevard past a stand publicising a new invention: Edison's phonograph. We wonder: what is to be the fate of the little work of art, the operetta, in the age of mechanical reproduction? A carriage crosses the screen and, leaving the sandwich man behind, the camera follows the vehicle, which comes to a halt. In the background we see theatrical posters bearing the names of Johann Strauss and Charles Lecocq. There is no mention of Offenbach. The camera peers into the carriage – an ageing, failing Offenbach is its lone occupant. We recognise that Offenbach's world has been transformed, that it is no longer his world at all. He and his operettas are anachronisms. What is important is that all this is conveyed not through captions or dialogue, but through the use of simple street images. The film's narrative is advanced as the camera pursues moving figures and objects – crowds, *flâneurs*, sandwich men, passers-by, carriages. Ideas are manifested as we follow the flow of life on the boulevard. Such scenes suggest that, had it been filmed, Kracauer's Offenbach motion picture would have been far more cinematic than one might initially imagine, and Kracauer the film theorist would probably have approved of *Offenbach: The Motion Picture*.

But I am rushing ahead. Let us rejoin Monsieur Offenbach in his salon. Above the hubbub of the assembled company we suddenly hear a snobbish young man recounting the latest rumours concerning the mutually destructive affair between Hortense Schneider and the ailing Duc de Gramont-Caderousse. Halévy, and the camera, are alone in noticing that Offenbach has stopped writing. In the next scene we encounter the unhappy lovers. The camera pans through the rooms and corridors of the rowdy Café Anglais,[47] just as it previously made its way into Offenbach's home. The camera struggles through crowds of dandies and demimonde, *bon viveurs* and bohemians, snobs and society figures, know-it-alls and non-entities. Gramont-Caderousse is at the head of a group of boisterous revellers, making a drunken spectacle of himself.[48] He finally collapses. In the grey morning light only Hortense Schneider has stayed to tend to him.

What is the significance of such images? Kracauer's *Motion Picture Treatment* presents imperial Paris as an empire of gossip, a cacophonous realm of chatter and conjecture, tittle-tattle and trivia, scandal and slander, speculation and rumour, a world of whispers delivered at full volume. It is depicted as a 'satiated society'[49] in which the 'sole object was to kill time, not to give it a meaning',[50] and in which the craving for novelty, intrigue and distraction ensured what it feared most: silence and boredom.[51] In the seamless movement between salon and street, café bar and boulevard, the opening scenes of the film suggest the blurring of public spaces and private lives, public lives and private spaces, the symbiosis of society figures and the popular press, artists and critics, celebrities and journalists, bourgeois and bohemian. These are all interlaced through the endless circulation of small talk, and in placing Offenbach so comfortably at the centre of this pantheon of prattle, this all too fragile house of *visiting cards*, Kracauer envisages him in a similar manner to Benjamin's 1929 image of Marcel Proust. For Benjamin, Proust's reflections offer incomparable insight into the aristocratic circles of *fin de siècle* Paris in which he moved. As Benjamin suggests, 'it was Proust's aim to design the entire inner structure of society as a physiology of chatter',[52] and this describes precisely the initial scenes of Kracauer's proposed movie, a filmic physiology of chatter represented by means of a panorama of chatterers.

This leads us to reconsider two aspects of Kracauer's Offenbach book. Firstly, it might suggest a new way of interpreting the key notion of a *Gesellschaftsbiographie*. Rather than understanding this as a 'biography of a society' in the broad sense, might this concept not better be seen as a depiction of 'society' in a narrower sense: as intimate social circles, as 'polite society'? In German the word *Gesellschaft* can mean 'company' and 'social gatherings' as well as society as a socioeconomic and political totality. Admittedly this would mean reading Kracauer against the grain of his own explicit intentions as outlined in the preface to the book, but this may be, as Benjamin claims, the most productive, indeed the only proper, way of reading anything.

Secondly, the notion of a physiology of chatter might lead us to revise our understanding of the style and structure of the Offenbach book, particularly Kracauer's reliance upon anecdotal material. Far from being mere 'old wives' tales', as Adorno complained, the anecdote serves as a device for the immanent unfolding of the innermost tendencies of a society nourished by hearsay and rumour. The use of anecdotes is, moreover, a critical technique for disclosing the distinctions and contradictions between surface appearances, perceptions and underlying realities, for unmasking the ideological pretensions and phantasmagoria of a particular era and class. This is central to Kracauer's *Motion Picture Treatment*: conceived as a series of anecdotal episodes highlighting the mismatch between what is and what seems to be, the film fosters the possibilities of comic debunking to the full.[53]

In this sense Halévy and Meilhac's initial encounter with Hortense sets the tone for everything that follows. After listening to the courtesan's histrionics, Meilhac is convinced that Schneider was born to play the role of *La Belle Hélène*, Helen of Troy, the epitome of female beauty. We laugh at the incongruity suggested by Meilhac's remark. We are reminded of Baudelaire's advocacy of the satirical images of the lithographer Honoré Daumier:

> Daumier pounced brutally upon antiquity and mythology and spit on them. The hot-headed Achilles, the prudent Ulysses, the wise Penelope, and that great ninny Telemachus, the beautiful Helen who ruined Troy, the ardent Sappho, patroness of hysterical women, were all portrayed with farcical homeliness that recalled those old carcasses of classical actors who take a pinch of snuff in the wings. [54]

Kracauer pounces and spits too: images of Hortense singing at the premier of *La Belle Hélène* are intercut with the final moments of the dissolute Gramont-Caderousse dying alone in a squalid hotel room. But is Meilhac really joking? His conviction is actually sound, but only because the operetta itself is a satirical form in which nothing is as it seems, and comic reversals and absurd contradictions are the order of the day. This is, after all, the topsy-turvy operetta world of the Second Empire. Hortense will indeed be the perfect Helen of Troy for Offenbach's operetta. Meilhac is telling us the truth, but we find it incredible. In one of Benjamin's radio scripts, a story concerned with bootlegging during the prohibition era in the United States, an African-American boy walks along a train waiting in a station selling 'iced tea' to the passengers. With a knowing smile and sly wink they readily pay exorbitant prices in the belief that the 'iced tea' is actually alcohol, but are mortified to discover that this 'iced tea' is in fact iced tea. As Melhman points out, notice is hereby given: everything is deception, trust nothing.[55] This warning also applies to Kracauer's *Motion Picture Treatment*.

The episode following the Café Anglais debauchery is the clearest example of this. Through the intercutting of two narratives the film presents an anecdote concerning Offenbach's 'reception' at a little town on the Rhine while en route to the fashionable spa resort of Bad Ems.[56] The composer's stopover at the unnamed location unfortunately coincides with a visit by the lieutenant governor (*Regierungspräsident*) from Wiesbaden, who has failed to keep the engagement. Coincidentally in the company of the deputation sent to bring the 'indisposed' dignitary, and with the local band coincidentally playing music from *Orpheus in the Underworld*, Offenbach steps off the Rhine steamer and greets the expectant crowds in the erroneous belief that he is being honoured. They in turn cheer the unwitting impostor and a festive procession ensues. Offenbach is duped and happy; the crowds are duped and happy. Everyone, save the embarrassed deputation, is duped and happy – temporarily at least. Offenbach finally realises his mistake when asked by the local mayor to say a few words to commemorate the town's new gas lighting system. The now-enlightened and humiliated composer makes a discreet exit from proceedings.[57]

These German provincials are not the only ones duped by appearances. Arriving in her cabriolet at the World Exhibition of 1867, Hortense is stopped by an attendant because she has inadvertently come to the entrance reserved for visiting royalty and nobility. The quick-thinking heroine immediately announces that she is none other than La Grande Duchesse de Gèrolstein – the character she was then playing – and the gate is opened with much servile bowing and scraping.[58]

While the film derives most of its comic force and critical edge from such ironic incidents, Kracauer is also not averse to inserting the occasional straightforward joke. On a Viennese street Offenbach comes to the aid of the aged and impoverished Rudolf Zimmer, the composer of eight bars of waltz music that have haunted Offenbach since childhood. When Zimmer has suitably recovered, Offenbach asks him to play the waltz in its entirety. Seated at the piano in a hotel room, Zimmer immediately plays the first eight bars but then stops – he has forgotten the rest too. This sequence is intercut with images of the Paris Commune, culminating in pictures of Paris ablaze and the Commune's bloody suppression.[59] In this intricate interlacing of narratives a smile is brought to, and immediately wiped off, our faces.

The juxtaposition of reality and illusion, ridiculous discrepancies and reversals, farcical impersonations and misunderstandings, ironies, parodies, and jokes that fall flat – these were the lifeblood of Offenbach's operetta. His stage works sought precisely the puncturing of pretension, the dispelling of phantasmagoria, through comedy. The Offenbach book was intended to partake of and correspond to this satirical spirit, this sense of the carnivalesque. So too was the Offenbach film. But we should be careful here. Although a comedy, and a musical one at that, Kracauer's *Motion Picture Treatment* should not be thought of as an attempt to turn Offenbach's life into some kind of operetta or to reproduce the operetta on screen. The reason for this is clear: Kracauer's film treatment is conceived as cinematic, not theatrical. In his *Theory of Film* Kracauer emphasises the antithetical moments of the operatic and the cinematic: 'Opera on the screen is a collision of two worlds detrimental to both'.[60] He explicitly rejects two common filmic strategies for managing this clash. The first is the 'canned operetta' or 'photographed theatre' approach, in which a stage performance is simply recorded by the cameras.[61] Not surprisingly, Kracauer condemns this as utterly uncinematic. Worse still is the attempt to fuse film and opera, to create a spurious synthesis of realistic (cinematic) and formative (artistic, theatrical) elements in order to produce some kind of *Gesamtkunstwerk*. 'As should be expected, this allegedly superior whole invariably reveals itself as an eclectic compromise between irreconcilable entities – a sham whole distorting either the opera or the film or both'.[62]

Neither canned operetta nor *Gesamtkunstwerk*, Kracauer's *Motion Picture Treatment* sought to incorporate Offenbach's music in multiple and complex ways, and this leads us to the very heart of his proposed film. The Offenbach movie is conceived as a musical film of a particular kind – not of the film musical variety, with characters suddenly bursting into song and launching themselves into dance in the midst of otherwise seemingly mundane activities, a genre which Kracauer surprisingly endorses, but rather of the kind which seeks 'to narrate the life of some virtuoso, composer or showman',[63] which integrates music as a 'component of the narrative' and, with its images of the composer at work and rehearsal scenes presents music as a 'product of real life processes'.[64] In the *Motion Picture Treatment* music weaves in and out of the drama, sometimes as background accompaniment linking together simultaneous or succeeding scenes; sometimes in the foreground as actual music, as when a fragment of a performance is presented; sometimes both. Beginning with actual, synchronous music in the theatre, the image of the stage dissolves and, with the music still playing, other images appear. The use of music from *La Belle Hélène* is typical. We start in Hortense Schneider's salon, where Offenbach is trying to persuade the reluctant diva to play the operetta's eponymous heroine.[65] With Offenbach at the piano playing an extract from the score to tempt her, Schneider instinctively begins to sing – cut to the premiere with her on stage in the closing bars before the interval. The camera then focuses on the intrigues of the celebrities and notables among the audience. The music begins playing again but instead of the stage we are presented with a montage of shots of European capitals and theatrical billboards suggesting the international success of the operetta[66] – cut to a Parisian hotel room and Gramont-Caderousse on his deathbed. Now actual music, now commentative, now synchronous, now asynchronous, now in parallel,[67] now in counterpoint[68] – Offenbach's music itself becomes the glue which holds the anecdotes together, creating a sense of the all-pervasiveness and ubiquity of this music and its embeddedness in this society.

This is not all. Offenbach's music is no mere cinematic soundtrack. Kracauer's *Motion Picture Treatment* makes continual use of one particular device: asynchronous, actual music. We hear music as it is being performed on stage, but we do not see the performance itself. *La Belle Hélène* is an excellent example, but there are many occasions when the camera seems wilfully to avoid focusing on the stage. At Bad Ems, for example, the singer Zulma Bouffar has scarcely opened her mouth before Offenbach leaves the theatre box – and we leave with him. Later, when Zulma sings a duet in the one-act *Lieschen und Fritzchen*, the camera is turned not on her but on the audience watching her. In the *Theory of Film* Kracauer argues that this deliberate inattention to actual musical performances has an important effect: it heightens the viewer's sensitivity to the music. In film one comes to an appreciation of music not so much through its direct representation, but rather through a series of digressions and diversions, through interruption and disturbance, above all through the alternation of concentration and distraction. Ironically, film does justice to music only when it seems to decentre it, only when it stays true to its own cinematic imperatives. Kracauer insists that

> loyalty to the medium might prove singularly rewarding. Precisely because it launches the spectator on visual pursuits, it might lead him to the kernel of the music he unavoidably neglects, so that he resembles the fairy-tale prince who, after a series of trials testing his devotion and steadfastness, ultimately finds his beloved at the most unexpected place.

Kracauer's example here is Rene Clair's:

> In the opera episode of *Le Million* the camera does not pay much
> attention to the fat singers and their love duet but literally turns its
> back on them, meandering through the painted stage world and
> focusing on the quarrelling lovers gone astray. We watch the recon-
> ciliation between the two, a pantomime to the sounds of the duet
> which ends with a fiery embrace; and we realise that the lovers are
> transported and driven into each other's arms by the enchanting
> voices and harmonies. And then something miraculous happens:
> absorbed in the sight of the lovers, we enter so completely into them
> that we are no longer aware of their presence but, as if we were they,
> yield to the impact of the duet. Having penetrated the images we find
> at their core, waiting for us, the very music we were forced to
> abandon.[69]

Benjamin would surely have appreciated this passage. After all, for
him digression was the only genuine mode of illuminating the 'truth content' of
the artwork, distraction the precondition for the technical mastery of the
everyday object world. In his essay on Goethe's *Elective Affinities*,[70] Benjamin
emphasises the necessity of circumspection and circumlocution in philosophical
enquiry. We cannot approach philosophical ideas and truths directly, but,
because they have their counterparts in the domain of art, we can approach
them indirectly through the analysis of artworks. Benjamin presents us with an
analogy here. Imagine encountering a shy, enigmatic, intriguing stranger. One
wishes to learn more, but direct questioning would be impolite and embar-
rassing. To avoid any such unpleasantness one asks friends, relatives and neigh-
bours, subtly, discreetly. From such sources we gain an impression of the
stranger. This is how we approach philosophical truths – not through rude inter-
rogation but rather through what others – artworks – have to say.

In this way Kracauer's vision of a world of chatter and gossip becomes
a metaphor for a method of critical analysis in which hearsay is transformed into a
heuristic device. This is how Offenbach's music is presented to us by Kracauer in
both book and film. We are to develop an ever more intimate acquaintance by
paying attention elsewhere, by exploring the social milieu from which it springs,
by understanding the circumstances attending its rise to popularity, by gauging
the responses it elicits from its admirers and detractors, by envisaging the conditions which will ensure its ultimate demise. In Kracauer's *Motion Picture Treatment* we discover something of the truth of Offenbach's music through another
medium. This is Kracauer's purpose. The camera turns its back on the music so
that we may discover it anew. Adorno's bitterest complaint regarding Kracauer's
book, that it ignores Offenbach's music, is characteristically myopic. It does ignore

the music, but for good reason: what Kracauer presents in his Offenbach studies is the operetta recaptured.[71] The *Motion Picture Treatment* brings this sharply and unmistakably into focus. The panoramic quality and continual, anecdotal digressions of *Jacques Offenbach and the Paris of His Time* owe their origins not to the dialectical imagination prized by Adorno, but rather to Kracauer's cinematic imagination.

Dialectics do have a place though. Kracauer's enthusiasm for the film musical genre derives from his understanding of its inner dialectic. For him the musical is to be valued not despite its theatricality but precisely because of it. By its very nature the film musical is concerned with juxtaposing realistic episodes with staged flights of fancy and artifice. In so doing the musical draws attention to and articulates the enduring tension at the heart of the film medium. 'Through its very structure,' Kracauer writes, 'it illustrates the eternal struggle for supremacy between the realistic tendency, suggested by the threadbare intrigue, and the formative tendency, which finds its natural outlet in the songs'.[72] The musical alternates between the intellectual demand for narrative coherence and consistency (characteristic of the non-story film), and the emotional participation generated by fictional contrivances (typical of the story film).

> The conflict between these two antinomic moves, which are natural outlets for the realistic and formative tendencies respectively, materialises in the very form of the musical. No sooner does the real-life intrigue of a musical achieve a certain degree of consistency than it is discontinued for the sake of a production number which often has been delineated at a prenatal stage, thereby corroding the intrigue from within. Musicals reflect the dialectic relation between the story film and the non-story film without ever trying to resolve it. This gives them an air of cinema. Penelope-fashion they eternally dissolve the plot they are weaving. The songs and dances they sport form part of the intrigue and at the same time enhance with their glitter its decomposition. [73]

Kracauer's *Motion Picture Treatment* is clearly more musical film than film musical. Nevertheless, his use of montage to counterpoise appearance and reality, anecdote and actuality, theatre stage and city street,[74] operetta and operetta world, is clearly concerned with maintaining this same tension, with setting in motion the dialectical play of enchantment and disenchantment. What emerges from and survives this process of dissolution – 'corroded intrigue' and 'glittering decomposition' – is the best possible description of the phantasmagoria of the Second Empire. 'All that is solid melts into air' – the air of the cinema. In this evaporation lay Kracauer's aspiration that the Parisian dream world of the nineteenth century might find its true expression in the Hollywood dream factory of the twentieth.

## Long shot

In his writings on Baudelaire and the image of the metropolitan crowd, Benjamin draws a simple but important distinction: the representation of confusion is not the same as confused representation. Similarly, when reading Kracauer's writings on Offenbach we should remember that the examination of the superficial is not to be mistaken for superficial examination. Adorno's rejection of Kracauer's book indicates his own confusion on this point. In refraining from critical aesthetic judgements on Offenbach's music itself, and in illuminating the circumstances which brought it to fleeting prominence, Kracauer did what Adorno was so conspicuously incapable of: treating light music lightly. It is Kracauer's deftness and wit that are so striking in the Offenbach writings, an irony and charm which correspond faithfully with his subject matter. In Kracauer, Offenbach's operetta finds not its abject apologist but its most astute sociologist. It also finds its would-be cinematographer. As the *Motion Picture Treatment* suggests, Kracauer's work was informed by cinematic imperatives: he sought to set images in motion, dancing to the merry rhythms of Offenbach's melodies. Such music was the accompaniment to the flow of life on the street, a score which found its perfect libretto in salon gossip. In envisaging the Parisian boulevard, in rendering the physiology of chatter, Kracauer understood the medium of film as a digression leading to the heart of the operetta. If it had been produced, his film would have constituted a popular entertainment portraying and penetrating the popular entertainment of the recent past. Perhaps it is as well that nothing came of it – imagine Adorno's scornful letter addressed to Kracauer in Beverley Hills!

If they are to become anything more than a dead book and forgotten screenplay, we must come to understand the depths of Kracauer's work on Offenbach. I will spare the reader the obvious Orpheus analogies here, but if we are to appreciate Offenbach and the Paris of his time afresh, if we are to stumble upon these writings anew, if Kracauer is also to be recaptured, we too must make a detour – via Hollywood.

## Notes

1   Kracauer (n.d.a:2).
2   In his brief correspondence with the Los Angeles-based film producer Max Laemmle in 1939, Kracauer asked Laemmle to speak with William Wyler, whose dance scenes in *Jezabel*

impressed Kracauer, and with the director of *The Great Ziegfeld*, Leonhard, in connection with the proposed Offenbach film (Deutsches Literaturarchiv 72.1531/1)

3   Published in England in 1937 as *Offenbach and the Paris of his Time* and in 1938 in the USA as *Orpheus in Paris: Offenbach and the Paris of His Time*. This translation has just been reissued by Zone Books (New York, 2002) under the full title *Jacques Offenbach and the Paris of His Time*. This new version includes a translation of Kracauer's original preface, absent from the 1937 translation. All references are to the 2002 edition.

4   Letter to Horkheimer of 16 September 1941 (Max Horkheimer archive, Mappe I 14, letter 161).

5   In the early autumn of 1941 Kracauer's hopes for the Offenbach film were briefly rekindled in New York. He wrote to Laemmle (24 September 1941) to ask for the return of his copy of the motion picture treatment, having left his own in Paris in the chaos of departure. With Horkheimer's encouragement, Kracauer made contact with the European Film Foundation. The film director and writer William Dieterle and his wife, the actress Charlotte Dieterle, expressed their enthusiasm. Kracauer realised, however, that there was no genuine interest in an Offenbach film and the subject was dropped (Max Horkheimer archive, Mappe I 14, letters 161–3).

6   Published in Volk (ed.) (1996:209–12).

7   See Deutsches Literaturarchiv 72.3620 and Max Horkheimer archive Mappe IX.

8   Letter to Laemmle, Deutsches Literaturarchiv 72.1531/3

9   See Kracauer (1997a(1960):151 and 200).

10  Letter to Horkheimer of 16 September 1941 (Max Horkheimer archive, Mappe I 14, letter 161).

11  Kracauer (2002:23).

12  Kracauer (2002:23).

13  Kracauer (2002:49).

14  In this sense, Kracauer notes, his study could be understood as 'a biography of a city' (Kracauer 2002:24).

15  Kracauer (2002:90).

16  'Only in Paris,' wrote Kracauer, 'were there all the elements, material and verbal, that made the operetta possible.' (Kracauer 2002:215-6). He later added 'All the ingredients of Offenbach's operettas existed in reality.' (Kracauer 2002:253).

17  Kracauer observed 'The operetta would never have been born had the society of the time not itself been operetta-like; had it not been living in a dream-world, obstinately refusing to wake up and face reality.' (Kracauer 2002:215).

18  Kracauer noted 'The music is also addressed to the infinite, and belongs to the never-never land to which Offenbach belonged, the only country in which he had any real roots.' (Kracauer 2002:109).

19  A 'peculiarly Parisian product' (Kracauer 2002:216).

20  Adorno and Benjamin (1999:183).

21  Adorno and Benjamin (1999:184).

22  Adorno and Benjamin (1999:184).

23  Adorno and Benjamin (1999:186).

24  Adorno and Benjamin (1999:186).

25  Max Horkheimer (archive Mappe VI:31–9) In this letter Adorno couches his critique in terms of a difficult but dutiful act to sustain a vital, meaningful friendship. 'I am attacking you,' he wrote, 'to defend you from yourself' (p. 39). Kracauer robustly defended his study in his reply of 25 May 1937, dismissing Adorno's critique as 'foolish' and misplaced. See: Deutsches Literaturarchiv 72.1119/18.

26  Adorno and Benjamin (1999:185).

27  Benjamin should have been far more wary of this high-handed view of supposedly misdirected intellectual endeavours. The next couple of years would see Adorno instructing him as to his own best interests with regard to the Baudelaire studies.

28  Published in the 6. Jahrgang, 1937, Nr. 3 pp. 697–8. Given the delicacy of the situation, Adorno proudly described this text as a successful 'dance on eggshells (*Eiertanz*)' (letter to Löwenthal of 1 October in Leo Löwenthal archive, Mappe A7:208).

29  In a letter to Adorno of 21 September 1937 Löwenthal praises the review of Kracauer's book and especially the care and subtlety of its devastating critique. (Leo Löwenthal archive Mappe A7:205).

30  The most notable exception being Frisby (1988). See also Grimstad in Kessler and Levin (eds.) (1990:59–76) and Koch in Kracauer (2002:11–21).

31  Jay (1986:166).

32  Jay (1986:167).

33  Kracauer (1997a (1960)).

34  Mehlman (1993).

35  Mehlman (1993:4).

36  See Kracauer (1997a (1960):52).

37  See Kracauer (1997a (1960):77–92).

38  Kracauer (1997a (1960):77).

39  Kracauer (1997a (1960):79).

40  Kracauer (1997a (1960):79).

41  As the definitive modern medium, film is a privileged mode for the penetration and representation of the definitive modern environment, the metropolis: this, rather than any naive advocacy of 'realism', is at the heart of Kracauer's much-misunderstood and falsely maligned theory of film. See for example Kracauer (1997a (1960):19, 31–2, 50, 62 and 72).

42  Kracauer (1997a (1960):62).

43  See Kracauer (1997a (1960):72 and 170).

44  Kracauer noted how 'In the midst of a crowded social gathering, Offenbach would suddenly become absorbed and begin to cover sheets of paper with innumerable little flies' feet. Others might have required deep peace, but the concentration and poise necessary for composition would come to him in the midst of a buzz of conversation.' (Kracauer 2002:196).

45  See Kracauer (2002:151).

46  Kracauer (n.d.a:20).

47  A particularly fashionable café after 1856 and the frequent setting for wild revelries and excesses. See Kracauer (2002:239–40).

48  The Duc apparently had a fondness for throwing the café's crockery out of the windows. See Kracauer (2002:240).

49  Benjamin (1973b:208).

50  Kracauer 2002:122).

51  Kracauer noted how those who chose 'to plunge headlong into a life of pleasure' were required to undergo 'the most extraordinary fatigues for its sake'. (Kracauer 2002:80).

52  Benjamin (1973b:208).

53  Kracauer emphasised how the cinematic medium, with its possibilities of intercutting narratives and of shifting vantage point to reveal something previously hidden or obscured, is particularly adept at comic debunking. See Kracauer (1997a (1960):306–8).

54  Benjamin (1999:743 convolute b2, 3).

55  See Mehlman (1993:7–12).

56  See Kracauer (2002:271–2).

57  See Kracauer (n.d.:4–5).

58 See Kracauer (n.d.:15) and Kracauer (2002:314). Kracauer noted that for Schneider 'the differences between operetta and real life were visibly obliterated'. (Kracauer 2002:313).

59 See Kracauer (n.d.:16).

60 Kracauer (1997a (1960):154).

61 Kracauer (1997a (1960):155).

62 Kracauer (1997a (1960):154). Interestingly, Kracauer's example of this is Michael Powell and Emeric Pressburger's 1951 film version of Offenbach's *Tales of Hoffmann* (Kracauer 1997a (1960):155).

63 Kracauer (1997a (1960):150).

64 Kracauer (1997a (1960):151). Kracauer's model here is Duvivier's Johann Strauss film *The Great Waltz*.

65 See Kracauer (2002:273).

66 As Kracauer observed, the success of this operetta was such that 'Europe was at Offenbach's feet'. (Kracauer 2002:280).

67 For example, panoramic images of the 1867 World Exhibition and of Schneider's mischievous visit merge into the gala of Offenbach's *La Grande-Duchesse de Gèrolstein.*

68 As when Paris burns to the accompaniment of Zimmer's tune.

69 Kracauer (1997a (1960):151–2).

70 Benjamin (1996:333).

71 Kracauer (1997a (1960):151).

72 Kracauer (1997a (1960):148).

73 Kracauer (1997a (1960):213).

74 Consider the following: A republican insurrection is taking place on the boulevards; from the side streets, mounted dragoons ride out to disperse the crowds; a street battle ensues; a woman screams.This scene of terror is viewed, Kracauer points out, from a bird's-eye perspective: from the balcony of the Théâtre des Variétés where Halévy is standing with two ballet dancers (See: Kracauer n.d.:15).

# Bibliography

Abrams, M.H. (1953) *The Mirror and the Lamp: Romantic Theory and the Critical Tradition*, Oxford: Oxford University Press.

Adorno, T.W. (1963) *Quasi una fantasia*, Frankfurt: Suhrkamp.

Adorno, T.W. (1969) *Minima Moralia*, Frankfurt: Suhrkamp.

Adorno, T.W. (1970) *Ästhetische Theorie*, Frankfurt: Suhrkamp.

Adorno, T.W. (1977) *Ohne Leitbild: Parva Aesthetica*, Frankfurt: Suhrkamp (Gesammelte Schriften 10.1).

Adorno, T.W. and Benjamin, W. (1999) (Lonitz, H., ed.; Walker, N., trans.) *The Complete Correspondence, 1928–1940. Theodor W. Adorno and Walter Benjamin*, Cambridge: Polity Press.

Anderson, S. (2000) *Peter Behrens and a New Architecture for the Twentieth Century*, Cambridge, Mass: MIT Press.

Anton, H. (1965) 'Modernität als Aporie und Ereignis', in Steffen, H. (ed.) *Aspekte der Modernität*, Göttingen: Vandenhoeck & Ruprecht.

Antonowa, I. and Merkert, J. (eds) (1995) *Berlin–Moskau, Moskau–Berlin 1900–1950*, Munich: Prestel.

Anz, T. and Stark, M. (1994) *Die Modernität des Expressionismus*, Stuttgart: Metzler.

Aristotle (1982) *Poetics*, New York: Norton.

Atkinson, W. C. (1934) *Spain: A Brief History*, London: Methuen.

Bahr, H. (1987) (Farkas, R., ed.) *Prophet der Moderne: Tagebücher 1888–1904*, Vienna: Böhlau.

Baldick, R. (1960) *The Goncourts*, London: Bowes and Bowes.

Band, H. (1999) *Mittelschichten und Massenkultur*, Berlin: Lukas Verlag.

Bann, S. (ed.) (1974) *The Tradition of Constructivism*, London: Thames and Hudson.

Barthes, R. (1994) *The Semiotic Challenge*, Berkeley: University of California Press.

Baudelaire, C. (1943) *Constantin Guys: le peintre de la vie moderne*, Geneva: La Palatine.

Baudelaire, C. (1968a) *Le peintre de la vie moderne*, Oeuvres Complètes Paris: du Seuil.

Baudelaire, C. (1968b) *Les Fleurs du Mal*, Oeuvres Complètes Paris: du Seuil.

Baudelaire, C. (1969) *The Painter of Modern Life and Other Essays*, London: Phaidon.

Baudelaire, C. (1970 and 1995) (Mayne, J., ed. and trans.) *The Painter of Modern Life and Other Essays*, London: Phaidon.

Bauer, H. (1963) 'Architektur als Kunst', in Bauer, H. (ed.) *Kunstgeshichte und Kunsttheorie im 19 Jahrhundert*, Berlin: De Gruyter.

Bauman, Z. (1989) *Modernity and the Holocaust*, Oxford: Blackwell.

Bauman, Z. (1991) *Modernity and Ambiguity*, Oxford: Blackwell.

Bauman, Z. (1995) *Moderne und Ambivalenz: Das Ende der Eindentigkeit*, Hamburg: Fisher.

Bauman, Z. (2000) *Liquid Modernity*, Cambridge: Polity.

Beaver, P. (1993) *The Crystal Palace: A Portrait of a Victorian Enterprise*, Chichester: Phillmore.

Beazley, E. and Lambert, S. (1964) 'The Astonishing City', *The Architects' Journal*, 1015.

Becket, D., (1990) *A Study of Le Corbusier's Poème de l'angle droit*, MPhil thesis, University of Cambridge.

Behler, E. (1988) *Studien zur Romantik und zur idealistischen Philosophie*, Paderborn: Schöning.

Behler, E. and Horisch, J. (eds) (1987) *Die Aktualität der Frühromantik*, Paderborn: Schöning.

Behne, A. (1919) *Wiederkehr der Kunst*, Leipzig: Kurt Wolff.

Behne, A. (1996) *The Modern Functional Building*, Santa Monica: Getty Center.

Behr, S., Fanning, D. and Jarman, D. (eds) (1993) *Expressionism Reassessed*, Manchester: Manchester University Press.

# Bibliography

Behrendt, C.W. (2000) *The Victory of the New Building Style*, Los Angeles: Getty Center.

Behrens, P. (1901) *Feste des Lebens und der Kunst: Eine Betrachtung des Theaters als höchsten Kultursymbols*, Leipzig: Diederichs.

Behrens, P. (1923) 'Die Dombauhütte', *Deutsche Kunst und Dekoration* **xxvi**, 220–30.

Behrens, P. (1932) 'Zeitloses und Zeitbewegtes', *Zentralblatt der Bauverwaltung*, **31**, 361–5.

Behrens, P. (1987) 'Die Form', in Fischer, W. (ed) *Zwischen Kunst und Industrie: Der Deutsche Werkbund*, Stuttgart: DVA, 181–4.

Benjamin, W. (1969a) *Charles Baudelaire: Ein Lyriker im Zeitalter des Hochkapitalismus*, Frankfurt: Suhrkamp.

Benjamin, W. (1969b) (Zohn, H., trans.) *Illuminations*, New York: Schocken.

Benjamin, W. (1973a) (Zohn, H., trans.) *Charles Baudelaire: A Lyric Poet in the Age of High Capitalism*, London: New Left Books.

Benjamin, W. (1973b) *Illuminations* (Introduced and edited by H Arendt), London: Fontana.

Benjamin, W. (1977) 'Paris, die Hauptstadt des XIX. Jahrhunderts', in *Illuminationen*, Frankfurt: Suhrkamp.

Benjamin, W. (1978) *Briefe*, Frankfurt: Suhrkamp.

Benjamin, W. (1979) *One-Way Street*, London: New Left Books.

Benjamin, W. (1980) *Gesammelte Schriften*, Frankfurt: Suhrkamp.

Benjamin, W. (1982) 'Das Passagen-Werk', in *Gesammelte Schriften V*, Frankfurt: Suhrkamp.

Benjamin, W. (1985) *One Way Street and Other Writings*, London: Verso.

Benjamin, W. (1996) *Selected Writings, Vol. 1*, Cambridge, Mass: Harvard University Press.

Benjamin, W. (1999) (Eiland, H. and McLaughlin, K., trans.) *The Arcades Project*, Cambridge, Mass: Harvard University Press.

Benson, T. (ed.) (2001) *Expressionist Utopias: Paradise, Metropolis, Architectural Fantasy*, Berkeley: University of California Press.

Berman, M. (1982) *All That Is Solid Melts Into Air*, London: Verso.

Bernstein, R. (ed.) (1985) *Habermas and Modernity*, Cambridge: Polity.

Bichet, P. (1890) *L'Art et le bien-être chez soi*, Paris: Marpon & Flammarion.

Billy, A. (1954) *Les Frères Goncourt*, Paris: Flammarion.

Billy, A. (1960) (Shaw, M., trans.) *The Goncourt Brothers*, London: Deutsch.

Bisanz, R. (1975) 'The Romantic Synthesis of the Arts: Nineteenth-century German Theories on a Universal Art', *Konsthistorisk Tidskrift*, **xxxxiv**, 38–46.

Blau, E. (1999) *The Architecture of Red Vienna, 1919–1934*, Cambridge, Mass: MIT Press.

Bloch, E. (1990) *Heritage of Our Times*, Berkeley: University of California Press.

Blomfield, R. (1904) '"Greek" Thomson, A Critical Note', *Architectural Review*, **xv**, 194.

Blumenberg, H. (1974) *Säkularisierung und Selbstbehauptung*, Frankfurt: Suhrkamp.

Blumenberg, H. (1983) *The Legitimacy of the Modern Age*, Cambridge: MIT Press.

Blumenberg, H. (1993) 'Nachahmung der Natur: Zur Vorgeschichte der Idee des schöpferischen Menschen', in *Wirklichkeiten, in denen wir leben*, Stuttgart: Reclam, 55–103.

Blundell-Jones, P. (1995) *Hans Scharoun*, London: Phaidon.

Boden, R.F (ed.) (1994) *Now Here Space, Time and Modernity*, Berkeley: University of California Press.

Boffrand, G. (2003) (Edited and introduced by Caroline van Eck, translated by David Britt) *A Book of Architecture*, Aldershot: Ashgate.

Böhringer, H. (1978) 'Avantgarde – Geschichte einer Metapher', *Archiv für Begriffsgeschichte*, **22**.

Borden, I. (1997) 'Space Beyond', *Journal of Architecture*, **2**, 313–35.

Bot, M. (1968) *Francis Picabia et la Crise de Valeurs Figuratives 1900–1925*, Paris: Klincksieck.

Bott, G. (1977) *Von Morris zum Bauhaus*, Hanau: Peters.

Bötticher, C.F. (1874) *Tektonik der Hellenen*, Berlin: Ernst & Korn.

Bourget, P. (1886) 'Mm. Edmond et Jules de Goncourt', in Bourget, P. (ed.) *Nouveaux essais de psychologie contemporaine*, Paris: Lemerre, 137–98.

Bradbury, M. and McFarlane, J. (1976) 'The Name and Nature of Modernism', in Bradbury and McFarlane (eds.) *Modernism*, Harmondsworth: Penguin.

Brett, D. (1992) *C.R. Mackintosh: The Poetics of Workmanship*, London: Reaktion Books.

Breuer, R. (1910) 'Häuser, die Künstler sich bauten', *Über Land und Meer*, **105**, 195–7.

Breuer, S. (1995) *Ästhetischer Fundamentalismus: Stefan George und der deutsche Antimodernismus*, Darmstadt: Wissenschaftliche Buchgesellschaft.

Brooks, H.A. (ed.) (1982–1984) *Le Corbusier Archive 8*, New York: Garland.

Brunner, O. (ed.) (1978) *Geschichtliche Grundbegriffe: Historisches Lexikon zur politisch-sozialen Sprache in Deutschland*, (Vol. IV) Stuttgart: Klett.

Bryant, G. (1999) 'Architecture as 'Precursor of Redemption'? Industrial Culture and the idea of the *Gesamtkunstwerk* in German modernism', *Macjournal*, **4**, 94–103.

Buchholz, K. *et al.* (eds) (2001) *Die Lebensreform: Entwürfe zur Neugestaltung von Leben und Kunst um 1900*, (2 vols.) Darmstadt: Häusser.

Buck-Morss, S. (1989) *The Dialectics of Seeing: Walter Benjamin and the Arcades Project*, Cambridge, Mass: MIT Press.

Buddensieg, T. and Rogge, H. (1980) *Industriekultur: Peter Behrens und die AEG 1907–1914*, Berlin: Gebr. Mann.

Buderath, B. (ed.) (1990) *Peter Behrens – Umbautes Licht: Das Verwaltungsgebäude der Hoechst AG*, Munich: Prestel.

Burkhauser, J. (ed.) (1990) *Glasgow Girls: Women in Art and Design 1880–1920*, Edinburgh: Canongate.

Bushart, M. (1990) *Der Geist der Gotik und die expressionistische Kunst*, Munich: Schreiber.

Bürger, P. (1974) *Theorie der Avantgarde*, Frankfurt: Suhrkamp.

Bürger, P. (1978) *Seminar Literatur und Kunstsoziologie*, Frankfurt: Suhrkamp.

Bürger, P. (1992) *The Decline of Modernism*, University Park, Penn.: Pennsylvania State University Press.

Cacciari, M. (1993) *Architecture and Nihilism*, New Haven: Yale University Press.

Calder, J. (ed.) (1986) *The Enterprising Scot – Scottish Adventure and Achievement*, Edinburgh: Royal Museum of Scotland.

Calinescu, M. (1977) *Faces of Modernity: Avant-Garde, Decadence, Kitsch*, Bloomington: Indiana University Press.

Calinescu, M. (1987) *Five Faces of Modernity*, Durham N.C.: Duke University Press.

Caramaschi, E. (1971) *Réalisme et Impressionisme dans l'oeuvre des frères Goncourt*, Paris: A.G. Nizet and Pisa, Libreria Goliardica.

Carl, P. (1988) 'Le Corbusier's Penthouse in Paris. 24 Rue Nungesser-et-Coli', *Daidalos*, **28**, 65–75.

Carr, R. (1982) *Spain 1808–1975*, Oxford: Clarendon Press.

Carswell, C. (1920) *Open The Door!*, London: Chatto and Windus.

Carlyle, T. (n.d.) *Sartor Resartus: The Life and Opinions of Herr Teufelsdrockh*, New York: Lovell, Coryell & Co.

Cascardi, A.J. (1987) 'Genealogies of Modernism', *Philosophy and Literature*, **11**, 207–25.

Cascardi, A.J. (1992) *The Subject of Modernity*, Cambridge: Cambridge University Press.

Cassirer, E. (1981) (Haden, J., trans.) *Kant's Life and Thought*, New Haven, Conn: Yale University Press.

Cassou, J. (1960) *Les Sources du XXe siècle: Les arts en Europe de 1884 à 1914*, Paris: Textes RMN.

Charbonnier, G. (1960) *Le Monologue du peintre: Entretien avec Le Corbusier*, Paris: Julliard.

Checkland, O. (1989) *Britain's Encounter with Meiji Japan, 1868–1912*, Edinburgh: Macmillan.

Chytry, J. (1989) *The Aesthetic State: A Quest in Modern German Thought*, Berkeley: University of California Press.

Clark, K. (1950) *The Gothic Revival*, London: Constabble.

Cohen, J.L. (1987) *Le Corbusier and the Mystique of the USSR. Themes and Projects for Moscow 1928–36*, Princeton: Princeton University Press.

# Bibliography

Collins, S. (1989) *From Divine Cosmos to Souvereign State: An Intellectual History of Consciousness and the Idea of Order in Renaissance England*, Oxford: Oxford University Press.

Colomina, B. (1994) *Privacy and Publicity*, Cambridge, Mass: MIT Press.

Colomina, B. (1994) 'Mies Not', in Mertins, D. (ed.) *The Presence of Mies*, New York: Princeton Architectural Press.

Commission of the Great Exhibition (1851 a) *Illustrated Catalogue of the Industry of All Nations*, London: Art Journal.

Commission of the Great Exhibition (1851 b) *The Official Descriptive and Illustrated Catalogue of the Great Exhibition*, London.

Constant, B. (1980) 'De la liberté des anciens comparée à celle des modernes – discours prononcé à l'Athénée Royale de Paris en 1819', in Gauchet, M. (ed.) *De la liberté chez les modernes*, Paris: Librairie Générale Française, 491–515.

Cook, E.T. and Wedderburn, A. (eds) (1903–12) *The Works of John Ruskin*, London.

Cooke, C. (1984) 'Fedor Osipovich Shekhtel', *AA Files*, 4–8.

Le Corbusier (1929) 'Tracés régulateurs', *L'Architecture Vivante*, Spring/Summer, 13–23.

Le Corbusier (1933a) 'Quel rôle joue l'esprit poétique?' *L'Architecture d'Aujourd'hui*, **10**, 63–4.

Le Corbusier (1933b) 'Discours d'Athènes', August 1933, *L'Architecture d'Aujourd'hui*, **10**, 80–89.

Le Corbusier (1933c) 'Le Pavillion Suisse à la Cité Universitaire', *L'Architecture Vivante*, **Winter**, 34–51.

Le Corbusier (1935) 'Les Arts dits primitifs dans la maison d'aujourd'hui', *L'Architecture d'Aujourd'hui*, **6**, 83–5.

Le Corbusier (1938) *Oeuvre plastique peintures et dessins architecture*, Paris: Albert Morancé.

Le Corbusier (1946a) *Towards a New Architecture*, London: The Architectural Press.

Le Corbusier (1946b) 'Programme of L'Esprit Nouveau, no. 1 (October, 1920)', in *Towards a New Architecture*, London: Architectural Press.

Le Corbusier (1948) *New World of Space*, New York: Raynal and Hitchcock.

Le Corbusier (1960) *Creation is a Patient Search*, New York: Praeger.

Le Corbusier (1964) (Boesiger, W., ed.) *Oeuvre complète II, 1929–34*, Zurich: Girsberger.

Le Corbusier (1967) *The Radiant City*, London: Faber.

Le Corbusier (1971) 'The Charter of Athens (1941)', in Conrads, U. (ed.) *Programs and Manifestos on 20th-Century Architecture*, Cambridge, Mass: MIT Press.

Le Corbusier (1987) *The Decorative Art of Today*, London: The Architectural Press.

Le Corbusier (1989) *Le Poème de l'angle droit*, Paris: Editions Connivences.

Crook, J.M. (1981) *William Burges and the High Victorian Dream*, London: Murray.

Crook, J.M. (1987) *The Dilemma of Style*, London: Murray.

Cumming, E. (1992) 'Industry and Art', in *Glasgow 1900: Art and Design*, Amsterdam: Van Gogh Museum, 9–14.

Curtis, W. (1986) *Le Corbusier: Ideas and Forms*, Oxford: Phaidon.

Dahlhaus, C. (1978) *Schönberg und Andere. Aufsätze*, Mainz: Schott.

Dahlhaus, C. (1982) *Esthetics of Music*, Cambridge: Cambridge University Press.

Daly, C. (1864) *L'Architecture privée au dix-neuvième siècle*, Paris: Morel.

Darwin, C. (1959) *The Origin Of Species*, Philadelphia: Pennsylvania University Press.

Donahue, N. (ed.) (1995) *Invisible Cathedrals: The Expressionist Art History of Wilhelm Worringe*, University Park: Pennsylvania State University Press.

Donald, J. (1992) 'Metropolis: the city as text', in Bocock, R. and Thompson, K. (eds) *Social and Cultural Forms of Modernity*, Milton Keynes: Open University Press, 417–61.

Donald, J. (1999) *Imagining the City*, London: Athlone.

Douglas, M. (1966) *Purity and Danger: an Analysis of the Concepts of Purity and Taboo*, London: Routledge and Kegan Paul.

Dreyfus, H.L. (1987) 'Misrepresenting Human Intelligence', in Born, R. (ed.) *Artificial Intelligence: The Case Against*, London and Sydney: Croom Helm.

Dubos, R. (1965) *Man Adapting*, New Haven and London: Yale University Press.

Durkheim, E. (1984) *The Division of Labour in Society*, Basingstoke: Macmillan.

Durkheim, E. (1995) *The Elementary Forms of Religious Life*, New York: Free Press.

Durning, L. and Wrigley, R. (eds) (2000) *Gender and Architecture*, Chichester: Wiley.

Eagleton, T. (1990) *The Ideology of the Aesthetic*, Oxford: Blackwell.

Eck, C.v. (1994) *Organicism in nineteenth-century architecture: an inquiry into its theoretical and philosophical background*, Amsterdam: Architectura & Natura.

Eco, U. (2002) *Baudolino*, London: Secker and Warburg.

Emmerich, W. and Wege, C. (1995) *Der Technikdiskurs in der Hitler-Stalin-Ära*, Stuttgart: Metzler.

Engels, F. (1962) 'The Housing Question', in K. Marx and F. Engels, *Selected Works*, Moscow: Progress Publishers, 557–635.

Enzensberger, H.M. (1962) 'Die Aporien der Avantgarde', in Enzensberger, H.M. (ed.) *Einzelheiten*, Frankfurt: Suhrkamp.

Ernst, M. (1948) *Beyond Painting, and Other Writings by the Artist and His Friends*, New York: Witterborn and Schulz.

Ernst, M. (1982) *Leaves Never Grow on Trees: Max Ernst's Histoire naturelle*, London: The Arts Council of Great Britain.

Evans-Pritchard, E.E. (1964) *The Position of Women in Primitive Society and Other Essays on Social Anthropology*, London and New York: The Free Press.

Farguell, I.M. and Grandval, V. (eds) (1998) *Hameaux, villas et cités de Paris*, Paris: Action Artistique de la Ville de Paris.

Flaubert, G. (1954) (Barzun, J., trans.) *Dictionary of Accepted Ideas*, Norfolk, Conn: New Directions.

Foucault, M. (1984) 'What is Enlightenment?', in Rabinow, P. (ed.) *The Foucault Reader*, New York: Pantheon, 35–50.

Foucault, M. (1992) *The Order of Things*, London: Routledge.

Fowle, F. (1991) 'The Hague School and the Scots: A Taste for Dutch Pictures', *Apollo*, **134**, 108–11.

Franciscono, M. (1971) *Walter Gropius and the Creation of the Bauhaus in Weimar: The Ideals and Artistic Theories of its Founding Years*, Urbana: University of Illinois Press.

Frazer, J.G. (1900) *Pausanias and Other Greek Sketches*, London and New York: Macmillan.

Friedland, R. and Boden, D. (eds) (1994) *Now Here Space, Time and Modernity*, Berkeley: University of California Press.

Friedrich, H. (1956) *Die Struktur der modernen Lyrik*, Hamburg: Rowohlt.

Frisby, D. (1985) *Fragments of Modernity*, Cambridge: Polity.

Frisby, D. (1988) *Fragments of Modernity: Theories of Modernity in the Works of Simmel, Kracauer and Benjamin*, Cambridge, Mass: MIT Press.

Frisby, D. (1990) 'Deciphering the Hieroglyphics of Weimar Berlin', in Suhr, C.W. (ed.) *Berlin, Culture and Metropolis*, Minneapolis: University of Minnesota Press, 152–65.

Frisby, D. (1992) *Simmel and Since*, London: Routledge.

Frisby, D. (2001) *Cityscapes of Modernity*, Cambridge: Polity.

Frisby, D. (2002) 'The City as Text', in Leach, N. (ed.) *Hieroglyphics of Space*, London: Routledge, 15–30.

Frisby, D. (2003) 'Straight or Crooked Streets: The Contested Rational Spirit of the Modern Metropolis', in Whyte, I.B. (ed.) *Modernism and the City of the Spirit*, London: Routledge, 57–64.

Frisby, D. (forthcoming) *Metropolitan Architecture and Modernity: Otto Wagner's Vienna*, Minneapolis: University of Minnesota Press.

Frisby, D. and Featherstone, M. (eds) (1997) *Simmel on Culture*, London: Sage.

Fritsche, P. (1996) *Reading Berlin 1900*, Cambridge, Mass: Harvard University Press.

Gadamer, H.G. (1975) *Truth and Method*, London: Sheed and Ward.

Gadamer, H.G. (1986a) *The Relevance of the Beautiful and Other Essays*, Cambridge: Cambridge University Press.

# Bibliography

Gadamer, H.G. (1986b) *Wahrheit und Methode: Grundzüge einer philosophischen Hermeneuti*, Tübingen: Mohr.

Gadamer, H.G. (1989) (Weinsheimer, J. and Marsh, D. G., trans.) *Truth and Method*, London: Sheed and Ward.

Gehlen, A. (1965) *Zeit-Bilder*, Frankfurt: Athenäum.

Germann, G. (1976) 'Gottfried Semper, über Konvention und Innovation', *Zeitschrift für Sweitzerische Archäologie und Kunstgeschichte*, **33**.

Geuss, R. (2001) *Public Goods, Private Goods*, Princeton: Princeton University Press.

Giddens, A. (1990) *The Consequences of Modernity*, Stanford: Stanford University Press.

Giddens, A. (1992) *Modernity and Self-Identity*, Cambridge: Polity.

Giedion, S. (1949) *Space, Time and Architecture*, Cambridge, Mass: Harvard University Press.

Gilloch, G. (1996) *Myth and Metropolis*, Cambridge: Polity.

Gilloch, G. (2002) *Walter Benjamin*, Cambridge: Polity.

Gleissner, R. (1988) *Die Entstehung der ästhetischen Humanitätsidee in Deutschland*, Stuttgart: Metzler.

Gombrich, E.H. (1960) *Art and Illusion*, London: Phaidon.

Gombrich, E.H. (1978a) *Kunst und Fortschritt: Wirkung und Wandlung einer Idèe*, Köln: du Mont.

Gombrich, E.H. (1978b) 'The Renaissance Conception of Artistic Progress and its Consequences', in Gombrich, E.H. (ed.) *Norm and Form*, London and New York: Phaidon.

de Goncourt, E. (1881) *La maison d'un artiste*, Paris: Charpentier.

de Goncourt, E. (1882) *La Faustin*, Paris: Charpentier.

de Goncourt, E. and de Goncourt, J. (1864) *Germinie Lacerteux*, Paris: Charpentier.

de Goncourt, E. and de Goncourt, J. (1879) *Les Frères Zemganno*, Paris: Charpentier.

de Goncourt, E. and de Goncourt, J. (1887) (preface by Zola) *Germinie Lacerteux – a realistic novel*, London: Vizetelly and Co.

de Goncourt, E. and de Goncourt, J. (1915) (West, J., ed. and trans.) *The Journal of the De Goncourts: Pages from a Great Diary*, London: Nelson.

de Goncourt, E. and de Goncourt, J. (1937) (Galantère, L., trans.) *The Goncourt Journals, 1851– 1870*, London: Cassell.

de Goncourt, E. and de Goncourt, J. (1955) (Turnell, M., trans.) *Germinie*, London: Weidenfeld and Nicholson.

de Goncourt, E. and de Goncourt, J. (1956) *Journal: mémoires de la vie littéraire*, Paris: Fasquelle et Flammarion.

de Goncourt, E. and de Goncourt, J. (1962) (Baldick, R., ed. and trans.) *Pages from the Goncourt Journal*, Oxford: Oxford University Press.

de Goncourt, E. and de Goncourt, J. (1966) (Clark, L. and Allan, I., trans.) *The Zemganno Brothers*, London: Mayflower-Dell.

de Goncourt, E. and de Goncourt, J. (1971) (Becker, G.J. and Philips, E., eds and trans.) *Paris and the Arts, 1851–1896: from the Goncourt Journal*, Ithaca: Cornell University Press.

Goodhart-Rendel, H.S. (1953) *English Architecture since the Regency*, London: Constable.

Gordon, J. (1979) 'Decadent Spaces: Notes for a Phenomenology of the Fin-de-Siècle', in Fletcher, I. (ed.) *Decadence and the 1890s*, London: Edward Arnold.

Grassi, E. (1962) *Die Theorie des Schönen in der Antike*, Köln: DuMont.

Greenberg, C. (1965) 'Modernist Painting', in *Art and literature*, no. 4/1965.

Greenberg, C. (1977) 'After abstract Expressionism', in Dickie and Sclafani (eds) *Aesthetics, A Critical Anthology*, New York: St Martin's.

Groys, B. (1987) 'Die totalitäre Kunst der 30er Jahre: Antiavantgardistisch in der Form und avantgardistisch im Inhalt', in Harten, J., Schmidt, H.-W. and Syring, M.L. (eds) *Die Art hat geblüht: Europäische Konflikte der 30er Jahre in Erinnerung an die frühe Avantgarde* (exhibition catalogue), Düsseldorf: Städtische Kunsthalle.

Groys, B. (1988) *Gesamtkunstwerk Stalin: Die gespaltene Kultur in der Sowjetunion*, Munich: Hanser.

Gumbrecht, H.V. (1978) 'Modern, Moderne, Modernität', in Brunner, O., Conze, W. and Kosellec, R. (eds) *Geschichtliche Grundbegriffe*, Vol. 4, Stuttgart: Klett-Cotta, 93–131.

Gusdorf, G. (1976) *Naissance de la conscience romantique au siècle des lumières*, Paris: Payot.

Haag Bletter, R. (1981) 'The Interpretation of the Glass Dream: Expressionist Architecture and the History of the Crystal Metaphor', *Journal of the Society of Architectural Historians*, **xi**, 20–43.

Habermas, J. (1962) *Strukturwandel der Öffentlichkeit*, Frankfurt: Suhrkamp.

Habermas, J. (1979) 'Modernity – An Incomplete Project', in Foster, H. (ed.) *Postmodern Culture*, London: Pluto, 3–15.

Habermas, J. (1981) *Theorie des kommunikativen Handelns*, Frankfurt: Suhrkamp.

Habermas, J. (ed.) (1985a) *Observations on the Spiritual Situation of the Age*, Cambridge, Mass: MIT Press.

Habermas, J. (1985b) *Der Philosophische Diskurs der Moderne*, Frankfurt: Suhrkamp.

Habermas, J. (1987) *The Philosophical Discourse of Modernity*, Cambridge, Mass: MIT Press.

Habermas, J. (1989) *The Structural Transformation of The Public Sphere*, Cambridge: Polity.

Hall, A. R. (1956) *The Scientific Revolution 1500–1800: The Formation of the Modern Scientific Attitude*, Boston: Beacon Press.

Hall, M. (2000) 'What do Victorian Churches Mean? Symbolism and Sacramentalism in Anglican Church Architecture, 1850–1870', *Journal of the Society of Architectural Historians*, **59/1**, 78–95.

Harbron, D. (1942) 'Thomas Harris', *Architectural Review*, **xcii**, 63–6.

Hardy, T. (1979) 'The Fiddler of the Reels', in *The Distracted Preacher and other Tales*, Harmondsworth: Penguin.

Harries, K. (1991) 'Theatricality and Re-presentation', *Perspecta*, **26**.

Harrison, C. and Wood, P. (eds) (1992) *Art in Theory, 1900–1990*, Oxford: Blackwell.

Harvey, D. (1985) *Consciousness and the Urban Experience*, Oxford: Blackwell.

Harvey, D. (1989) *The Condition of Postmodernity*, Oxford: Blackwell.

Haxthausen, C.W. and Suhr, H. (eds) (1990) *Berlin, Culture and Metropolis*, Minneapolis: University of Minnesota Press.

Häusler, G., (1990) 'In the Artwork we become One: The Problem of the *Gesamtkunstwerk* in the Visual Arts in the Early 20th Century', MPhil thesis, University of Cambridge, Department of Architecture and History of Art.

Hegel, G.W.F. (1972) (Reichelt, H., ed.) *Grundlinien der Philosophie des Rechts*, Frankfurt: Ullstein.

Hegel, G.W.F. (1991) (Wood, A.W., ed.; Nisbet, H.B., trans.) *Elements of the Philosophy of Right*, Cambridge: Cambridge University Press.

Heidegger, M. (1992) *Being and Time*, Oxford: Blackwell.

Helland, J. (1996) *The Studios of Frances and Margaret Macdonald*, Manchester: Manchester University Press.

Henrich, D. and Iser, W. (eds) (1982) *Theorien der Kunst*, Frankfurt: Suhrkamp.

Herbert, R. (ed.) (1964) *Modern Artists on Art*, Englewood Cliffs: Prentice-Hall.

Herf, J. (1984) *Reactionary Modernism: Technology, Culture and Politics in Weimar and the Third Reich*, Cambridge: Cambridge University Press.

Hermand, J. (ed.) (1971) *Jugendstil*, Darmstadt: Wissenschaftliche Buchgesellschaft.

Herrmann, W. (ed.) (1992) *'In What Style should we build?': The German Debate on Architectural Style*, Santa Monica: Getty Center.

Hessel, F. (1984) *Ein Flaneur in Berlin*, Berlin: Arsenal.

Hessel, F. (1999a) 'Das andre Berlin', in *Sämtliche Werke* (Vol. 3), Oldenburg: Igel.

Hessel, F. (1999b) 'Von den schwierigen Kunst spazieren zu gehen', in *Sämtliche Werke* (Vol. 5), Oldenburg: Igel.

Hessisches Landesmuseum (1977) *Ein Dokument Deutscher Kunst 1901–1976* (5 vols.) Darmstadt: Roether.

# Bibliography

Heynen, H. (1999) *Architecture and Modernity: A Critique*, Cambridge, Mass: MIT Press.

Heynen, H. (1999) 'Architecture as Critique of Modernity', in H. Heynen, *Architecture and Modernity*, Cambridge, Mass: MIT Press, 148–174.

Hillebrand, B. (ed.) (1978) *Nietzsche und die deutsche Literatur* (2 vols.), Tübingen: Niemeyer.

Hofmann, W. (1970) *Von der Nachahmung zur Erfindung der Wirklichkeit: Die schöpferische Befreiung der Kunst 1890–1917*, Cologne: DuMont.

Honey, S. (1986) *Mies van der Rohe: European Works*, London: Academy Editions.

Horkheimer, M. and Adorno, T.W. (1973) *Dialectic of Enlightenment*, London: Allan Lane.

Horn-Oncken, A. (1967) *Über das Schickliche*, Göttingen: Akademie der Wissenschaften.

Howard, E. (1945) *Garden Cities of To-Morrow*, London: Faber and Faber.

Hübsch, H. (1828) *In welchem Stil sollen wir bauen?*, Karlsruhe: C.F. Müller.

Hvattum, M. (2001) 'Methods of Invention: Gottfried Semper between Poetics and Practical Aesthetics', *Zeitschrift für Kunstgeschichte*, **64**, 537–46.

Institut Mathildenhöhe Darmstadt (1990) *Aufbruch zur Moderne: Die Darmstädter Künstlerkolonie zwischen Tradition und Innovation*, Darmstadt: Roether.

Jaeschke, W. and Holzhey, H. (eds) (1990) *Früher Idealismus und Frühromantik: Der Streit um die Grundlagen der Ästhetik 1795–1805*, Hamburg: Meiner.

Jameson, F. (1991) *Postmodernism, or the Cultural Logic of Late Capitalism*, London: Verso.

Jameson, F. (2002) *A Singular Modernity*, London: Verso.

Janofske, E. (1984) *Architektur-Räume*, Braunschweig: Vieweg.

Jauss, H.R. (1970) 'Literarische Tradition und gegenwärtiges Bewusstsein der Modernität', in Jauss, H.R. (ed.) *Literaturgeschichte als Provokation*, Frankfurt: Suhrkamp.

Jauss, H.R. (1989) *Studien zum Epochenwandel der ästhetischen Moderne*, Frankfurt: Suhrkamp.

Jay, M. (1986) *Permanent Exiles: Essays on the Intellectual Migration from Germany to America*, New York: Columbia University Press.

Jefferies, M. (1995) *Politics and Culture in Wilhelmine Germany: The Case of Industrial Architecture*, Oxford: Berg.

Jencks, C. (1989) *What is Postmodernism?*, London: Academy Editions.

Jenkins, I. (1992) *Archaeologists and Aesthetes in the Sculpture Galleries of the British Museum 1800–1939*, London: British Museum.

Jennings, H.F. (1902) *Our Homes and How to Beautify Them*, London: Harrison & Sons.

Johnson, W. (1865–7) *Proceedings of the Glasgow Architectural Society, 1865–7*, Glasgow: The Glasgow Architectural Society.

Johnson, P. (1947) *Mies van der Rohe*, New York: Museum of Modern Art.

Kalas, E. (1933) 'The Art of Glasgow', in *Charles Rennie Mackintosh, Margaret Macdonald: Memorial exhibition catalogue*, Glasgow: McLellan Galleries.

Kamphausen, A. (1952) *Gotik ohne Gott: Ein Beitrag zur Deutung der Neugotik und des 19. Jahrhunderts*, Tübingen: Matthiesen.

Kaplan, W. (ed.) (1996) *Charles Rennie Mackintosh*, New York and Glasgow: Abbeville Press.

Kelly, M. (ed.) (1996) *Critique and Power. Recasting the Foucault/Habermas Debate*, Cambridge, Mass: MIT Press.

Kessler, M. and Levin, T. (eds) (1990) *Siegfried Kracauer. Neue Interpretationen*, Tübingen: Stauffenburg Verlag.

Kinchin, J. (1996) 'Interiors: nineteenth-century essays on the "masculine" and the "feminine" room', in Kirkham, P. (ed.) *The Gendered Object*, Manchester: Manchester University Press, 12–29.

Kinchin, J. (1998) 'Art and History into Life: the Revival of Pageantry in Scotland', *Journal of the Scottish Society for Art History*, **2**, 42–51.

Kinchin, P. and Kinchin, J. (1988) *Glasgow's Great Exhibitions 1888, 1901, 1911, 1938, 1988*, Wendlebury: White Cockade Publishing.

Klein, R. (1952) 'Sur la signification de la synthèse des arts', *Revue d'esthetique*, **V**, 289–311.

Klinger, C. (1995) *Flucht – Trost – Revolte: Die Moderne und ihre ästhetischen Gegenwelten*, Munich: Hanser.

Klotz, H. (1988) *The History of Postmodern Architecture*, Cambridge, Mass: MIT Press.

Kockelmans, J.J. and Kisiel, T.J. (eds) (1970) *Phenomenology and the Natural Sciences*, Evanston: Northwestern University Press.

Kohlschmidt, W. *et al.* (eds) (1958) *Reallexikon der deutschen Literaturgeschichte*, Berlin: de Gruyter.

Koselleck, R. (ed.) (1977) *Studien zum Beginn der modernen Welt*, Stuttgart: Klett-Cotta.

Koselleck, R. (1985) (Tribe, K., trans.) *Futures Past; on the Semantics of Historical Time*, Cambridge, Mass: MIT Press.

Koselleck, R. (1989) *Vergangene Zukuft: Zur Semantik geschichtlicher Zeiten*, Frankfurt: Suhrkamp.

Koselleck, R. and Widmer, P. (eds) (1980) *Niedergang: Studien zu einem geschichtlichen Thema*, Stuttgart: Klett-Cotta.

Kostka, A. and Wohlfarth, I. (eds) (1999) *Nietzsche and an 'Architecture of our Minds'*, Los Angeles: Getty Center.

Kracauer, S. (1947) *From Caligari to Hitler: A Psychological Study of the German Film*, Princeton, NJ: Princeton University Press.

Kracauer, S. (1974) *Kino: Essays, Studien, Glossen zum Film*, Frankfurt: Suhrkamp.

Kracauer, S. (1979) *Von Caligari zu Hitler: Eine psychologische Geschicthe des deutschen Films*, Frankfurt: Suhrkamp.

Kracauer, S. (1987) *Strassen in Berlin und anderswo*, Berlin: Das Arsenal.

Kracauer, S. (1990) *Schriften* Vol. 5, 1–3, Frankfurt: Suhrkamp.

Kracauer, S. (1995) *The Mass Ornament: Weimar Essays*, Cambridge, Mass: Harvard University Press.

Kracauer, S. (1996) *Berliner Nebeneinander: Ausgewählte Feuilletons 1930–33*, Zürich: Edition Epoca.

Kracauer, S. (1997a (1960)) *Theory of Film: The Redemption of Physical Reality*, Princeton, NJ: Princeton University Press (first pub. Oxford University Press 1960).

Kracauer, S. (1997b) *Frankfurter Turmhäuser: Ausgewählte Feuilletons 1906–30*, Zürich: Edition Epoca.

Kracauer, S. (1998) *The Salaried Masses: Duty and Distraction in Weimar Germany*, London: Verso.

Kracauer, S. (2002) (David, G. and Mosbacher, E., trans.) *Jacques Offenbach and the Paris of his Time*, New York: Zone Books.

Kracauer, S. (n.d.a) *Jaques Offenbach: Motion Picture Treatment*, Marbach am Neckar: Deutsches Literaturarchiv 72.3529.

Kracauer, S. (n.d.b) *Siegfried Kracauer Nachlass*, Marbach am Neckar: Deutsches Literaturarchiv.

Krawietz, G. (1995) *Peter Behrens im Dritten Reich*, Weimar: VDG.

Krustrup, M. (1991) *Le Corbusier: Porte Email; Palais d'Assemblée de Chandigarh*, Copenhagen: Arkitektens Forlag.

Kuhn, T. (1962) *The Structure of Scientific Revolutions*, Chicago, Ill: University of Chicago Press.

Lachtermann, D.R. (1989) *The Ethics of Geometry: A Genealogy of Modernity*, London: Routledge.

Lampugnani, V.M. and Schneider, R. (eds) (1994) *Moderne Architektur in Deutschland – 1900 bis 1950: Expressionismus und Neue Sachlichkeit*, Stuttgart: Hatje.

Lang, J. (1987) *Creating an Architectural Theory*, New York: Van Nostrand Reinhold.

Lankheit, K. (1976) 'Gottfried Semper und die Weltausstellung, London 1851', in Vogt, Reble and Fröhlich (eds) *Gottfried Semper und die Mitte des 19. Jahrhunderts*, Basel: Birkhäuser.

Lash, S. and Friedman, J. (eds) (1992) *Modernity and Identity*, Oxford: Blackwell.

Leach, N. (ed.) (1997) *Rethinking Architecture*, London: Routledge.

Leck, R.M. (2000) *Georg Simmel and Avant-Garde Sociology*, Amherst: Humanity Books.

Lefebvre, H. (1984) *Everyday Life in the Modern World*, New Brunswick: Transaction.

Lefebvre, H. (1991) *The Production of Space*, Oxford: Blackwell.

Lefebvre, H. (1995) *Introduction to Modernity*, London: Verso.

# Bibliography

Lepenies, W. (1978) *Das Ende der Naturgeschichte*, Frankfurt: Suhrkamp.

Lewis, W.S. (ed.) (1955) *Horace Walpole's Correspondence* (Vol. 28), London and New Haven: Oxford University Press and Yale University Press.

Libeskind, D. (1980) *End Space*, London: Architectural Association.

Logan, G. (1905) 'A Colour Symphony', *The Studio*, **36**, 118.

Long, R.-C.W. (ed.) (1995) *German Expressionism: Documents from the End of the Wilhelmine Empire to the Rise of National Socialism*, Berkeley: University of California Press.

Loos, A. (1962) 'Die Geschichte des armen reichen Mannes', in Loos, A. (ed.) *Sämtliche Schriften*, Vienna: Herold. Also in English translation, Adolf Loos, *Spoken into the void: Collected Essays 1897–1900*, Cambridge, Mass., MIT Press, 1982, 125–7.

Love, N.S. (1986) *Marx, Nietzsche and Modernity*, New York: Columbia University Press.

Lovejoy, A.O. (1974) *The Great Chain of Being*, Cambridge, Mass: Harvard University Press.

Lovelace, A. (1991) *Art for Industry: The Glasgow Japan Exchange of 1878*, Glasgow: Glasgow Museums.

Löwith, K. (1949) *Meaning in History*, Chicago: Chicago University Press.

Löwith, K. (1967) *Weltgeschichte als Heilsgeschichte*, Stuttgart: Kohlhammer.

Lynch, K. (1960) *The Image of the City*, Cambridge, Mass: MIT Press.

Lyotard, J.F. (1984) *The Postmodern Condition*, Minneapolis: University of Minnesota Press.

Macdonald, M. (1903) *Letter from Margaret Macdonald to Anna Muthesius*, private collection.

Macdonald, J. (1992) 'Metropolis: the City as Text', in Thompson, R.B. (ed.) *Social and Cultural Forms of Modernity*, Cambridge: Polity.

Mackay, T. (1900) *The Life of Sir John Fowler, Engineer, Bart., K.C.M.G., etc.*, London.

Mackintosh, C.R. (1990) 'Seemliness (1902)', in Robertson, P. (ed.) *Charles Rennie Mackintosh: the Architectural Papers*, Wendlebury: White Cockade Publishing.

Macready, S. and Thompson, F.H. (eds) (1985) *Influences in Victorian Art and Architecture*, London: Society of Antiquaries of London.

Mallgrave, H.F. (1985) 'Gustav Klemm and Gottfried Semper, the meeting of ethnological and anthropological theory', *RES, Journal of Anthropology and Aesthetics*, **9**.

Mallgrave, H.F. (ed.) (1993) *Otto Wagner. Reflections on the Raiment of Modernity*, Santa Monica: Getty Center.

Mallgrave, H.F. (1996) *Gottfried Semper, Architect of the Nineteenth Century*, New Haven and London: Yale University Press.

Mallgrave, H.F. and Herrmann, W. (eds) (1989) *Gottfried Semper: The Four Elements of Architecture and other writings*, Cambridge and London: Cambridge University Press.

Man, P.d. (1984) *The Rhetorics of Romanticism*, New York: Columbia University Press.

Mandelbaum, M. (1971) *History, Man and Reason: A Study in Nineteenth-Century Thought*, Baltimore: Johns Hopkins.

Manuel, F.E. and Manuel, F.P. (1979) *Utopian Thought in the Western World*, Cambridge: Harvard University Press.

Martin, D. (1897) *The Glasgow School of Painting*, London: George Bell & Son.

Marx, K. (1978) 'Speech at the Anniversary of the People's Paper (1856)', in Tucker, R.C. (ed.) *The Marx–Engels Reader*, New York: Norton.

Marx, K. and Engels, F. (1973) *The Communist Manifesto*, New York: Labor Publications.

Mattenklott, G. (1985) *Bilderdienst: Ästhetische Opposition bei Beardsley und George*, Frankfurt: Rowohlt.

Measom, G. (1859) *The Official Illustrated Guide to the Lancaster and Carlisle, Edinburgh and Glasgow, and Caledonian Railways*, Glasgow: Murray and Sons.

Mehlman, J. (1993) *Walter Benjamin for Children: An Essay on his Radio Years*, Chicago: University of Chicago Press.

Meier-Graefe, J. (1905) 'Peter Behrens = Düsseldorf', *Dekorative Kunst*, **viii**, 381–428.

Meier-Graefe, J. (1987) *Entwicklungsgeschichte der modernen Kunst*, Munich: Piper.

Meinecke, F. (1972) (Anderson, J. E., trans.) *Historism: The Rise of a New Historical Outlook*, London: Routledge and Kegan Paul.

Merleau-Ponty, M. (1962) (Smith, C., trans.) *Phenomenology of Perception*, London: Routledge and Kegan Paul.

Mertins, D. (2002) 'Living in a Jungle: Mies, Organic Architecture, and the Art of City Building', in Lambert, P. (ed.) *Mies in America*, New York: Abrams.

Mézières, N. Le Camus de (1992) *The Genius of Architecture; or, the Analogy of that Art with our Sensations*, Santa Monica: Getty Center.

Michelson, A. (1998) 'Where is your rupture? Mass Culture and the Gesamtkunstwerk', in Krauss, R. *et al.* (eds) *October: The Second Decade*, Cambridge, Mass: MIT Press, 95–115.

Middleton, R. (ed.) (1996) *The Idea of the City*, London: The Architectural Association.

Mitchel, W. (1995) *City of Bits: Space, Place, and the Infobahn*, Cambridge, Mass: MIT Press.

Mix, Y.-G. (ed.) (2000) *Naturalismus, Fin de Siecle, Expressionismus 1890–1918*, Munich: dtv.

Monzie, A. de (1925) 'Eloge de Charcot', *Revue neurologique*, **1**, 1159–62.

Moore, R. (1977) *Le Corbusier, Myth and Meta-Architecture: The Late Period 1947–65*, Atlanta: Georgia State University.

Moos, S.V. (1979) *Le Corbusier: Elements of a Synthesis*, Cambridge: MIT Press.

Moravánszky, A. (1996) *Competing Visions. Aesthetic Innovation and Social Imagination in Central European Architecture, 1867–1918*, Cambridge. Mass: MIT Press.

Morris, W. (1892) *Preface to The Nature of Gothic: A Chapter of The Stones of Venice*, Hammersmith: Kelmscott Press.

Motycka Weston, D., (1994) *The Problem of Space in Early Twentieth-Century Art and Architecture*, PhD thesis, University of Cambridge.

Motycka Weston, D. (2002) 'The Hour of the Enigma: The Phenomenal Temporality in the Metaphysical Painting of Giorgio de Chirico', in Heck, C. and Lippincott, K. (eds) *Symbols of Time in the History of Art*, London: Brepols.

Mountjoy, D. (1910) *A Creel of Peat: Stray Papers*, London: Adelphi Press.

Muir, J.H. (1901) *Glasgow in 1901*, Glasgow and Edinburgh: John Murray.

Muthesius, H. (1902) 'Die Glasgower Kunstbewegung: Charles R Mackintosh und Margaret Macdonald-Mackintosh', *Dekorative Kunst*, 193–221.

Muthesius, H. (1979) (Sharp, D., ed.) *The English House*, New York: Rizzoli.

Neat, T. (1994) *Part Seen Part Imagined*, Edinburgh: Canongate.

Nerdinger, W. *et al.* (eds) (1990) *Revolutionsarchitektur: Ein Aspekt der europäischen Architektur um 1800*, Munich: Hirmer.

Neumann, F.v. (1899) 'Die Moderne in der Architeketur und im Kunstgewerbe', *Zeitschrift des Österreichischen Ingenieur-und Architekten-Vereins*, **51**, 145–9.

Neumeyer, F. (1986) *Mies van der Rohe: Das kunstlose Wort*, Berlin: Siedler.

Neumeyer, F. (1991) *The Artless World, Mies van der Rohe on the Building Art*, Cambridge, Mass: MIT Press.

Newbery, F. (1909) *The Birth and Growth of Art*, Glasgow.

Nietzsche, F. (1906) 'Die fröhliche Wissenschaft', in *Werke*, Leipzig: Naumann.

Nietzsche, F. (1980) *Sämtliche Werke*, Vol. 6, Munich: dtv.

Nietzsche, F. (1988) *Kritische Studienausgabe*, Munich: dtv.

Nipperdey, T. (1988) 'Wehlers Gesellschaftsgeschichte', *Geschichte und Gesellschaft*, **14**.

Nisbet, H.B. (1972) *Goethe and the Scientific Tradition*, London: Institute of Germanic Studies, University of London.

Nochlin, L. (1968) 'The Invention of the Avant-Garde', in Hess, T.B. and Ashbery, J. (eds) *Avant-Garde Art*, New York: Collier.

Nochlin, L. (1971) *Realism*, Harmondsworth: Penguin.

de Noussane, H. (1896) *Le goût dans l'ameublement*, Paris: Firmin-Didot.

Oettermann, S. (1980) *Das Panorama: Die Geschichte eines Massenmediums*, Frankfurt: Syndikat.

Osborne, H. (1979) *Abstraction and Artifice in Twentieth Century Art*, Oxford: Clarendon.

Osborne, P. (1992) 'Modernity is a Qualitative, Not a Chronological Category', *New Left Review*, **192**, 65–8.

Osborne, P. (1995) *The Politics of Time: Modernity and Avant-Garde*, London: Verso.

Pagan, J. (1847) *Sketches of the History of Glasgow*, Glasgow: Stuart.

Page, D. and Stamp, G. (2001) 'Interpreting the Egyptian Halls', *Alexander Thomson Society Newsletter*, **28**, 6–10.

Pausanias (1971) (Levi, P., trans.) *Guide to Greece*, Harmondsworth: Penguin.

Payne, A. (1999) *The Architectural Treatise in the Italian Renaissance: Architectural Invention, Ornament, and Literary Culture*, Cambridge: Cambridge University Press.

Paz, O. (1976) *Point de convergence*, Paris: Gallimard.

Pehnt, W. (1998) *Die Architektur des Expressionismus*, Ostfildern-Ruit: Hatje.

Percy, C. and Ridley, J. (eds) (1985) *The Letters of Edwin Lutyens to his Wife Lady Emily*, London: Collins.

Petit, J. (ed.) (1970) *Le Corbusier lui-même*, Geneva: Rousseau.

Pevsner, N. (1972) *Some Architectural Writers of the Nineteenth Century*, Oxford: Clarendon.

Platz, G. (1975) 'Elements in the Creation of a New Style', in Benton, T. C. and Sharp, D. (eds) *Form and Function*, London: Open University Press.

Podro, M. (1982) *The Critical Historians of Art*, New Haven: Yale University Press.

Poggioli, R. (1968) *The Theory of the Avant-Garde*, Cambridge, Mass: Belknap.

Pope, A. (1921) 'Epitaph intended for Sir Isaac Newton', in *The Poetical Works of Alexander Pope*, London: Macmillan.

Posener, J. (1981) 'Vorlesungen zur Geschichte der neuen Architektur III: Das Zeitalter Wilhelms II', *Arch+*.

Posener, J. (1995) *Berlin auf dem Wege zu einer neuen Architektur*, Munich: Prestel.

Prange, R. (1991) *Das Kristalline als Kunstsymbol: Bruno Taut und Paul Klee: Zur Reflexion des Abstrakten in Kunst und Kunsttheorie der Moderne*, Hildesheim: Olms.

Preziosi, D. (ed.) (1998) *The Art of Art History: A Critical Anthology*, Oxford: Oxford University Press.

Pye, D. (1968) *The Nature and Art of Workmanship*, Cambridge: Cambridge University Press.

Rabinow, P. (ed.) (1984) *The Foucault Reader*, New York: Pantheon.

Raeburn, M. and Wilson, V. (1987) *Le Corbusier, Architect of the Century*, London: The Arts Council of Great Britain.

Rasch, W. (ed.) (1970) *Bildende Kunst und Literatur: Beiträge zum Problem ihrer Wechselbeziehungen im 19. Jahrhundert*, Frankfurt: Klostermann.

Richardson, A.E. (1914) *Monumental Classic Architecture in Great Britain and Ireland*, London and New York: Batsford.

Ricoeur, P. (1965) 'Universal Civilization and National Cultures', in Ricoeur, P. (ed.) *History and Truth*, Evanston: Northwestern University Press.

Riehl, W. (1851–69) *Die Naturgeschichte des deutschen Volkes als Grundlage einer deutschen Sozialpolitik*, Stuttgart: Cotta, 4 vols.

Robson-Scott, W.D. (1965) *The Literary Background of the Gothic Revival in Germany*, Oxford: Clarendon.

Rosemont, F. (ed.) (1978) *What is Surrealism? Selected Writings*, London: Pluto Press.

Rosen, C. and Zerner, H. (1982) 'What is, and is not, Realism?', *New York Review of Books*, 18 Feb and 4 March.

Rowe, C. and Koetter, F. (1978) *Collage City*, Cambridge, Mass: MIT Press.

Rowe, C. (1994) *The Architecture of Good Intentions*, London: Academy Editions.

Ruskin, J. (1907a) 'The Nature of Gothic', in *The Stones of Venice*, Vol. 2, London: Dent.

Ruskin, J. (1907b) 'Unto This Last', in *Sesame and Lilies, Unto This Last, and the Political Economy of Art*, London: Cassell.

Russel, D.A. and Winterbottom, M. (eds) (1972) *Ancient Literary Criticism: The Principal Texts in New Translations*, Oxford: Oxford University Press.

Russell, J.B. (1887) *The House in relation to Public Health*, Transactions of the Insurance and Actuarial Society of Glasgow, 2nd series, No. 5.

Salmon, F. (2000) *Building on Ruins: The Rediscovery of Rome and English Architecture*, Aldershot: Ashgate.

Sant'Elia, A. (1973) 'Manifesto of Futurist Architecture', in: Apollonio, V. (ed.) *Futurist Manifestos*, London: Thames and Hudson.

Sayer, D. (1992) *Capitalism and Modernity*, London: Routledge.

Scaff, L. (1995) 'Social Theory, Rationalism and the Architecture of the City', *Theory, Culture and Society*, **12, 2**, 63–86.

Scharoun, H. (1967) 'Bauen und Leben', *Bauwelt*, **58**.

Scheerbart, P. (1889) *Das Paradies, die Heimat der Kunst Berlin*, Berlin: Verlag Deutscher Fantasten.

Schelling, F.W.J. (1966) *Philosophie der Kunst*, Darmstadt: Wissenschaftliche Buchgesellschaft.

Schinkel, K.F. (1979) *Architektonisches Lehrbuch*, Munich and Berlin: Deutsche Kunstverlag.

Schlegel, F. (1958) *Kritische Friedrich–Schlegel–Ausgabe*, Paderborn: Schöning.

Schorske, C.E. (1963) 'The Idea of the City in European Thought: Voltaire to Spengler', in Handlin, O. and Burchard, J. (eds) *The Historian and the City*, Cambridge, Mass: MIT Press.

Schrade, H. (1936) *Schicksal und Notwendigkeit der Kunst*, Leipzig: Armanen.

Schuster, K.P. (ed.) (1980) *Peter Behrens und Nürnberg: Geschmackswandel in Deutschland, Historismus, Jugendstil und die Anfänge der Industrieform*, Munich: Prestel.

Schwarzer, M. (1995) *German Architectural Theory and the Search for Modern Identity*, Cambridge: Cambridge University Press.

Scott, G.G. (1857) *Remarks on Secular and Domestic Architecture, Present and Future*, London: John Murray.

Scott, G.G. (1879) *Personal and Professional Recollections by the late Sir George Gilbert Scott, R.A.*, London.

Scott, G.G. (junior) (1881) *An Essay on the History of English Church Architecture*, London: Simpkin, Marshall and Co.

Seckler, E. and Curtis, W. (1978) *Le Corbusier at Work*, Cambridge, Mass: Harvard University Press.

Semper, G. (1852) *Practical Art in Metals and Hard Materials; its Technology, History and Styles*, London.

Semper, G. (1853) 'Outline for a System of Comparative Style-Theory, MS 122 ff.1–37 and MS 124 ff.5–28', *RES, Journal of Anthropology and Aesthetics*, **6**, 5–32.

Semper, G. (1856) 'The Attributes of Formal Beauty', in Herrmann, W. (ed.) *Gottfried Semper: In Search of Architecture*, Cambridge, Mass: MIT Press.

Semper, G. (1860) *Der Stil in den technischen und tektonischen Künsten oder praktische Ästhetik: ein Handbuch für Techniker, Künstler und Kunstfreunde*, Frankfurt: Verlag für Kunst und Wissenschaft.

Semper, G. (1884) *Kleine Schriften*, M. und H. Semper, Berlin and Stuttgart.

Semper, G., (1977) *Der Stil in den technischen und tektonischen Künsten oder praktische Ästhetik*, Mittenwaldt: Mäander.

Senkevitch, A. (1982) 'Introduction: Mosei Ginzburg and the Emergence of a Constructivist Theory of Architecture', in Ginzburg, M. (ed.) *Style and Form*, Cambridge, Mass: MIT Press.

Sert, J.L., Léger, F. and Giedion, S. (1993) 'Nine Points on Monumentality (1943)', in Ockmann, J. (ed.) *Architecture Culture 1943–1968*, New York: Rizzoli.

Seuphor, M. (1971) *Cercle et Carré*, Paris: Belfond.

Silverman, D. (1989) *Art Nouveau in Fin-de-Siècle France*, Berkeley: University of California Press.

# Bibliography

Simmel, G. (1902) 'Tendencies in German Life and Thought since 1870', *International Monthly,* **5**.

Simmel, G. (1909) 'Die Kunst Rodins und das Bewegungsmotiv in der Plastik', *Nord und Süd,* **129 II**, 189–96.

Simmel, G. (1997) 'The Metropolis and Mental Life', in Frisby, D. and Featherstone, M. (eds) *Simmel on Culture*, London: Sage, 174–85.

Simmel, G. (2004) *The Philosophy of Money*, London: Routledge.

Sitt, M. (1994) 'Jacob Burckhardt as Architect of a New Art History', *Journal of the Warburg and Courtauld Institutes*, **57**, 227–43.

Smith, J.M. (1875) 'The Architecture of the Future', *Building News*, **xxxv**, 425.

Sparling, T.A. (1982) *The Great Exhibition: A Question of Taste*, New Haven: Yale Center for British Art.

Spengler, O. (1991) (Atkinson, C. F., trans.) *The Decline of the West*, Oxford: Oxford University Press.

Stackhouse, M.L. (1972) *Ethics and the Urban Ethos: An Essay in Social Theory and Theological Reconstruction*, Boston: Beacon.

Stamp, G. (1992) 'Mackintosh, Burnet and Modernity', *Architectural Heritage III* (journal of the Architectural Heritage Society of Scotland), 8–31.

Stamp, G. (1999) *Alexander 'Greek' Thomson*, London: L. King.

Stamp, G. (ed.) (1999) *The Light of Truth and Beauty: the Lectures of Alexander 'Greek' Thomson Architect 1817–1875*, Glasgow: The Alexander Thomson Society.

Stamp, G. (2002) *An Architect of Promise: George Gilbert Scott junior and the Late Gothic Revival*, Donington: Shaun Tyas.

Stamp, G. (2003) 'Gothic Thomson', *The Alexander Thomson Society Newsletter*, **33**, 10–18.

Stamp, G. (2004) 'An Architect of the *Entente Cordiale* Eugène Bourdon (1870–1916)' *Architectural Heritage XV* (Journal of the Architectural Society of Scotland), forthcoming.

Stapleton, A. (1996) 'John Moyr Smith 1839–1912', *Decorative Arts Society Journal*, **20**, 18–28.

Stern, F. (1961) *The Politics of Cultural Despair: A Study in the Rise of the Germanic Ideology*, Berkeley: University of California Press.

Stewart, R. (ed.) (1962) *The English Notebooks of Nathaniel Hawthorne* (Vol. 13), New York: Modern Languages Association of America.

Stocking, G.W.J. (1987) *Victorian Anthropology*, New York: Free Press.

Stuckenschmidt, H.H. (1974) *Schönberg. Leben–Umwelt–Werk*, Zurich: Atlantis.

Summerson, J. (1994) 'On discovering 'Greek' Thomson', in Stamp, G. and McKinstry, S. (eds) *'Greek' Thomson*, Edinburgh: Edinburgh University Press.

Swingewood, A. (1998) *Cultural Theory and the Problem of Modernity*, Houndmills: Macmillan.

Szeemann, H. (ed.) (1983) *Der Hang zum Gesamtkunstwerk*, Aarau: Sauerländer.

Szondi, P. (1991) *Poetik und Geschichtsphilosophie*, Frankfurt: Suhrkamp.

Taut, B. (1914a) *Glashaus: Werkbundausstellung Köln 1914*, Cologne.

Taut, B. (1914b) 'Eine Notwendigkeit', *Der Sturm*, **4**, 174–5.

Taut, B. (1918) 'Architektur-Programm', *Mitteilungen des deutschen Werkbundes*, **4**, 16–19.

Taut, B. (1919a) *Alpine Architektur*, Hagen: Folkwang.

Taut, B. (1919b) *Die Stadtkrone*, Jena: Diederichs.

Taut, B. (1920) *Die Auflösung der Städte*, Hagen: Folkwang.

Taut, B. (1982) 'Letter to Max Taut, 8 June 1904', in Whyte, I.B. (ed.) *Bruno Taut and the Architecture of Activism*, Cambridge: Cambridge University Press.

Taylor, J. (1916) 'The Glasgow School of Art', *The Studio*, 124–5.

Taylor, C.H. (1989) *The Sources of the Self: The Making of the Modern Identity*, Cambridge: Cambridge University Press.

Tegethoff, W. (1989) 'From Obscurity to Maturity', in Schultze, F. (ed.) *Mies van der Rohe: Critical Essays*, New York: Museum of Modern Art.

Teige, K. (1966) 'Konstruktivismus a likvidace umění (Constructivism and the liquidation of art)', in *Svět stavby a Básně* (*The World of Building and Poetry*), Prague:Československý Spisovatel.

Thomson, A. (1859) 'On Masonry, and How it may be Improved'. Glasgow: Lecture given to the Glasgow Architectural Society, 22 February 1859. Reprinted in Stamp, G. (ed.) (1999).

Thomson, A. (1866) 'A Protest against Gothic' (letter), *Building News*, 25 May, 384. Reprinted in Stamp, G. (ed.) (1999).

Thomson, A. (1871) 'How is it that there is no modern style of architecture?' *Glasgow Herald*, 8 April 1871, reprinted in Stamp, G. (ed.) (1999).

Thomson, A. (1874) *Art and Architecture: A Course of Four Lectures ... delivered at the Glasgow School of Art and Haldane Academy*, Manchester. Reprinted in Stamp, G. (ed.) (1999).

Thomson, J.E.H. (1881) *Memoir of George Thomson, Cameroon Mountains, West Aftrica, by one of his nephews*, Edinburgh.

Tipps, D. (1973) 'Modernization theory and the study of national societies: A critical view', *Comparative Studies in Society and History*, **15**.

Toulmin, S. (1990) *Cosmopolis: The Hidden Agenda of Modernity*, Chicago: University of Chicago Press.

Tönnies, F. (1955) *Community and Association*, London: Routledge.

Van der Woud, A. (2001) *The Art of Building*, Aldershot: Ashgate.

Vattimo, G. (1988) (Suyder, J., trans.) *The End of Modernity*, Cambridge: Polity Press.

Vesely, D. (1987) 'Architecture and the Conflict of Representation', *AA Files*, **8**, 20–39.

Vidler, A. (2000) 'Spaces of Passage: The Architecture of Estrangement', in A. Vidler, *Warped Spaces*, Cambridge, Mass: MIT Press, 65–80.

Volk, A. (ed.) (1996) *Siegfried Kracauer: Zum Werk des Romanciers, Feuilletonisten, Architekten, Filmwissenschaftlers und Soziologen*, Zurich: Seismo Verlag.

Vosskamp, W. (ed.) (1985) *Utopieforschung: Interdisziplinäre Studien zur neuzeitlichen Utopie*, (3 vols.) Frankfurt: Suhrkamp.

Wagner, O. (1988) *Modern Architecture*, Santa Monica: Getty Center.

Wagner, P. (1993) *A Sociology of Modernity*, London: Routledge.

Wagner, R. (1983) *Kunst und Revolution*, Frankfurt: Insel.

Walden, R. (ed.) (1982) *The Open Hand: Essays on Le Corbusier*, Cambridge: MIT Press.

Warburg, A. (1999) (Britt, D., trans.) *The Revival of Antiquity*, Los Angeles: Getty Center.

Ward, J. (2002) *Weimar Surfaces: Urban Visual Culture in 1920s Germany*, Berkeley: University of California Press.

Watkin, D. (1977) *Architecture and Morality*, Oxford: Clarendon Press.

Webber, M.M. (1964) 'The Urban Place and the Nonplace Urban Realm', in Webber, M.M., Dyckman, J.W. and Foley, D.L. (eds) *Explorations into Urban Structure*, Philadelphia: University of Pennsylvania Press.

Weber, M. (1963) *Gesammelte Aufsätze zur Religionssoziologie*, Tübingen: Mohr.

Weightman, J. (1973) *The Concept of the Avant-Garde: Explorations in Modernism*, London: Alcove Press.

Weinstein, J. (1990) *The End of Expressionism: Art and the November Revolution in Germany 1918–1919*, Chicago: University of Chicago Press.

Wells, A. (1902) 'The late Daniel Cottier', *Journal of Decorative Art and the British Decorator*, **XXII**, 145.

Welter, V.M. (2002) *Biopolis. Patrick Geddes and the City of Life*, Cambridge, Mass: MIT Press.

Whewell, W. (1852) 'The general bearing of the Great Exhibition on the progress of art and science', in *Lectures on the result of the Great Exhibition of 1851 delivered before the Society of Arts*, London: Society of Arts, Manufacturers and Commerce, 3–34.

White, C. (1989) *The Fairy Tale World of Jessie M. King*, Edinburgh: Canongate.

Whitfield, S. (1992) *Magritte* (exhibition catalogue), London: Brepols.

## Bibliography

Whyte, I.B. (1982) *Bruno Taut and the Architecture of Activism*, Cambridge: Cambridge University Press.

Whyte, I.B. (ed.) (2003) *Modernism and the Spirit of the City*, London: Routledge.

Windsor, A. (1985) *Peter Behrens: Architekt und Designer*, Stuttgart: DVA.

Winfried Nerdinger *et al.* (1990) *Revolutionsarchitektur: Ein Aspekt der europäischen Architektur um 1800*, Munich: Hirmer.

Wingler, H.M. (1966) *Gottfried Semper: Wissenschaft, Industrie und Kunst und andere Schriften über Architektur, Kunsthandwerk und Kunstunterricht*, Mainz and Berlin: Kupferberg.

Wood, C.H. (ed.) (1992) *Art in Theory, 1900–1990*, Oxford: Blackwell.

Worringer, W. (1927) *Form in Gothic*, London: Putnams.

Yeats, W.B. (1937) *A Vision*, London: Macmillan.

Zinser, H. (ed.) (1986) *Der Untergang von Religionen*, Berlin: Reimer.

# Index

Page numbers in italics refer to illustrations

Acton Burnell Church, Shropshire *106*
AEG *see* Allgemeine Elektrizitäts-Gesellschaft (AEG)
aesthetic fundamentalism: antimodern modernism 70; and modern architecture 68–80, 159
aesthetic state 72–3
aesthetics: and modernity 7, 30
alienation: cities 16, 234, 295
Allgemeine Elektrizitäts-Gesellschaft (AEG): and Peter Behrens 160, 162
antimodern modernism: art 70
Arcades Project: Walter Benjamin 13, 271–90
architectural history: academic discipline 57, 63, 65, 69; and modernity 56–8
architectural modernism 69, 73, 77, 78; and socialism 48; and theology 46–7; and totalitarianism 48–50
architectural style 58, 83, 107; and historical laws 133–4
architecture: as art 60–1; art forms (*Kunstformen*) 58–9; Christian *106*, *111*, 115–16; and crafts 62–3; and creativity 93–4, 96, 97; cultural context 81–99; dictator- ships 48–50; expressionism 256–70; fragments 174–89; *Gestalt* 93–4; Glasgow 237–9; use of glass 260–2, *261*, *263*, 264–7, 279–80; gothic revival 107, 108, 109–16, 120; and historicism 45–6; and idealism 73–5; meaning 60–3, 173–5; and modernity 3–4, 18, 42–55; plate glass 103, 104, 110, 112; and political ideology 48; public attitudes 81; Queen Anne style 112–15; as revolution 71; and romanticism 92–3; style 61–3; and technology 83–6; totalitarianism 48–50; universality 95–6; work forms (*Werkformen*) 58–9; and *Zeitgeist* 45–6, *see also* classical architecture; modern architecture; Victorian architecture
art: architecture as 60–1; autonomy 34–7; impressionism 32; as institution 35; internationalism 83; modernism and isolation 38–9; and perspectivity 174; progress 35–6, 37; realism 33–4; representation 32–3; and revolution 74–6
artist: as redemptive spirit 259–60
Aspdin, Joseph: Portland cement 195
Athens: Choragic Monument of Thrasyllus *106*, *see also* Greece
avant-garde 94, 98, 178, 267, 270; concrete 198–9, *207*, *211*; inwardness 94–6, 98–9; nature of 29, 30, 31–2; and representation 87, 89–90; Russia 198–200, *207*, *211*

Barcelona: nineteenth-century extension 221–2, *221*, *222*, *see also* Spain
barricades: Paris 279, 282–3, 284
Baudelaire 24, 52, 283–4; and modernity 7, 26–34, 222–3, 274–6
Behrens, Peter 52, 74, 159–62, 167; and Allgemeine Elektrizitäts-Gesellschaft (AEG) 162, 164, 166; Dombauhütte 164, *165*; Gesamtkunstwerk 159–68; Hoechst Administration Building 156, *157*, 163–7, *163*; *Jugendstil* 161; Mathildenhöhe, Darmstadt 160–1, *160*
Benjamin, Walter 6, 10, 16, 17, 19, 39, 166, 222, 292–3, 297–8, 310–19; Arcades Project 271–90; definition of modernity 13–14; philosophical enquiry 318; prehistory of modernity 315, 271–90; radio scripts 311, 315
Berlin 12, 19, 48, 243, 255, 258, 267, 278, 291–303; as capital of Germany 256–8; entertainment 294–8, 301; housing 256; industrialisation 257; Kulturforum 96, *97*; as metropolis 294–5, 298; Meyers Hof, Wedding *257*; modernity 294–303; Philharmonie 94, *95*; population increase 256; suburbanisation 258

# Index

Bötticher, Carl Gottlieb Wilhelm: art forms
(*Kunstformen*) 58–9; and classical
architecture 58–61; *Die Tektonik der
Hellenen* 58–61; work forms
(*Werkformen*) 58–9
Britain 51, 52, 104, 107, 108, 234, 240; use of
concrete in 196
Britannia Bridge, Menai Strait 116
buildings: mass production 199, 201–2, 206

capitalism 44, 199, 232, 272, 280; dynamism
of 7–8; and modernity 6, 7–9, 17
Celticism: Glasgow 252
cement, Portland: invention of 195
Cerdá, Ildefonso: early life 219–20; extension
of Barcelona 221–2, *222*; and modernity
217–30; portrait *220*; and 'rurisation'
229; theory of urbanism 222–30
chain of being 99
Cheryomushki *see* Noviye Cheryomushki
Choragic Monument of Thrasyllus: Athens *106*
Christianity: architecture 47, *106*, *111*, 115–16
cities: alienation 295; home as refuge from
151–3, 285–7; ideas of 53, 262–4;
industrial 51, 232, 236–7, 243, 257;
modernity 51–3, 271–90; reading of 19–20
civic pride: Glasgow 240–1
classical architecture: historical study 56–67
collage 174, 175, 178, 180, 189; situational
187
collective consciousness 272; and society 9–10
Cologne: Glashaus at Werkbund Exhibition
260–2, *261*, *263*
commentary: and modern art 39
commodity fetishism 8
communication: difficulties in 90–1
communications: and urbanism 227
concrete 117, 180, 183, 188; attitudes to 196;
and avant-garde 198–9, *207*, *211*; Britain
196; and mass production 199, 201–2,
206; and modernity 195–214; origins
195; Russia 195–214, *207–14*; Soviet
Union 195–214, *207–14*, *see also*
reinforced concrete
constructivism: deconstructive 88, *89*; Russia
199, 204
conurbation: nature of 249
craft skills: Glasgow 242–3
crafts: and architecture 62–3
Cranston *see* Miss Cranston's Tearooms

creativity: and architecture 93–4, 96, 97;
nature of 87; and production 87
Crystal Palace: and the Great Exhibition
(1851) 124–35; main nave *124*;
mediaeval court 130–1, *130*; plan *126*,
*see also* Great Exhibition (1851)
cultural life: and natural periodicities 91
cultural rationalisation 37–8
culture: and architecture 81–99; discontinuity
96–9; and expressionism 93; inwardness
93–6; mass 294–8, 300–1; and
romanticism 92–3
Cuvier: evolutionary theories 227

design skills: Glasgow 242–3
development: and technology 91–2
dictatorships: architecture 48–50

emptiness: modern architecture 90
entertainment: Berlin 294–8, 301
Ernst, Max: *frottage* 186, 187–8; In the Stable
of the Sphinx 186
evolutionary theories: Cuvier 227; Lamarck 227
expressionism: and culture 93; German
architecture 93, 256–70; inwardness 93–4

film: and music 306; and Siegfried Kracauer
298, 299, 301, 307–20
fragments 10, 13, 38, 81, 90, 140, 200, 277–8,
293, 311, 317; and architecture 174–89
France *see* Paris

garden cities: Germany 258, 260; philosophy
of 51, 52, 229
Germany: Berlin as capital 256–8;
expressionist architecture 256–70; garden
cities 258, 260; *Gläserne Kette* 265, 267;
modernity 4, 14–15; Nazi architecture
48–9, 50, *see also* Berlin; Cologne
*Gesamtkunstwerk* xii, 83; Behrens, Peter
159–68; Hoechst Administration Building
156, *157*, 163–7, *163*; modern
architecture 156–68; nature of 158–9,
260, 316
*Gestalt*: architecture 93–4
*Gläserne Kette*: Germany 265, 267
Glasgow: architectural internationalism
237–9; Celticism 252; civic pride 240–1;
craft skills 242–3; design 241–3, 246–7;
expansion 234–6, 239–40, 245; filthiness

233, 245, 246; gothic revival *113*, *121*;
International Exhibition (1901) 237, *238*;
international trade 236; lower classes
233, 245; middle classes 240–1, 244,
245; Miss Cranston's Tearooms 239, 241,
244; modernity 232–55; and
Protestantism 251–2; Style pageants 241–2
Glasgow Style 247, 250, 252
glass: in architecture 260–2, *261*, *263*, 264–7,
279–80; Glashaus at Werkbund Exhibition,
Cologne 260–2, *261*, *263*; Swiss Pavilion
183, 184–5, 186, *see also* plate glass
Gombrich, Ernst: progress and modernity
30–1, 35, 37
Goncourt, Edmond de: 53 boulevard
Montmorency 139, *143*, *144*, *145*, *147*,
148–50, *149*; home decoration 142–51;
*La maison d'un artiste* 139–40, 142–7
Goncourt, Edmond and Jules de: as novelists
140–2
gothic: and Hoechst Administration Building
164; secular 164–5; Victorian architecture
107, 108, 109–16, 120
Great Exhibition (1851) 124–35; axis of
progress 127–8; and the Crystal Palace
124–35, *124*, *126*, *130*; expanding
figure of a man composed of 7000
working parts *125*; purpose 126; steam
machine in the shape of an Egyptian
temple *132*
Greece: ancient ruins 273–4, *see also* Athens
Gropius, Walter 42, 43, 49, 72, 160, 164;
Bauhaus Manifesto 73–4; spiritual
revolution 75–6
*Gründerzeit*: Germany 256–7

Habermas, Jürgen 6, 7, 16, 38, 44, 72; and
modernity 14–15
Hablik, Wenzel August: *Fantasy* 266; *The Path
of Genius* 259, 260
Haeckel, Ernst: *Kunstformen in der Natur* 268
Haussmann, Baron: rebuilding of Paris 277,
279, 281, 282–3
Hawthorne, Nathaniel: on Glasgow 232–3
historical laws: and architectural style 133–4
historical periodisation: and modernity 6
historicism: and architecture 45–6; definition 45
history: concepts of 127, 131, 133–4
Hoechst Administration Building 156, *157*,
163–7, *163*

home: as artistic collection 151; expressing
character of occupant 138, 142–51, 153;
and liberty 137, 153; as refuge from city
151–3, 285–7; tasteful decoration 138–9
housing: Berlin 256; and mass production
199, 201–2, 206; Moscow 201, *213*
Howard, Ebenezer 48, 229; New Towns 51,
229

idealism: and architecture 73–5
impressionism: in art 32
individualism: and modernity 9–10
industrialisation: cities 51, 232, 236–7, 243,
256–7; Germany 256–7
internationalism: of art 83
iron: Victorian architecture 103, 104, 110–12

*Jugendstil*: Mathildenhöhe, Darmstadt 160–1,
*160*; Peter Behrens 161

Kant 6, 14, 35, 138; and taste 138–9
Kracauer, Siegfried 14, 16, 19, 222, 278, 287:
and film 298, 299, 301, 307–20; *Jacques
Offenbach and the Paris of his Time* 308–
11; and modernity
294–303; Offenbach film 307–20; piano
players 291, 296, 301–3, 305; *Strassen in
Berlin und anderswo* 291–306; *Theory of
Film* 298, 312, 316, 317–18
*Kunstformen*: architecture 58–9

Lamarck: evolutionary theories 227
Le Corbusier 42, 46, 47, 48, 49, 52, 71;
*Composition with the Moon* (1929) 177–8,
*177*; fragments 178–89; ideas of the city
53; *objets à réaction poétique* 175–8, 181–
2; Radiant City 179; *Sculpture and Nude*
184; Swiss Pavilion, Paris *176*, 178–89,
*179*, *181*, *182*, *185*, *187*
liberty: and home 137, 153; ideas of 137, 153;
and urbanism 229–30
Libeskind, Daniel: 'deconstructive
construction' 88, *89*

Mackintosh, Charles Rennie: domestic
interiors 246; house *235*
Magritte, René: *Memory of a Voyage* 180–1
Marx, Karl 13, 16, 17, 206, 224, 265, 271,
276, 285; *Communist Manifesto* 7–9; on
modernity 44

# Index

mass production: and concrete 199; housing 199, 201–2; problems 206

mathematics: and reality 90

Mendelsohn, Erich: Einsteinturm, Potsdam *269*; garden pavilion project *269*

metropolis: Berlin 294–5, 298; and modernity 3–4, 10–11, 52–3; self-destruction 258

Mies van der Rohe, Ludwig 42, 49, 77, 90 95, 117, 160; architecture and technology 84, 86; ideas of the city 53; Lake Shore Drive *85*; universality of architecture 95–6

*mikrorayon*: Moscow 205, *212*

Miss Cranston's Tearooms: Glasgow 239, 241, 244

modern architecture: and aesthetic fundamentalism 68–80; cultural discontinuity 96–9; emptiness 90; *Gesamtkunstwerk* 156–68; meaning 173–5; visibility 97–8

modern culture: inwardness 93–9

modernism: antimodern 70; and architecture 46–7, 48–9, 50; and artistic isolation 38–9; concepts of 50–1, 54; death 42–3; definition 44; historical construction 23–5; and novelty 27–8; and politics 48–50; and theology 46–7, *see also* modernity

modernity: aesthetic 7, 30; Arcades Project 271–90; and architectural history 56–8; and architecture 3–4, 18, 42–55; Baudelaire 7, 26–7, 222–3, 275–6; Berlin 294–303; and capitalism 6, 7–9, 17; Cerdá, Ildefonso 217–30; cities 51–3, 271–90; concepts of 29–30, 43–4, 50–1, 77; and concrete 195–214; cultural 30; definitions 10–11, 13–14; diversity 16; Germany 4, 14–15; Glasgow 232–55; and historical periodisation 6; history of 24–5, 44–6; and individualism 9–10; and Kracauer, Siegfried 294–303; and the metropolis 3–4, 10–11, 52–3; and modernisation 5–6; and monumentalism 4; nature of 4–5; Paris 271–90; prehistory 271–90; and progress 29–31; and rationalisation 15; Russia 197; and society 5–13; spatial dimensions 16; transitoriness 7; and urbanism 222–30; Vienna 4; Wagner, Otto 3–4, *see also* modernism

monumental neoclassicism: as international style 49

monumentalism: and modernity 4

Moscow: concrete 197–8, *207–14*; housing 201, *213*; metro 200–1, *208*; *mikrorayon* 205, *212*; Noviye Cheryomushki 204, *210*; rebuilding 196–7, *see also* Russia; Soviet Union

movement: and urbanism 226–7

museum: ideal 128–30, *129*, 132–4

music: and film 306; progress in 30–1

Natural History Museum, London *120–1*

Nazi-ism: architecture 48–9, 50

New Towns: philosophy of 51, 52, 229

Nietzsche, Friedrich 14, 17, 69, 70, 71–6, 159, 161, 165, 167, 169, 259, 262; critique of modernity 12–13

novelty: and modernism 27–8

Noviye Cheryomushki: Moscow 204, *210*

*objets à réaction poétique*: Le Corbusier 175–8, 181–2

Offenbach, Jacques: biography 308–11; operetta 308, 309–10, 312–13, 315

Paris: Arcades Project 271–90; barricades 279, 282–3, 284; modernity 271–90; rebuilding by Baron Haussmann 277, 282–3; Second Empire 309, 312–14, 319–20; Swiss Pavilion *176*, 178–89, *179*, *181*, *182*, *185*, *187*

Pausanias: ancient Greek ruins 273–4

periodicities, natural: and cultural life 91

perspectivity: and art 174

philosophical enquiry 318

piano players: Siegfried Kracauer 291, 296, 301–3, 305

plate glass: Victorian architecture 103, 104, 110, 112

political ideology: and architecture 48

politics: and modernism 48–50

population increase: Berlin 256

Portland cement: invention of 195

postmodernity: nature of 17

production: and creativity 87; nature of 87–8

progress: and autonomy in art 35–6, 37; Great Exhibition (1851) 127–8; and modernity 29–31; music 30–1

Protestantism: Glasgow 251–2

protozoa *268*

public: attitudes to architecture 81

Pueblo Indians: rituals 64–5
Pugin, A.W.N. 110, 115, 117, 121; mediaeval court, Crystal Palace 130–1, *130*

Queen Anne style: Victorian architecture 112–15

Radiant City: Le Corbusier 179
radio scripts: Benjamin, Walter 311
rationalisation: cultural 37–8; and modernity 15
realism: art 33–4
reality: and mathematics 90; nature of 81–2, 86; and technology 91
reinforced concrete 117; invention 195; mass production of housing 205, *see also* concrete
representation: art 32–3; and the avant-garde 87, 89–90; nature of 82–3; purpose of 86–7
revolution: aesthetic 71–2; through architecture 71; and art 74–6
romanticism: architecture 92–3; and culture 92–3
Ruskin, John 51, 108, 115, 177, 250; architecture and political ideology 48
Russia: avant-garde 198–200, *207*, *211*; concrete 195–214, *207–14*; constructivism 199, 204; and modernity 197; Noviye Cheryomushki 204, *210*, *see also* Moscow; St Petersburg; Soviet Union

St Petersburg: development 197
Scharoun, Hans 199, 267; Berlin Philharmonie 94, *95*; expressionism 93; Gestalt in architecture 93–4; Volkshaus *92*
Schiller, Friedrich 35, 74, 76; aesthetic state 72
Scott, Sir George Gilbert: architectural style 104, 109–15, 119–22; Hunterian Museum, Glasgow *113*; Kelham Hall, Newark-on-Trent *114*; portrait *105*; St George's Church, Doncaster *111*; University of Glasgow *121*
Semper, Gottfried 16, 46, 222, 243, 286; architectural style 61–3, 133–5; and the Great Exhibition (1851) 124–35; ideal museum 128–30, *129*, 132–4; 'method of inventing' 133–4
Simmel, Georg: definition of modernity 10–11
sociability: and urbanism 229–30
socialism: and architectural modernism 48
society: collective consciousness 9–10; and modernity 5–13

Soviet Union: concrete 195–214, *207–14*; concrete production figures 204, 205; hero workers 203–4; mass production of buildings 199, 201–6; *piati-etashniye* 204, 206, *see also* Moscow; Russia
space: and modernity 16
Spain: nineteenth century history 217–19, *see also* Barcelona
steel: use in building 60
*Strassen in Berlin und anderswo*: Siegfried Kracauer 291–306
suburbanisation: Berlin 258
Swiss Pavilion, Paris *176*, 178–89, *179*, *181*, *182*, *185*, *187*

taste: domestic decoration 138–9
Taut, Bruno: *Alpine Architektur 263*; Cathedral of Socialism 74–5; *Die Auflösung der Städte* 265, *265*, 268; *Die Weltbaumeister* 265; garden cities 260; *Gläserne Kette* 265, *267*; Glashaus at Werkbund Exhibition, Cologne 260–2, *261*, *263*; glass in architecture 260–2, *261*, *263*, 264–7; *Monument to the New Law 266*; portrait *261*
technology: and architecture 83–6; and development 91–2; and reality 91
theology: and architectural modernism 46–7
*Theory of Film*: Siegfried Kracauer 298, 312, 316, 317–18
Thomson, Alexander ('Greek'): architectural style 103–22; Holmwood House, Cathcart *114*; Howard Street warehouse, Glasgow *113*; use of iron 103, 104, 110–12; Natural History Museum, London *120–1*; use of plate glass 102, 103, 104, 110; portrait *105*; St Vincent Street Church, Glasgow *111*, *235*
totalitarianism: and architectural modernism 48–50
transport: and urbanism 226–7

urbanism: and Cerdá, Ildefonso 222–30; and communications 227; etymology 225–6; form and function 227; and liberty 229–30; and modernity 222–30; and movement 226–7; and 'rurisation' 229; and sociability 229–30; and transport 226–7
USSR *see* Soviet Union

# Index

Victorian architecture 103–22; Christian *111*, 115–16; gothic 107, 108, 109–16, 120; iron 103, 104, 110–12; plate glass 103, 104, 110, 112; Queen Anne style 112–15; Scott, Sir George Gilbert 104, 109–15, 119–22; Thomson, Alexander ('Greek') 103–22

Vienna 46, 52, 238; and modernity 4

visibility: modern architecture 97–8

Wagner, Otto 20, 46, 52; architecture 20; *Modern Architecture* 3–4; modernity 3–4, 18

Wagner, Richard 68, 70, 74, 75, 161, 166, 168, 262; *Gesamtkunstwerk* 158–9, 168; progress in music 30–1

Warburg, Aby: art history 63–5

Weber, Max 15, 51, 70; cultural rationalisation 37; theory of modernity 11–12

work forms (*Werkformen*): architecture 58–9

*Zeitgeist*: and architecture 45–6